BIOTECHNOLOGY IN AGRICULTURE SERIES

General Editor: Gabrielle J. Persley, International Service for National Agricultural Research (The Hague, Netherlands), and Project Manager, World Bank/ISNAR/ACIAR/AIDAB Biotechnology Study.

For a number of years, biotechnology has held out the prospect for major advances in agricultural production, but only recently have the results of this new revolution started to reach application in the field. The potential for further rapid developments is however immense.

The aim of this new book series is to review advances and current knowledge in key areas of biotechnology as applied to crop and animal production, forestry and food science. Some titles will focus on individual crop species or groups of species, others on specific goals such as plant protection or animal health, with yet others addressing particular methodologies such as tissue culture, transformation or immunoassay. In some cases, relevant molecular and cell biology and genetics will also be covered. Issues of relevance to both industrialized and developing countries will be addressed, and social, economic and legal implications will also be considered. Most titles will be written for research workers in the biological sciences and agriculture, but some will also be useful as textbooks for senior-level students in these disciplines.

Editorial Advisory Board:

P.J. Brumby, formerly of the World Bank, Washington DC, USA.
E.P. Cunningham, FAO, Rome, Italy.
P. Day, Rutgers University, New Jersey, USA.
J.H. Dodds, International Potato Center (CIP), Peru.
J.J. Doyle, International Laboratory for Research on Animal Diseases, Nairobi, Kenya.
S.L. Krugman, United States Department of Agriculture, Forest Service.
W.J. Peacock, CSIRO, Division of Plant Industry, Australia.

Titles Available:

1: Beyond Mendel's Garden: Biotechnology in the Service of World Agriculture *G.J. Persley*
2: Agricultural Biotechnology: Opportunities for International Development *Edited by G.J. Persley*
3: The Molecular and Cellular Biology of the Potato *Edited by M.E. Vayda and W.D. Park*
4: Advanced Methods in Plant Breeding and Biotechnology *Edited by D.R. Murray*

Advanced Methods in Plant Breeding and Biotechnology

Edited by

David R. Murray

Senior Research Associate,
Department of Biological Sciences,
University of Sydney,
Sydney NSW 2006
Australia

C·A·B *International*

C·A·B International   Tel: Wallingford (0491) 32111
Wallingford   Telex: 847964 (COMAGG G)
Oxon OX10 8DE   Telecom Gold/Dialcom: 84: CAU001
UK   Fax: (0491) 33508

British Library Cataloguing-in-Publication Data
 A catalogue record for this book is available from the British Library

ISBN 0 85198 706 0
ISSN 0960 202X

Typeset by Enset (Photosetting), Midsomer Norton, Bath, Avon
Printed and bound in the UK by Redwood Press Ltd, Melksham

Contents

Preface

Agricultural practices that rely heavily on applications of expensive nitrogenous fertilizers and pesticides are no longer sustainable. There are some ecologically sound alternative approaches to solving agricultural problems, one of which is to breed plants with improved defences against a variety of pests and pathogens, and with better capacities to utilize nutrients economically, or withstand environmental stresses. The integration of rapidly developing 'molecular' techniques with conventional methods of plant breeding can speed up the process of developing new cultivars with desirable characteristics. However, developments have taken place so rapidly that it has become difficult for plant breeders to be aware of all the techniques currently available. The choice of procedures with the best chances of success for a given purpose may well be overlooked.

My objective in the planning of this book was to gather together contributions that would illustrate the application of all the methods broadly termed 'genetic engineering' to the protection and improvement of cultivated plants, especially those that provide food. The introductory chapter provides a historical setting and considers how the goals of plant breeding may change in the future. The following chapters then fall approximately into two groups. The first group is technique oriented, considering both indirect (vector-mediated) and direct methods of gene transfer, and the ways in which successful gene transfer can be monitored.

The second group of chapters is goal oriented, and considers how progress is being made towards the selective transfer of desirable characteristics such as disease resistance, pest resistance and stress tolerance. Central to this book is a review of somatic embryogenesis (Chapter 7), since plant cells with modified genomes must ultimately be encouraged to differentiate in such a way that new and fertile plants are readily obtained.

The coverage is comprehensive, including electroporation (Chapter 4) and biolistics (Chapter 5). More than 1000 references provide quick access to the scientific literature. Accordingly, the book should serve as a valuable guide for potential research students, as well as for those already engaged in applied research.

It is a pleasure to record my thanks to the contributing authors, to everyone who supplied information and material for inclusion in the book, and to copyright owners, for permission to reproduce figures already published. I am also grateful to Mr Tim Hardwick and the staff of CAB International for their co-operation and support during publication.

David R. Murray

Contributors

K. C. Armstrong, *Plant Research Centre, Central Experimental Farm, Ottawa, Ontario, Canada K1A 0C6.*

Robert G. Birch, *Department of Botany, University of Queensland, Queensland 4072, Australia.*

A. K. Chakravorty, *Department of Biochemistry, University of Queensland, Queensland 4072, Australia.*

M. C. Christey, *DSIR Crop Research, Private Bag, Christchurch, New Zealand.*

A. J. Conner, *DSIR Crop Research, Private Bag, Christchurch, New Zealand.*

Christopher A. Cullis, *Dean of Mathematics and Natural Sciences, 7080 Crawford Hall, Case Western Reserve University, Cleveland, Ohio 44106, USA.*

E. M. Dommisse, *DSIR Crop Research, Private Bag, Christchurch, New Zealand.*

Tricia Franks, *Department of Botany, University of Queensland, Queensland 4072, Australia.*

John A. Gatehouse, *Department of Biological Sciences, University of Durham, South Road, Durham DH1 3LE, UK.*

J. E. Grant, *DSIR Crop Research, Private Bag, Christchurch, New Zealand.*

S. Hinnisdaels, *Université de Bruxelles, Laboratoire Génétique des Plantes, Institut Biologie Moléculaire, Paardenstraat 65, B-1640 Rode–St Genèse, Belgium.*

S. A. Merkle, *School of Forest Resources, University of Georgia, Athens, Georgia 30602, USA.*

A. Mouras, *Université de Bordeaux II, Laboratoire Biologie Cellulaire, Avenue des Facultés, 33405 Talence Cedex, France.*

David R. Murray, *Senior Research Associate, Department of Biological Sciences, University of Sydney, Sydney NSW 2006, Australia.*

W. A. Parrott, *Department of Agronomy, University of Georgia, Athens, Georgia 30602, USA.*

G. Pelletier, *Laboratoire de Biologie Cellulaire, INRA, CNRA, Route de St Cyr, 78026 Versailles Cedex, France.*

Carl Rathus, *Bureau of Sugar Experiment Station, PO Box 86, Indooroopilly, Queensland 4068, Australia* (Present address: *CSIRO Division of Tropical Crops and Pastures, 306 Carmody Road, St Lucia, Queensland 4067, Australia*).

Ray J. Rose, *Department of Biological Sciences, University of Newcastle, Newcastle, New South Wales 2308, Australia.*

K. J. Scott, *Department of Biochemistry, University of Queensland, Queensland 4072, Australia.*

C. Taylor, *Centre for Cereal Biotechnology, Waite Agricultural Research Institute, University of Adelaide, Glen Osmond, South Australia 5064.*

Gail M. Timmerman, *DSIR Crop Research, Private Bag, Christchurch, New Zealand.*

Norman F. Weeden, *Department of Horticultural Sciences, New York State Agricultural Experiment Station, Cornell University, Geneva, NY 14456, USA.*

Maureen C. Whalen, *Department of Biology, Colby College, Waterville, Maine 04901, USA.*

E. G. Williams, *Department of Botany, University of Georgia, Athens, Georgia 30602, USA.*

Chapter 1
Breeding Plants for the Twenty-first Century

David R. Murray

Department of Biological Sciences, University of Sydney,
Sydney NSW 2006, Australia

Introduction

The science of genetics and the discipline of plant breeding have emerged only in the past 200 years. But plants have been selected and bred from prehistoric times – from the stage of deliberate cultivation. The non-shattering rachis of cereal stalks (Zohary, 1969; Harlan *et al.*, 1973) and the pod that remains closed while it dries out (Waines, 1975; Ladizinsky, 1979a) were two of the earliest selected alterations of wild-type seed dispersal mechanisms. These recessive characteristics facilitate harvesting, and became generally established in cultivated cereals and legumes more than 10,000 years ago.

Other characteristics have been selected inadvertently, e.g. aluminium tolerance of certain wheat cultivars growing in acid soils (Foy *et al.*, 1974). Each major crop species has its own long history, and for some, hundreds or even thousands of distinct cultivars have been generated. The modern plant breeder has unparalleled opportunities to take advantage of existing genetic variability in closely allied taxa, and to consider strategies for incorporating genes from distant and alien sources when endeavouring to modify a plant's genome for a particular purpose. This chapter serves to question the goals of plant breeding for the coming century, to illustrate the pitfalls and potential benefits of adopting 'molecular' methods, and to address the need for regulation of the release of transgenic plants.

Genetic resources

The reserves of genomic variability available to plant breeders include collections of cultivars, and more importantly, collections of wild or weedy

1

species that share an immediate common ancestry with the crop plant under consideration. Tentative identifications of wild progenitors based on cytogenetic observations can in many instances be confirmed by comparative electrophoretic studies of seed proteins (Ladizinsky, 1979b; Murray, 1984; Krishna & Mitra, 1988).

Harlan (1969, 1984), Heiser (1976) and Hawkes (1990), among others, have long advocated the conservation of wild relatives of cultivated plants to establish a comprehensive gene pool. In addition to national collections in many countries, there are now significant international collections. A volume marking the tenth anniversary of the formation of the International Board for Plant Genetic Resources (IBPGR) describes progress in conservation, evaluation and utilization of genetic resources for crop improvement (Holden & Williams, 1984). The IBPGR is one of 13 centres established by the Consultative Group on International Agricultural Research (CGIAR) under the auspices of the Food and Agriculture Organization of the United Nations. Those centres with primary concerns for the conservation and evaluation of crop plant diversity are listed in Table 1.1.

Table 1.1. International Agricultural Research Centres with plant genetic resource programmes.

AVRDC	Asian Vegetable Research and Development Centre, Republic of China (Taiwan)
CIAT	Centro Internacional de Agricultura Tropical, Colombia
CIMMYT	Centro Internacional de Mejoramiento de Maiz y Trigo, Mexico
CIP	International Potato Centre, Lima, Peru
ICARDA	International Centre for Agricultural Research in the Dry Areas, Aleppo, Syria
ICRISAT	International Crops Research Institute for the Semi-Arid Tropics, Hyderabad, India
IITA	International Institute of Tropical Agriculture, Ibadan, Nigeria
IRRI	International Rice Research Institute, Republic of the Philippines

At present there is growing financial pressure on both national and international agencies to cut back on the numbers of genotypes actively maintained. Such streamlining is premature. It is not yet possible to judge which minority genotypes are superfluous to future breeding requirements, and which not. Many biochemical and physiological variations within species are only now being recognized (Krishna & Murray, 1988; Murray, 1988). Major collections have never been surveyed for these emerging differences, and their implications for agronomic performance have not yet been assessed.

Defining objectives – what are desirable attributes?

In order to effect improvements in any plant's productivity or performance, the breeder must be fully aware of the attributes and breeding history of the plant in question. As well as more conventional concerns such as disease resistance and day-length control of flowering, some destabilized environmental factors must now be given more attention. The predominant genotypes for each crop may in future need to be reassessed and modified in light of their responses to:

- increasing mean global atmospheric CO_2 concentration (Tucker, 1981);
- climate change affecting precipitation patterns and temperature range (Chapter 14; Daly, 1989; Schneider, 1990);
- increasing penetration of damaging ultraviolet B radiation to the Earth's surface as a consequence of thinning of the stratospheric ozone layer (Worrest & Caldwell, 1986).

Some examples of possible interactions and trade-offs are considered later in this section.

Yield

The physiological basis of differences in performance, and hence yield, has formed an important focus for research in the past (Evans, 1975). It would be fair to judge that although considerable progress has been made in understanding the way assimilates are translocated within the plant and delivered to developing fruits and seeds (Murray, 1987, 1988), we do not fully understand the way partitioning of resources is regulated. Yield of the desired plant part is still the best empirical measure of plant performance.

The genetic basis of yield is elusive, although there is now a valuable reservoir of past experience to draw upon. It must nevertheless be recognized that sometimes any further real improvement in yield by genetic manipulation might not be feasible, or the size of the likely gain might not be worth the effort that would need to be committed.

Sunflower

The cultivated sunflower, *Helianthus annuus*, is generally regarded as one of the most productive C_3 plants (Rawson & Constable, 1980; Rawson *et al.*, 1980; Diepenbrock, 1988), although its consumption of water per unit dry matter produced is greater than in C_4 plants such as maize (Blanchet *et al.*, 1981). In common with other oilseed crops, there is an inverse relation between seed oil content and seed protein content – both cannot be increased simultaneously (Alba & Greco, 1981; Röbbelen, 1981).

For the oil type of sunflower it does not appear possible to further improve water use efficiency, seed (achene) yield, or the oil plus protein content of the seed, either by continued breeding or by liberal application of nitrogenous fertilizer. The futility of adding nitrogen above a certain optimal application is illustrated by the data of Table 1.2. The most important external factors limiting yield in sunflower are usually potassium supply, and water supply, which must be adequate throughout vegetative growth and flowering (Blanchet *et al.*, 1981; Diepenbrock, 1988). The response to excess nitrogen is delayed flowering and fruiting. This is risky if maturation is extended into a rainy period, when soaked heads will become infected by fungi such as grey mould (*Botrytis cinerea*) before they can be harvested.

Table 1.2. Response of sunflower to applied nitrogenous fertilizer.

N^a (kg/ha)	Head diameter (cm)	Achene yield (kg/ha)	Oil content (%, w/w)	Oil yield (kg/ha)
0	14.7	2193	48.1	1054
56	16.5	2841	46.3	1320
112	17.0	3043	45.3	1381

Data of Zubriski & Zimmerman (1974).
[a]As NH_4NO_3.

Annual legumes

When yield factors are being studied in legumes, which are capable of forming N_2-fixing root nodules inhabited by *Rhizobium* spp., the genotypes of both the host plant and the strain of *Rhizobium* have a bearing on the extent and effectiveness of nodulation (e.g. Miller *et al.*, 1986). For ease of interpretation, comparisons among cultivars of a given legume might best be performed with a single strain of inoculant *Rhizobium* (Fernandez & Miller, 1985a, 1987). However, under field conditions, comparisons are more realistic if conducted with mixed strain inocula, so that the optimal combinations of bacterial strain with host cultivar have the opportunity to develop (Fernandez & Miller, 1985b; Miller *et al.*, 1986; Fernandez *et al.*, 1989).

Seed yields under field conditions were compared for four indeterminate cultivars of cowpea (*Vigna unguiculata* subsp. *unguiculata*) of low or high nitrogen-fixing capacity (Table 1.3). A soil pretreatment of methyl bromide was given, which controlled soil pathogens and weeds. Three nutritional regimes were established: inoculation with cowpea 'EL' mixed strain peat inoculum; application of 100 kg/ha nitrate N (as calcium nitrate); and un-modified control, depending on a small amount of residual soil nitrogen and endogenous strains of *Rhizobium*.

Plants were harvested at full maturity, which varied as shown in Table 1.3. Total plant mass, seed yield, mean seed mass, number of seeds per pod, and number of pods per plant were recorded (Table 1.4). A harvest index was calculated as the ratio of seed yield to total plant mass.

Table 1.3. Characteristics of five cultivars of cowpea analysed for yield components by Fernandez & Miller (1985b).

Cultivar	N_2 fixation	Growth habit	Time to flowering (days)	Time to maturity (days)
A. Bush Purple Hull	Low	Determinate	32	48
B. Mississippi Silver	Low	Indeterminate	42	71
C. California Blackeye	High	Indeterminate	49	81
D. Lady	Low	Indeterminate	49	92
E. Brown Crowder	High	Indeterminate	49	85

Of the seed yield components, the number of seeds per pod and the mean seed mass were strongly genotype-dependent, as in other legumes, and not appreciably influenced by treatment. The number of pods per plant was least responsive to the nitrogen regime in cultivar Mississippi Silver (5 ± 1), but practically doubled for inoculated California Blackeye plants (Table 1.4). Variability in this parameter largely explains the observed differences in seed yield. All the indeterminate cultivars showed greater seed yield when grown symbiotically with inoculum compared with growth with added nitrate. Compared with the control, both treatments gave superior seed yields by a factor approaching two. One cultivar, Brown Crowder, stood out as providing superior seed yields to all other cultivars under both treatments (Table 1.4).

In annual legumes, the extent of redistribution of previously assimilated nitrogen from within the plant to seeds varies strongly with cultivar and species (Pate, 1984; Murray, 1988). African cultivars of cowpea also show marked differences in seed protein profiles (Carasco *et al.*, 1978; Khan *et al.*, 1980; Murray *et al.*, 1983). It would be useful in future breeding programmes for cowpea and other legumes to relate differences in seed protein content and composition to other yield parameters.

Table 1.4. Influence of nitrogen nutrition on yield parameters of cowpea.

Cultivar[a]	Plant mass (kg/ha)	Seed yield (kg/ha)	Seed mass (mg)	Seeds per pod	Pods per plant	Harvest index[b]
Control						
A	2080	910	87	8	6	0.44
B	5070	780	106	12	4	0.19
C	5070	740	116	6	9	0.15
D	5070	760	45	13	13	0.17
E	5460	910	96	11	8	0.18
Inoculated						
A	1950	1170	85	9	8	0.58
B	5850	1500	110	11	6	0.29
C	7280	1490	111	7	17	0.21
D	7540	1400	44	14	18	0.19
E	6760	2020	99	12	11	0.31
+ Nitrate						
A	1820	1110	92	9	8	0.64
B	7150	1430	120	12	5	0.21
C	7160	1300	117	7	12	0.18
D	5460	1170	43	13	13	0.21
E	8190	1820	104	12	11	0.22

Data of Fernandez & Miller (1985b).
[a]Cultivar identities are as in Table 1.3.
[b]Seed yield/plant mass.

Nitrogen harvest index in cereals

Nitrogen harvest index is the proportion of total nitrogen in the mature plant recovered in the seeds. It reflects the ability of the plant to mobilize previously assimilated nitrogen. The concept was first developed explicitly for cereals, which depend on external sources of nitrogen (Austin *et al.*, 1977; Desai & Bhatia, 1978).

Within comparable groups of plants, nitrogen harvest index values can differ by almost a factor of two: from 57 to 83% for fifteen cultivars of durum wheat (Desai & Bhatia, 1978), and from 35 to 65% among wild and cultivated barleys (Corke *et al.*, 1989). However, a high nitrogen harvest index does not necessarily mean that seeds have a high protein content. In this study of durum wheat, 'grain protein concentration was neither correlated significantly to plant nitrogen nor to nitrogen harvest index' (Desai & Bhatia, 1978). Similar observations were made for bread wheat by Johnson *et al.* (1967). Hence there is little point in using high nitrogen harvest index as a selection criterion in the absence of other critical information, such as the distribution of seed nitrogen between protein and nucleic acids, and the quality of the protein.

The potential to rearrange the genetic components of seed yield using both cultivated and wild forms of the major cereal crops has not yet been fully realized (Bhatia, 1975; Desai & Bhatia, 1978; Corke *et al.*, 1989). A breeding objective of fundamental importance in the future will be to enhance both grain yield and grain protein yield without depending on high input of nitrogenous fertilizer. Plants that do well with limited applications of fertilizers are economic in every sense. They are likely to have a high nitrogen harvest index as a consequence of translocation effectiveness (Johnson *et al.*, 1967) in combination with other attributes.

Scale of agriculture

The scale of agriculture and the method of harvesting will often govern what constitutes an improvement in yield. For pea plants in New Zealand, subject to a once-only mechanical harvest, a semi-leafless determinate variety may yield 4 tonnes per hectare, twice as much as a conventional garden pea (cv. Melbourne Market). In physiological terms, this is a misleading comparison. The substantial difference comes about because of the indeterminate sequential flowering habit of the conventional variety. If pea plants are grown on a market garden scale and hand-picked periodically, then the yields of traditional cultivars are equal to or superior to those of the newer determinate types. A full discussion of the morphogenetic effects of the *det* mutation is provided by Singer *et al.* (1990).

Time of harvest

Response to photoperiod

The time from sowing to harvest in annuals depends on temperature regime and day-length versus night-length. Flowering is often strictly controlled by the duration of the uninterrupted dark period in the daily cycle (Evans, 1969). Such plants are still referred to as 'short-day' or 'long-day'. Short day-length control has sometimes proved undesirable, inhibiting the spread of crops from the tropics to temperate regions, e.g. soybean from Asia to Europe (Schuster & Böhm, 1981).

For cultivars with a weak response to photoperiod alone ('day-neutral' plants), the simple multiple of mean diurnal temperature and mean day-length can often be used to predict flowering time. Use of this parameter has a long history, reviewed by Aitken (1974). Given the arbitrary nature of both components and the complexities of genetic control of plant reproductive growth (Marx, 1983), this equation works surprisingly well.

Refinements, such as correction for a base temperature, are discussed by Fernandez & Chen (1989), who applied several different formulae in studies of photoperiod-insensitive mung bean (*Vigna radiata*). Several such genotypes were identified following screening of more than 1000 mung bean accessions at the AVRDC. Representative genotypes were assessed under field conditions in nine countries, ranging in latitude (north) from 4°07′ (Malaysia) to 31°19′ (Pakistan), to test predictions based on information gained at invariant photoperiods of 11.5 or 13.5 h and in the mean diurnal temperature range of 22–30°C. These authors concluded that the flowering dates of photoperiod-insensitive mung beans can be predicted and adjustments to sowing time made according to latitude.

Modified harvest maturity

Faster-maturing plants can provide two or even three harvests per year instead of one. Increased frequency of harvesting was the chief advantage of the new rice introduced from Vietnam to China almost 1000 years ago. In areas with diminished rainfall, such as Kano State, Nigeria, varieties of sorghum (*Sorghum bicolor*) with a shorter growing season have proved invaluable in avoiding total crop failures (Tomlins, 1987). Normally the local sorghum would take about 120 days to mature, stand 2–3 m tall and, with adequate water, produce about 1 tonne of grain per hectare. Improved varieties were developed from US and Indian genotypes. These mature in 80–90 days, reach only 1–2 m in height, and yield 1.5–2.5 tonnes of grain per hectare.

Locally the sorghum straw is used as a building material. Does the reduced yield of straw matter? In this situation, the answer is no. To have a yield of grain at all is the over-riding consideration. Some plots of the traditional sorghum are still planted to provide a continued supply of building straw, conditions permitting.

The importance of being able to predict flowering time, and hence harvest time, is appreciated by fruit growers attempting to spread out their production. The benefits are twofold: more efficient utilization of harvesting and packing resources, including labour, and avoidance of a glut in the market, which has the effect of depriving growers of adequate financial returns.

An over-reliance on post-harvest storage and bureaucratic interference in constraining choice of cultivar have marred the Australian apple-producing industry in the recent past. The advantages of existing cultivars that are not widely grown, such as Lady Williams and Gala, are being rediscovered (Miller, 1990), and a more even spread of apple production is being encouraged.

Stone-fruit growers now have available new cultivars of 'low-chill' peaches and nectarines developed in Florida, USA (Geyle, 1990). These will mature 7–14 days before, or after, standard varieties such as Prince, Gold and King (peaches) or Sundown (nectarines). These varieties will prove popular for exactly the same reasons as 'off-peak' apples – and there is certainly interest in the development of new cultivars of other fruits with modified harvest maturity.

Quality

Organoleptic criteria

Flavour and aroma constituents are major determinants of organoleptic quality in fruits, vegetables and grain crops, but this aspect of quality has often been ignored in past breeding programmes. The tasteless but tough commercial tomato is a prime example. The incorporation of superior flavour into improved cultivars does sometimes happen, as in the novel cowpeas referred to by Tomlins (1987).

Adjustments to flavour can be made deliberately, by either selecting for increased contents of desirable compounds, or selecting against the presence of significant amounts of undesirable compounds. Gas–liquid chromatography techniques for assaying the volatile compounds contributing to aroma and flavour have been available for more than 20 years. A refinement such as 'headspace' condensation allows compounds created or destroyed by steam distillation (or cooking) to be identified (Schreyen *et al.*, 1976a, b; Dirinck *et al.*, 1981). Such methods allow an objective measure of quality that would assist selection of genotypes. There is considerable scope for the improvement of flavour by conventional breeding techniques, or by identifying appropriate genes for modification or transfer once biosynthetic pathways have been fully elucidated.

Nutritional value

Assessment of nutritional quality also requires some kind of analysis so that cultivars can be compared. In wheat and triticale, nutritional quality and functional parameters related to milling and baking are closely linked to seed protein composition. Recent progress in this area is reviewed elsewhere (Murray, 1989; Simmonds, 1989).

The often quoted 'need' to improve the protein quality of legume seeds is usually more apparent than real. This impression is based on the false premise that the total seed protein content of legume seeds is invariably deficient in the sulphur-amino acids, cysteine and methionine. This may be true of some legumes, such as cowpea (Murray *et al.*, 1983), but it is certainly not true of chick pea (Jaya & Venkataraman, 1980; Bhatty, 1982; Murray & Roxburgh, 1984). Similarly, the amounts of tryptophan recorded in a recent

Table 1.5. Tryptophan content of total protein extracted from seeds of cultivated legumes and sunflower.

Species	Tryptophan (%, w/w)
Cicer arietinum (chick pea)	0.91
Vicia faba (broad bean)	1.00
Lens culinaris (lentil)	1.02
Vigna unguiculata (cowpea) Vita 3	1.10
Phaseolus vulgaris (common bean)	1.10
Pisum sativum (pea) Telephone	1.26
Vigna mungo (black gram)	1.33
Helianthus annuus (sunflower)	1.36
Cajanus cajan (red gram)	1.56
Pisum sativum, cv. Melbourne Market	1.62
Phaseolus vulgaris, cv. Streamline	1.66
Phaseolus vulgaris, cv. Seafarer	1.82
Vigna radiata (green gram)	1.99

Data of Sastry and Murray (1986).

survey (Table 1.5) indicate that staple legume seeds are already sufficient sources of this essential amino acid.

Tryptophan, lysine, threonine and methionine are likely to be limiting in vegetarian diets reliant on cereal grains or starchy roots or tubers as main sources of protein and energy. Provided a compensating mixture of legume seeds and seedlings complements such a diet, adequate quantities of all the essential amino acids can be obtained (Sastry & Murray, 1986, 1987).

Plants like cassava (*Manihot utilissima* [*M. esculenta*]) cannot be modified to produce roots with high protein content simply by incorporating the gene for a seed protein from some other species. Many different aspects of transport physiology and metabolism would require alteration simultaneously in order to engineer a cassava plant with high-protein roots. Such a breeding objective must still be regarded as fanciful and unlikely to be attained in the near future.

Disease and pest resistance

The genetic basis of resistance

Earlier breeding programmes have often focused on improving resistance to one or more major pathogens, e.g. wheat and various rusts (Johnson *et al.*, 1967; Appels & Lagudah, 1990). Breeding for resistance is still the least contentious area of endeavour for plant breeders, because all that is being contemplated is the reinstatement of natural resistance still found in wild forms but no longer expressed in some or all cultivars. The development of multiple-disease resistance has become a major priority for CGIAR centres (Table 1.1), since this is a cost-effective way 'to assist resource-poor farmers in developing countries' (Nene, 1988).

The assessment of genetic resources available from major collections is a normal prerequisite for breeding programmes aimed at improving disease resistance. Such surveys have produced interesting results. For example, a survey of 23 accessions of nine Portuguese cabbages and kales (*Brassica oleracea*) revealed that all are resistant to downy mildew (*Peronospora parasitica*) and most are resistant to 'cabbage yellows' (*Fusarium oxysporum* f. sp. *conglutinans* race 1), but none is resistant to bacterial blackrot (*Xanthomonas campestris* pv. *campestris*) (Monteiro & Williams, 1989). Susceptibility to downy mildew is common among crop plants, but in *Brassica*, resistance is conferred by production of allyl isothiocyanate from the glucoside sinigrin (Harborne, 1982). Cultivars of *Brassica oleracea* lacking resistance appear to have resulted from selection for milder flavour.

In contrast to constitutive compounds like sinigrin, other plant products conferring resistance to pathogens are not formed until after infection or contact. These are a heterogeneous group termed phytoalexins (see Harborne, 1982; Dixon, 1986; Chapter 11). Isoflavonoids comprise one major group of phytoalexins. Their synthesis following fungal invasion has been monitored inside infected cells (Snyder & Nicholson, 1990). The genotypic difference between susceptibility and resistance is expressed in the rate at which phytoalexins are synthesized – the more rapid the response, the greater the likelihood of successful resistance.

It is an over-simplification to classify resistance as purely qualitative or purely quantitative as Carson & Carson (1989) proposed. Ultimately it will be possible to explain all disease resistance in genetic terms, to isolate genes responsible for components of resistance and to transfer these genes select-ively into a recipient genome. Progress in this direction is a recurring theme throughout this volume.

Avoidance of pesticides

The chief advantage of disease- or pest-resistant crop plants is the elimina-tion of the need, either real or apparent, to apply pesticides. Whether insecticides, miticides (acaricides) or nematicides are considered, similar intractable problems have arisen repeatedly.

Lack of specificity for the target organism means that beneficial organ-isms are also killed. Selection pressure on the target organism means that a few resistant survivors generate resistant progeny. In a short time, resistant genotypes predominate. The working life of each newly devised pesticide is typically 4–5 years (Seymour, 1982), but may be as little as one season. With hindsight, we are told that such and such a compound is now known to be carcinogenic and therefore hazardous to those who applied the compound to the crop or to those who ate contaminated plant or animal products.

All of these negative aspects of intensive agriculture are well illustrated by what has happened with nematicides (Feldmesser *et al.*, 1985). Often the pest nematodes lie deep within the soil, making access the first obstacle for control. Twenty-three chemical nematicides have been tried out in place of steam treatment (Russell, 1961) or crop rotation. Of these, fewer than half are currently registered. 'DBCP', the most widely used nematicide in 1977, was deregistered for all uses in the USA except pineapple when shown to cause sterility and cancer. Methyl bromide, another carcinogen, is restricted to specialty uses. Aldicarb was suspended in April 1990 (*American Vegetable Grower*, May 1990, p. 25). Other effective nematicides, of which there are very few, face restrictions in the future as evidence for mammalian toxicity mounts. Besides these inherent problems, nematicides are expensive both to purchase and to apply.

Breeding for nematode resistance is clearly desirable, but conventional breeding methods have not always proved adequate. For example, potatoes are susceptible to cyst nematodes, including *Globodera pallida* and *G. rostochiensis*. Several diploid wild species of *Solanum* have been shown to be resistant to two pathotypes, P_4A and P_5A, and to transmit resistance to hybrids with diploid *Solanum tuberosum* at better than 50% of progeny (Chavez *et al.*, 1988a, b). However, many cultivated potatoes are tetraploid, and not all the transmitted attributes from wild types are desirable, e.g. excessive solanine production (Fenwick *et al.*, 1990). Selective incorporation of genetic material governing nematode resistance should be possible through the application of molecular techniques.

Difficulties with infertile hybrids hinder the transfer of root-knot nematode resistance from *Cucumis anguria* and *Cucumis metuliferus* to muskmelon, *Cucumis melo* (Fassuliotis & Nelson, 1988). Again, advanced selective techniques could ensure the incorporation and expression of genes conferring nematode resistance into cultivars of *Cucumis melo*.

Conflicting priorities

Natural resistance to aphids is often conferred by glandular hairs (trichomes) present on the epidermis. The secretions of trichomes contain plant products that act as toxins or deterrents (Chapter 10; Turner *et al.*, 1980). Cultivars of plants whose wild relatives possess trichomes may lack these in sufficient density to protect the plant, e.g. tomato (Goffreda *et al.*, 1990).

Introduction of the genetic capacity for producing trichomes on stems and leaves would not have an adverse effect on the fruit surface – smooth-skinned tomatoes would remain that way. Hence the acquisition of this trait by cultivars should have solely beneficial effects, including faster growth and the avoidance of aphid-borne viral diseases without the application of insecticides. However, the same kind of breeding objective applied to sunflower would result in a reduction in yield – not of the seeds, but of the stems, which are often used as animal feed. Smooth stems are preferred for this purpose (Diepenbrock, 1988).

Complex interactions

Will the yields of crop plants in the future be depressed in a synergistic fashion by a combination of damage from exposure to ultraviolet B radiation (280–320 nm) and susceptibility to disease? Genotype-dependent variation in resistance to the growth-suppressing effects of ultraviolet light has been well documented for such plants as soybean (Biggs *et al.*, 1981; Murali *et al.*, 1988) and cucumber (Tevini & Teramura, 1989).

A recent study with several cultivars of cucumber (*Cucumis sativus*) indicates that the effect of ultraviolet radiation on the severity of fungal infection varies according to cultivar, timing of irradiation (pre- or post-spore inoculation), duration of irradiation, strength of inoculum and plant age (Orth *et al.*, 1990). Anthracnose (*Colletotrichum lagenarium*) and cucumber scab (*Cladosporium cucumerinum*) were the two fungal diseases investigated. One cultivar, Straight 8, is susceptible to both diseases, whereas Calypso Hybrid and Poinsette are resistant.

All three cultivars showed increased severity of disease symptoms when seedlings were irradiated prior to inoculation (Fig. 1.1). However, for the normally resistant cultivars, the damage was confined to the cotyledons. The results suggest that cucumbers should indeed be bred for resistance to fungal diseases, and in addition, be protected from ultraviolet exposure by screening, at least during seedling establishment. Increasing intensities of ultraviolet B may mean that cucumbers will need to be grown entirely in glasshouses in the future. How many other field crops will also need to be enclosed?

Genetic engineering – problems and potential

The term 'genetic engineering' has been widely adopted to apply to the deliberate alteration of the genome by biochemical methods, possibly involving incorporation of DNA from prokaryotes, or from any eukaryote, including yeasts, other fungi, algae and mammals. Despite more than 15 years of debate (see Ada, 1979; Langley, 1979), the public perception of this process is often one of mistrust (Hallen, 1990; Hindmarsh, 1990; Phelps, 1990).

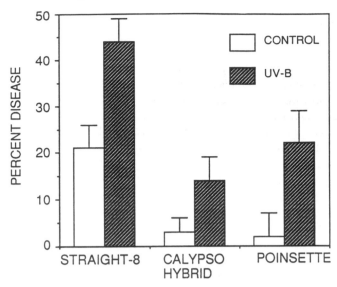

Fig. 1.1. The severity of symptoms resulting from *Colletotrichum lagenarium* infection with and without prior irradiation for 7 days with UV-B (11.6 effective kJ/m^2). Cucumber seedlings were inoculated with 1.0×10^5 spores per ml at the end of the UV exposure and rated 4 days later (18 days after sowing). The main effect of UV treatment is significant with $P \leqslant 0.0001$. From Orth *et al.* (1990).

So far as plants are concerned, I have no doubt that the application of recombinant DNA techniques could be beneficial – if directed towards appropriate breeding objectives. However, in practice, genetic engineering of plants has been directed largely towards goals that fuel public prejudice against the technology. At best, many of these goals seem trivial; at worst, misguided.

Because genes will now be mixed into combinations that could not have occurred without human intervention, the degree of confinement that can be guaranteed for newly incorporated genetic information becomes an important issue. Yet the most frequent application of genetic engineering to plants, that of conferring herbicide resistance, involves the greatest probability of genetic escape.

The problem of herbicide resistance

Herbicides are chemical compounds that kill plants – weeds and crop plants alike. Selectivity and rapid breakdown are not always obtained. Left-over herbicide applied to weeds before a crop is planted can persist in the soil and decrease yield (e.g. soybeans, Bowman *et al.*, 1989).

Crop species 'engineered' to be resistant to one or more herbicides might cope better with pre-sowing herbicide applications and suffer no loss of yield. Oxtoby & Hughes (1989) state that if crop plants were resistant to herbicides, 'weed control by chemicals would be more effective, and this could lead to increased yields and improved agricultural practice'.

This is putting the cart before the horse. Improved agricultural practice must in any event come first. Weed control by chemicals is economically justified only if weed cover is likely to reach about 10% of the crop area (e.g. Wahmhoff & Heitefuss, 1988). Either it is effective, or it is not. The real question at issue is whether engineering crop plants for herbicide resistance will lead to increased or decreased rates of application of synthetic herbicides, compounds whose long-term effects on human health are largely unexplored.

If genetically engineered crop plants are known to have increased herbicide resistance, the temptation for farmers to over-treat with herbicide during pre-sowing applications will be very strong. This has occurred in the past with the organochlorines 2,4-D and 2,4,5-T, leading to problems with dioxin, spread as a contaminant of commercial preparations of 2,4,5-T (Buist, 1989).

There is no guarantee that genes conferring herbicide resistance will remain confined to the crop species in which they are placed. This will depend on the identity of the crop plant and its degree of relatedness to attendant weeds. Almost every field crop has at least one related weed form (Harlan, 1969). In some instances, interbreeding between crop plants and closely related weeds happens routinely. An example is the carrot (*Daucus carota*), which interbreeds readily with the wild carrot. Evidence for this is the occurrence of white carrots with objectionable flavour in crops of orange carrots growing in Europe (Wijnheijmer *et al.*, 1989). Gene transfer occurs in both directions, hence some modifications for herbicide resistance introduced to carrot cultivars could readily be transmitted to the weedy wild carrot.

A field trial of transgenic potatoes resistant to chlorsulfuron provided evidence of low but significant interbreeding and transmission of genetic resistance (Tynan *et al.*, 1990). Among the progeny of wild-type potato plants growing within the trial area, the frequency of chlorsulfuron-resistant seedlings was 1.14%. Among the progeny from wild-type potatoes growing up to 4.5 m from the trial area, the frequency of resistant seedlings fell to 1 in 2000, and between 4.5 and 10 m from the trial, their frequency was zero. The wild potato is not a weed of the cultivated potato, but this demonstration does prove the point that pollen cannot be confined, even though the transgenic plants serving as a source of pollen can be. It must be noted that low rates of cross-pollination are usually found in potatoes, and the degree of 'escape' may well be higher for other species.

Herbicide-resistant weeds are an undesirable consequence of any programme aimed at improving crop plants. Already the useful 'life' of individual herbicides is governed by the rate at which herbicide resistance spreads in target weed populations. Intensive use of a given herbicide can act as a selective agent in favour of herbicide-resistant genotypes, until the herbicide is no longer useful. There is a distinct possibility that breeding for herbicide-resistant crop plants could undermine the use of herbicides altogether.

Unpredictable effects of gene transfer

Techniques for determining the base sequences of DNA or RNA are reliable (Howe & Ward, 1989). In terms of base sequence, it is possible to know exactly what has been incorporated into a recipient genome. However, all the consequences of introducing a novel gene (or genes) into a plant cannot be predicted with absolute certainty. Beversdorf *et al.* (1980) report a study on *Brassica*, where triazine resistance from *Brassica campestris* (turnip) was introduced to *Brassica napus* (oil-seed rape) by sexual crossing and then backcrossing for eight generations with *B. napus* as the male (pollen) parent. In this way the maternally inherited (chloroplast) character of triazine resistance from *B. campestris* was retained in the progeny, but, as an unexpected consequence, the yield of the new plant was 30% less than that of the original *B. napus*.

In another instance, resistance to the herbicide glyphosate, involving a single enzyme, was successfully transferred from the bacterium *Salmonella typhimurium* to tomato (Comai *et al.*, 1985). However, the degree of resistance was insufficient to allow the modified tomato plants to tolerate glyphosate spraying at commercial concentrations without marked reduction in growth. It is not only the incorporation of a new gene into a genome that is important, but exactly where it is located, since this will govern the timing, the tissue specificity and the degree of its expression.

Given that the unpredictability associated with gene modification is more likely to involve under-expression than over- (or even adequate) expression, how valid are fears that genetically engineered crop plants might turn out to be 'super weeds', competing successfully against members of surrounding plant communities and taking over? On scientific grounds, this possibility must be considered to be extremely remote. There are many characters that disadvantage cultivated plants in the wild. Merely accumulating all known pest resistances into one genotype would not overcome serious reproductive disadvantages, such as the lack of seedcoat-imposed dormancy (Murray, 1984), and the lack of effective seed dispersal mechanisms. Without these attributes a seedbank of dormant seeds could not be built up in the soil.

Any genetically engineered crop plant that did revert to wild-type for mechanism of seed dispersal would be identified during trials and never released. The degree of care required is no different from that which is normally exercised when weeds are maintained in a collection (Harlan, 1984).

Regulations and variety rights – who benefits?

The regulations required for the oversight of recombinant DNA technology and the introduction to agriculture of suitably modified plants are still being refined in several countries. In Australia a Federal Government Inquiry is in progress at present. Procedures for risk assessment in advance of any release will need to be established, along with an administration responsible for recording and monitoring released organisms. Such procedures are well established in several OECD countries.

The protection of genetically engineered plants by plant variety rights legislation is supposed to encourage investment in plant improvement on the part of the private sector in industrialized countries. Brumby *et al.* (1990) have identified intellectual property rights as a major constraint to the transfer of biotechnology to developing countries. Many of these countries refuse to recognize the alienation of a common heritage implicit in plant 'patenting' and variety rights. It may be argued that the world's plants and their genetic diversity should represent a communal resource, a shared inheritance that should be transmitted without impediment to future generations. Attempts to protect 'intellectual property' are also perceived as a mechanism for perpetuating economic inequality.

This is not a debate that will be settled quickly. The breeding priorities of the transnational agrichemical and seed companies in the private sector are obviously different from those of the publicly funded plant breeding institutions. It is the latter that must be relied upon to cater for genuine improvements in agriculture to benefit the greatest numbers of people in industrialized and developing countries alike.

Acknowledgements

It is a pleasure to record my thanks to those authors who sent reprints in response to my requests for information, and to those who gave permission for reproduction of data.

References

Ada, G. L. (1979) Genetic engineering research: evaluation and containment of potential risks. In: White, F. G. (ed.), *Scientific Advances and Community Risk*. Australian Academy of Science, Canberra, ACT, pp. 7–20.

Aitken, Y. (1974) *Flowering Time, Climate and Genotype*. Melbourne University Press, Melbourne.

Alba, E. & Greco, I. (1981) Problems in simultaneous breeding for oil and protein content in sunflower. In: Bunting, E. S. (ed.), *Production and Utilization of Protein in Oilseed Crops*. M. Nijhoff, The Hague, Boston and London, pp. 43–9.

Appels, R. & Lagudah, E. S. (1990) Manipulation of chromosomal segments from wild wheat for the improvement of bread wheat. *Australian Journal of Plant Physiology* 17, 253–66.

Austin, R. B., Ford, M. A., Edrich, J. A. & Blackwell, R. D. (1977) The nitrogen economy of wheat. *Journal of Agricultural Science* 88, 159–67.

Beversdorf, W. D., Weiss-Lerman, J., Erickson, L. R. & Souza-Machado, Z. (1980) Transfer of cytoplasmically inherited triazine resistance from bird's rape to cultivated oil seed rape (*Brassica campestris* and *B. napus*). *Canadian Journal of Genetics and Cytology* 22, 167–72.

Bhatia, C. R. (1975) Criteria for early generation selection in wheat breeding programmes for improving protein productivity. *Euphytica* 24, 789–94.

Bhatty, R. S. (1982) Albumin proteins of eight edible grain legume species: electrophoretic patterns and amino acid composition. *Journal of Agricultural and Food Chemistry* 30, 620–2.

Biggs, R. H., Kossuth, S. V. & Teramura, A. H. (1981) Response of 19 cultivars of soybeans to ultraviolet B irradiance. *Physiologia Plantarum* 53, 19–26.

Blanchet, R., Marty, J.-R., Merrien, A. & Puech, J. (1981) Main factors limiting sunflower yield in dry areas. In: Bunting, E. S. (ed.), *Production and Utilization of Protein in Oilseed Crops*. M. Nijhoff, The Hague, Boston and London, pp. 205–26.

Bowman, J. E., Hartman, G. L., McClary, R. D., Sinclair, J. B., Hummel, J. W. & Wax, L. M. (1989) Effect of row spacing, tillage and herbicides on seed quality in rotated and continuous soybeans. *Seed Science and Technology* 17, 531–42.

Brumby, P., Pritchard, A. J. & Persley, G. J. (1990) In: Persley, G. J. (ed.), *Agricultural Biotechnology: Opportunities for International Development*. CAB International, Wallingford, Oxon, pp. 429–36.

Buist, R. (1989) *Food Chemical Sensitivity*. Collins, Sydney, Australia.

Carasco, J. F., Croy, R., Derbyshire, E. & Boulter, D. (1978) The isolation and characterization of the major polypeptides of the seed globulin of cowpea (*Vigna unguiculata* L. Walp.) and their sequential synthesis in developing seeds. *Journal of Experimental Botany* 29, 309–23.

Carson, S. D. & Carson, M. J. (1989) Breeding for resistance in forest trees – a quantitative genetic approach. *Annual Review of Phytopathology* 27, 373–95.

Chavez, R., Jackson, M. T., Schmiediche, P. E. & Franco, J. (1988a) The importance of wild potato species resistant to the potato cyst nematode, *Globodera pallida*, pathotypes P_4A and P_5A, in potato breeding. I. Resistance studies. *Euphytica* 37, 9–14.

Chavez, R., Jackson, M. T., Schmiediche, P. E. & Franco, J. (1988b) The importance of wild potato species resistant to the potato cyst nematode, *Globodera pallida*, pathotypes P_4A and P_5A, in potato breeding. II. The crossability of resistant species. *Euphytica* 37, 15–22.

Comai, L., Facciotti, D., Hiatt, W. R., Thompson, G., Rose, R. E. & Stalker, D. M. (1985) Expression in plants of a mutant *aroA* gene from *Salmonella typhimurium* confers tolerance to glyphosate. *Nature* 317, 741–4.

Corke, H., Avivi, N. & Atsmon, D. (1989) Pre- and post-anthesis accumulation of dry matter and nitrogen in wild barley (*Hordeum spontaneum*) and in barley cultivars (*H. vulgare*) differing in final grain size and protein content. *Euphytica* 40, 127–34.

Daly, J. L. (1989) *The Greenhouse Trap*. Bantam Books, Transworld Publishers (Australia) Pty Ltd, Sydney.

Desai, R. M. & Bhatia, C. R. (1978) Nitrogen uptake and nitrogen harvest index in durum wheat cultivars varying in their grain protein concentration. *Euphytica* 27, 561–6.

Diepenbrock, W. (1988) Yield development in the sunflower – a survey. In: *Plant Research and Development*, vol. 27. Institute for Scientific Co-operation, Tübingen, Germany, pp. 38–58.

Dirinck, P. J., De Pooter, H. L., Willaert, G. A. & Schamp, N. M. (1981) Flavor quality of cultivated strawberries: the role of the sulfur compounds. *Journal of Agricultural and Food Chemistry* 29, 316–21.

Dixon, R. A. (1986) The phytoalexin response: elicitation, signalling and control of host gene expression. *Biological Reviews* 61, 239–91.

Evans, L. T. (1969) *The Induction of Flowering*. Macmillan, South Melbourne.

Evans, L. T. (1975) *Crop Physiology*. Cambridge University Press, Cambridge, United Kingdom.

Fassuliotis, G. & Nelson, B. V. (1988) Interspecific hybrids of *Cucumis metuliferus* × *C. anguria* obtained through embryo culture and somatic embryogenesis. *Euphytica* 37, 53–60.

Feldmesser, J., Kochansky, J., Jaffe, H. & Chitwood, D. (1985) Future chemicals for control of nematodes. In: Hilton, J. L. (ed.), *Agricultural Chemicals of the Future*, BARC Symposium 8, Rowman and Allanheld, Totowa, pp. 327–44.

Fenwick, G. R., Johnson, I. T. & Hedley, C. L. (1990) Toxicity of disease-resistant plant strains. *Trends in Food Science and Technology* July, 23–5.

Fernandez, G. C. J. & Chen, H. K. (1989) Temperature and photoperiod influence reproductive development of reduced-photoperiod-sensitive mungbean genotypes. *Journal of the American Society for Horticultural Science* 114, 204–9.

Fernandez, G. C. J. & Miller, J. C. Jr (1985a) Estimation of heritability by parent–offspring regression. *Theoretical and Applied Genetics* 70, 650–4.

Fernandez, G. C. J. & Miller, J. C. Jr (1985b) Yield component analysis in five cowpea cultivars. *Journal of the American Society for Horticultural Science* 110, 553–9.

Fernandez, G. C. J. & Miller, J. C. Jr (1987) Analysis of host genotype × rhizobial strain interaction in N₂ fixation. *Euphytica* 36, 903–11.

Fernandez, G. C. J., Chen, H. K. & Miller, J. C. Jr (1989) Adaptation and environmental sensitivity of mungbean genotypes evaluated in the international mungbean nursery. *Euphytica* 41, 253–61.

Foy, C. D., Lafever, H. N., Schwartz, J. W. & Fleming, A. L. (1974) Aluminium tolerance of wheat cultivars related to region of origin. *Agronomy Journal* 66, 751–7.

Geyle, A. (1990) Growing stonefruit. *Commercial Horticulture* Winter, 27–8.

Goffreda, J. C., Szymkowiak, E. J., Sussex, I. M. & Mutschler, M. A. (1990) Chimeric tomato plants show that aphid resistance and triacylglucose production are epidermal autonomous characters. *Plant Cell* 2, 643–9.

Hallen, P. (1990) Genetic engineering: 'miracle of deliverance' or 'destroyer of worlds'? *Habitat* 18(1), 9–12.

Harborne, J. B. (1982) *Introduction to Ecological Biochemistry,* 2nd edn. Academic Press, London.

Harlan, J. R. (1969) Evolutionary dynamics of plant domestication. *Japanese Journal of Genetics* 44, Supplement 1, 337–43.

Harlan, J. R. (1984) Evaluation of wild relatives of crop plants. In: Holden, J. H. W. & Williams, J. T. (eds), *Crop Genetic Resources: Conservation and Evaluation.* George Allen and Unwin, London, pp. 212–22.

Harlan, J. R., de Wet, J. M. J. & Price, E. G. (1973) Comparative evolution of cereals. *Evolution* 27, 311–25.

Hawkes, J. G. (1990) *The Potato – Evolution, Biodiversity and Genetic Resources.* Belhaven, London.

Heiser, C. B. Jr (1976) *The Sunflower.* University of Oklahoma Press, Norman.

Hindmarsh, R. (1990) Biotechnology – a challenge for the green movement. *Habitat* 18(5), 9–12.

Holden, J. H. W. & Williams, J. T. (1984) *Crop Genetic Resources: Conservation and Evaluation.* George Allen and Unwin, London.

Howe, C. J. & Ward, E. S. (1989) *Nucleic Acid Sequencing – A Practical Approach.* Oxford University Press, Oxford, New York, Tokyo.

Jaya, T. V. & Venkataraman, L. V. (1980) Effect of germination on the nitrogenous constituents, essential amino acids, carbohydrates, enzymes and anti-nutritional factors in chickpea and greengram. *Indian Food Packer* 34, 3–11.

Johnson, V. A., Mattern, P. J. & Schmidt, J. W. (1967) Nitrogen relations during spring growth in varieties of *Triticum aestivum* L. differing in grain protein content. *Crop Science* 7, 664–7.

Khan, M. R. I., Gatehouse, J. A. & Boulter, D. (1980) The seed proteins of cowpea (*Vigna unguiculata* L. Walp.). *Journal of Experimental Botany* 31, 1599–611.

Krishna, T. G. & Mitra, R. (1988) The probable genome donors to *Arachis hypogaea* L. based on arachin seed storage protein. *Euphytica* 37, 47–52.

Krishna, T. G. & Murray, D. R. (1988) Effects of cycloheximide and actinomycin D on glycosidase activities in the cotyledons of legume seeds following imbibition. *Journal of Plant Physiology* 132, 745–9.

Ladizinsky, G. (1979a) Seed dispersal in relation to the domestication of middle east legumes. *Economic Botany* 33, 284–9.

Ladizinsky, G. (1979b) Species relationships in the genus *Lens* as indicated by seed-protein electrophoresis. *Botanical Gazette* 140, 449–51.

Langley, B. W. (1979) Genetic engineering research: an industrial viewpoint. In: White, F. W. G. (ed.), *Scientific Advances and Community Risk.* Australian Academy of Science, Canberra, ACT, pp. 21–32.

Marx, G. A. (1983) Developmental mutants in some annual seed plants. *Annual Review of Plant Physiology* 34, 389–417.

Miller, J. C. Jr, Zary, K. W. & Fernandez, G. C. J. (1986) Inheritance of N_2-fixation efficiency in cowpea. *Euphytica* 35, 551–60.

Miller, P. (1990) Apple extra. *Commercial Horticulture* Winter, 6–8.

Monteiro, A. A. & Williams, P. H. (1989) The exploration of genetic resources of Portuguese cabbage and kale for resistance to several *Brassica* diseases. *Euphytica* 41, 215–25.

Murali, N. S., Teramura, A. H. & Randall, S. K. (1988) Response differences between two soybean cultivars with contrasting UV-B radiation sensitivities. *Photochemistry and Photobiology* 48, 653–7.

Murray, D. R. (1984) The seed and survival. In: Murray, D. R. (ed.), *Seed Physiology*, vol. 1, *Development*. Academic Press, Sydney, pp. 1–40.

Murray, D. R. (1987) Nutritive role of seedcoats in developing legume seeds. *American Journal of Botany* 74, 1122–37.

Murray, D. R. (1988) *Nutrition of the Angiosperm Embryo*. Research Studies Press, Taunton, United Kingdom.

Murray, D. R. (1989) *Biology of Food Irradiation*. Research Studies Press, Taunton, United Kingdom.

Murray, D. R. & Roxburgh, C. McC. (1984) Amino acid composition of the seed albumins from chickpea. *Journal of the Science of Food and Agriculture* 35, 893–6.

Murray, D. R., Mackenzie, K. F., Vairinhos, F., Peoples, M. B., Atkins, C. A. & Pate, J. S. (1983) Electrophoretic studies of the seed proteins of cowpea, *Vigna unguiculata* (L.) Walp. *Zeitschrift für Pflanzenphysiologie* 109, 363–70.

Nenc, Y. L. (1988) Multiple-disease resistance in the grain legumes. *Annual Review of Phytopathology* 26, 203–17.

Orth, A. B., Teramura, A. H. & Sisler, H. D. (1990) Effects of ultraviolet-B radiation on fungal disease development in *Cucumis sativus*. *American Journal of Botany* 77, 1188–92.

Oxtoby, E. & Hughes, M. A. (1989) Breeding for herbicide resistance using molecular and cellular techniques. *Euphytica* 40, 173–80.

Pate, J. S. (1984) The carbon and nitrogen nutrition of fruit and seed – case studies of selected grain legumes. In: Murray, D. R. (ed.), *Seed Physiology*, vol. 1, *Development*, Academic Press, Sydney, pp. 41–82.

Phelps, R. (1990) Genetic engineering – a leap into the unknown. *Consuming Interest* No. 45, 4–8.

Rawson, H. M. & Constable, G. A. (1980) Carbon production of sunflower cultivars in field and controlled environments. I. Photosynthesis and transpiration of leaves, stems and heads. *Australian Journal of Plant Physiology* 7, 555–73.

Rawson, H. M., Constable, G. A. & Howe, G. N. (1980) Carbon production of sunflower cultivars in field and controlled environments. II. Leaf growth. *Australian Journal of Plant Physiology* 7, 575–86.

Röbbelen, G. (1981) Potential and restrictions of breeding for protein improvement in rapeseed. In: Bunting, E. S. (ed.), *Production and Utilization of Protein in Oilseed Crops*. M. Nijhoff, The Hague, Boston and London, pp. 3–16.

Russell, E. J. (1961) *The World of the Soil*. Fontana, Collins, London and Glasgow.

Sastry, M. C. S. & Murray, D. R. (1986) The tryptophan content of extractable seed protein from cultivated legumes, sunflower and *Acacia*. *Journal of the Science of Food and Agriculture* 37, 535–8.

Sastry, M. C. S. & Murray, D. R. (1987) The contribution of trypsin inhibitors to the nutritional value of chick pea seed protein. *Journal of the Science of Food and Agriculture* 40, 253–61.

Schneider, S. (1990) Prudent planning for a warmer planet. *New Scientist* 128 (no. 1743), 39–41.

Schreyen, L., Dirinck, P., Van Wassenhove, F. & Schamp, N. (1976a) Volatile flavor components of leek. *Journal of Agricultural and Food Chemistry* 24, 336–41.

Schreyen, L., Dirinck, P., Van Wassenhove, F. & Schamp, N. (1976b) Analysis of leek volatiles by headspace condensation. *Journal of Agricultural and Food Chemistry* 24, 1147–52.

Schuster, W. & Böhm, J. (1981) Experience in soyabean breeding in middle Europe. In: Bunting, E. S. (ed.), *Production and Utilization of Protein in Oilseed Crops.* M. Nijhoff, The Hague, Boston and London, pp. 158–76.

Seymour, J. (1982) Integrated control of orchard mites. *Rural Research* 116 Spring, 15–19.

Simmonds, D. H. (1989) *Wheat and Wheat Quality in Australia.* CSIRO, Melbourne.

Singer, S. R., Hsiung, L. P. & Huber, S. H. (1990) Determinate (*det*) mutant of *Pisum sativum* (Leguminosae, Papilionoideae) exhibits an indeterminate growth pattern. *American Journal of Botany* 77, 1330–5.

Snyder, B. A. & Nicholson, R. L. (1990) Synthesis of phytoalexins in *Sorghum* as a site-specific response to fungal ingress. *Science* 248, 1637–9.

Tevini, M. & Teramura, A. H. (1989) UV-B effects on terrestrial plants. *Photochemistry and Photobiology* 50, 479–87.

Tomlins, G. (1987) A volunteer's experiences in Africa. *Biologist* 34(1), 49–51.

Tucker, G. B. (1981) *The CO₂–Climate Connection.* Australian Academy of Science, Canberra, ACT.

Turner, J. C., Hemphill, J. K. & Mahlberg, P. G. (1980) Trichomes and cannabinoid content of developing leaves and bracts of *Cannabis sativa* L. (Cannabaceae). *American Journal of Botany* 67, 1397–406.

Tynan, J. L., Williams, M. K. & Conner, A. J. (1990) Low frequency of pollen dispersal from a field trial of transgenic potatoes. *Journal of Genetics and Breeding* (in press).

Wahmhoff, W. & Heitefuss, R. (1988) Studies on the use of economic injury thresholds for weeds in winter barley. In: *Plant Research and Development,* vol. 27. Institute for Scientific Co-operation, Tübingen, Germany, pp. 59–91.

Waines, J. G. (1975) The biosystematics and domestication of pea (*Pisum* L.). *Bulletin of the Torrey Botanical Club* 102, 385–95.

Wijnheijmer, E. H. M., Brandenburg, W. A. & Ter Borg, S. J. (1989) Interactions between wild and cultivated carrots (*Daucus carota* L.) in the Netherlands. *Euphytica* 40, 147–54.

Worrest, R. C. & Caldwell, M. M. (1986) *Stratospheric Ozone Reduction, Solar Ultraviolet Radiation and Plant Life.* Springer, Berlin.

Zohary, D. (1969) The progenitors of wheat and barley in relation to domestication and agricultural dispersal in the Old World. In: Ucko, P. J. & Dimbleby, D. W. (eds), *The Domestication and Exploitation of Plants and Animals.* Duckworth, London, pp. 47–66.

Zubriski, J. C. & Zimmerman, D. C. (1974) Effects of nitrogen, phosphorus and plant density on sunflower. *Agronomy Journal* 66, 798–801.

Chapter 2
Chromosomal Organization and Gene Mapping

Norman F. Weeden

Department of Horticultural Sciences, New York State
Agricultural Experiment Station, Cornell University,
Geneva, New York 14456, USA

Introduction

The association of the heritable material in germ cells with chromosomes (Sutton, 1903) marked an important milestone in the science of genetics. Biologists had thus identified a structure within which Johannsen's abstract concept of a gene (Johannsen, 1909) could begin to take on physical reality. The term 'gene' came to mean the 'ultimate unit of inheritance' and much of the science of genetics was soon dedicated to understanding the structure of these units and their organization within the chromosomes. Numerous models were proposed and tested. The genetic material turned out to be deoxyribonucleic acid rather than the initial suspect, protein. Astoundingly, the information was encoded by just four bases in a delightfully efficient mechanism that was not at all obvious at the beginning of the century.

As genes began to be assembled into linkage groups and the structure of individual genes began to be probed, unanticipated results were frequently obtained. Initial models such as beads on a string (Morgan, 1926) or the one gene, one enzyme hypothesis (Beadle & Tatum, 1941) have required significant modifications. Serious workers in the field might have suspected the existence of regulatory as well as structural genes before Jacob & Monod (1961) developed the operon model, but it is doubtful whether anyone predicted the presence of introns before the sequencing of the haemoglobin gene in rabbit (Jeffreys & Flavell, 1977). The current model of eukaryotic genome organization is not as aesthetically pleasing or comprehensible as the genetic code. Furthermore, the concept of a gene has had to be modified to such an extent, even since Watson & Crick (1953) gave a molecular construct to Johannsen's concept, that a simple definition is now impossible.

Although the structure and organization of the eukaryote genome is much more complex than originally envisioned, techniques for analysing the heritable material continue to improve. The field of genetics is progressing rapidly, particularly in the investigation of the human genome. Fortunately, new insights gained through studies of human DNA are usually directly applicable to plant genetics, and many techniques (such as controlled crosses with large F_2 populations) are impossible in human genetics but are taken for granted in plant breeding.

Species such as *Arabidopsis thaliana* are serving as the *Drosophila* of the plant kingdom. The capabilities for culturing mutated tissues and transforming single cells permit many experiments on *Arabidopsis*, tobacco and other model species to be designed similar to procedures used with fungi and bacteria. Progress, therefore, can be made at a comparable rate. Although detailed information on plant genome organization is presently available from relatively few species, much more data will be reported in the near future. We may expect that in many areas plant genetics will continue to play a major role in the advancement of scientific knowledge.

The genome of angiosperms

Repetitive DNA

One of the most startling features of the eukaryotic genome is its tremendous diversity in size. Even if we restrict our sample to diploid plant taxa the DNA content can vary 100-fold. Significant changes in DNA content per haploid complement can occur within a genus, and species with low DNA contents such as *Arabidopsis thaliana*, *Vigna radiata*, *Rosa wichuriana* and *Rumex sanguineus* are found in widely disparate taxa. Increases in DNA content at the diploid level are not usually accompanied by changes in the number of genes being expressed. For instance, *Vicia faba* and *Vicia sativa* express the same number of isozyme loci despite a sevenfold change in haploid DNA content (C value) (Rees *et al.*, 1966). The C value can change even without a significant alteration in karyotype as exemplified by two species of onion, one with twice the C value of the other (Jones & Rees, 1969).

Such variation in DNA content among diploid plants is caused by differences in the amount of repetitive DNA sequences in the genome. This repetitive DNA is usually present in the eukaryotic genome in both a clustered and a 'short-period' interspersed arrangement (Davidson *et al.*, 1975; Flavell *et al.*, 1977; Murray *et al.*, 1978). The classes are not mutually exclusive, but different types of DNA repeats tend to predominate in each class. Clustered repeats include the ribosomal genes and centromeric and telomeric sequences, whereas most of the interspersed repeats are referred to as variable-number tandem repeats (VNTR) (Nakamura *et al.*, 1987) or 'minisatellite' DNA (Jeffreys *et al.*, 1985a).

Both the 45S and 5S ribosomal genes have been well characterized in a number of plant species (Mascia *et al.*, 1981; Choumane & Heizmann, 1988; Kavanaugh & Timmis, 1988; Zimmer *et al.*, 1988; Dobrowolski *et al.*, 1989; Cordesse *et al.*, 1990; and reviews by Appels & Honeycutt, 1986, and Rogers & Bendich, 1987). The internal structure of the 45S repeat unit consists of highly conserved and relatively variable regions, making this repeat one of the most useful for mapping studies. The highly conserved region ensures that the clone from garden pea or maize will hybridize to virtually all other angiosperm rDNA repeat units. The highly variable spacer region will usually reveal polymorphisms within experimental or wild populations (Saghai-Maroof *et al.*, 1984; Zimmer *et al.*, 1988; Cordesse *et al.*, 1990). Thus, rDNA clusters are often the first genes investigated when one is developing a linkage map based on DNA markers (Saghai-Maroof *et al.*, 1984; Polans *et al.*, 1986; Vallejos *et al.*, 1986). Locating these clusters on a linkage map has the additional advantage that linkage groups containing the rDNA genes can immediately be placed on cytogenetic maps because they represent those chromosomes containing the nucleolar organizer regions.

Telomeric DNA has presented a puzzle to molecular biologists ever since the demonstration that DNA polymerases have an absolute requirement for a 3' primer. How do the very ends of the chromosomes get primed if there is no upstream sequence to act as a primer? The fascinating answer to this question has been gradually elucidated by Blackburn and co-workers (1989). The structure of the telomere is, as one might guess, unique. One DNA strand is rich in G residues, and this strand forms a 3' overhang of 12–16 nucleotides (Zakian, 1989). The high G content may permit these nucleotides to fold back on themselves, the structure being stabilized by inter-base hydrogen bonds. Replication of this strand involves the addition of entire repeat units by a special enzyme called a 'telomerase'. This enzyme appears to function much like a reverse transcriptase because it has an RNA template as part of its structure (Greider & Blackburn, 1989). The telomerase may not be present in all cells, and the chromosomes in some somatic tissues may actually shorten with age (Hastie *et al.*, 1990).

A particularly interesting type of repetitive DNA is the moderately repetitive class distributed throughout the genome. Sequences in this category are found as short segments interspersed between stretches of single-copy DNA. Such sequences played havoc with early DNA mapping techniques using genomic libraries because a cloned piece of DNA containing such a sequence hybridizes to hundreds of restriction fragments on a Southern blot, making interpretation of the pattern impossible. Many large (> 1 kb) genomic clones will contain such a sequence, and a genomic library must be prescreened to eliminate such clones. This prescreening also removes other highly repetitive DNA such as ribosomal and cpDNA inserts, but often means that one is discarding 90% of the recombinant clones being analysed.

To get around this tedious and frustrating procedure researchers have used certain enzymes that will not cut at restriction sites where certain bases (particularly cytosine in a 'CpG' nucleotide sequence) have been methylated. Such methylation is presumed to occur primarily on non-transcribed or 'inactive' DNA. Thus, by restricting plant DNA with an enzyme such as *Pst*I, and selecting only the 500–3000 bp size fraction for ligation into a vector, one can theoretically eliminate most of the larger, uncut repetitive sequences from the library. This approach has worked well in tomato (Zamir & Tanksley, 1988), potato (Gebhardt *et al.*, 1989), soybean (Keim & Shoemaker, 1988) and common bean (C. E. Vallejos, unpublished data).

Interspersed repetitive DNA evolves rapidly (Flavell, 1982; Evans *et al.*, 1983), and for many species, repetitive sequences specific to that taxon have been isolated (Sakowicz *et al.*, 1986; Koukalova *et al.*, 1989). Indeed, genome-specific clones have been developed for four of the six genome types of rice (Zhao *et al.*, 1989), for the A, B and D genomes of hexaploid wheat (Chao *et al.*, 1990) and for the C genome of *Avena* (Fabijanski *et al.*, 1990).

The rapid evolution of these sequences also permits an alternative method for circumventing problems associated with clones containing them. Although the clone may not be useful for mapping work on the particular species from which the DNA was derived, it can make an excellent probe for a related species or genus in which the interspersed repetitive DNA has diverged significantly. Assuming the clone also contains a conserved low-copy sequence, such as a portion of a structural gene, it should hybridize to only this more conserved sequence in the genome of the related organism. We have found that many genomic probes from the garden pea (*Pisum sativum*) containing repetitive DNA will give simple 1–3 banded patterns in lentil (*Lens culinaris*) when high-stringency washes are used (unpublished results).

Organization of coding sequences

The eukaryotic haploid genome can be envisioned as consisting of extremely long strands of DNA covered with histones and other proteins, the entire structure being observable as chromosomes. Usually repetitive sequences are present on the strand, including those defining the centromeres and telomeres as well as the interspersed repeats characteristic of the species. A cluster of tandemly repeated ribosomal repeat units is usually associated with a restriction on a chromosomal arm. Much of the remainder of the genome consists of single- or low-copy sequences coding for the protein machinery required for cellular function. Are these genes organized in an interpretable pattern, allowing the prediction of the location of genes based on their function, DNA composition or any other characteristic?

Work by Bernardi and co-workers has shown that both animal and plant genomes include long regions (> 300 kb) of remarkably uniform GC content (Bernardi *et al.*, 1985; Salinas *et al.*, 1988; Montero *et al.*, 1990). These regions are called isochores, and genes present in these regions display a very similar GC content to the flanking sequences. The isochores are much larger than individual genes and are not affected by the differences in GC composition characteristic of introns and exons (see further discussion below). In other words, either both introns and exons have higher GC contents in isochores of high GC content or more coding sequences are clustered in high GC regions because coding sequences are richer in G and C nucleotides than non-coding segments (Montero *et al.*, 1990).

When the linkage map for tomato was beginning to take shape the eminent tomato geneticist and breeder, C. M. Rick, noted that the mapped loci appeared to fall into clusters (Rick, 1971). Lima-de-Faria (1983) further championed the idea of genes being distributed non-randomly along chromosomes, citing evidence in tomato, garden pea and barley, as well as certain non-plant genomes such as *Drosophila*. Could the presence of isochores provide an explanation for this apparent clustering of genes?

Unfortunately, isochores probably are not large enough to account for a gene cluster much over 0.5 cM[1] in size, much smaller than the clusters identified in tomato. One could speculate about 'clusters' of isochores, but as yet there are insufficient data concerning isochore distribution along chromosomes. More pertinent to this question is the extensive mapping of the tomato, pea and barley genomes done since Rick and Lima-de-Faria offered their hypotheses. Working with random cDNA clones as well as sequences known to code for specific genes, geneticists have developed much finer scale linkage maps for tomato (Bernatzky & Tanksley, 1986b; Helentjaris *et al.*, 1986; Vallejos *et al.*, 1986). These maps do not reveal an obvious clustering of structural genes. Similarly, work with isozyme and DNA sequences encoding known proteins in garden pea failed to demonstrate any clear clustering of these loci (Casey *et al.*, 1986; Weeden & Wolko, 1990). At present we must conclude that there is very little evidence suggesting that certain regions (5–20 cM) of the plant genome are packed with structural genes or other important loci while other areas are devoid of such genes.

This conclusion should not be interpreted to mean that minor clusters of genes do not exist in the plant genome. The data argue much to the contrary. Many proteins, such as the major chlorophyll *a*/*b*-binding proteins, the small subunit of ribulose bisphosphate carboxylase, actin, seed proteins, esterases

---

[1] For the remainder of this chapter the centiMorgan (cM) is used as the standard unit of linkage distance. The centiMorgan is the preferred unit for indicating map distances in sexually reproducing organisms, and for short distances, is equal to the frequency of recombination between markers.

and leghaemoglobin, have been shown to be coded by small families of genes, and usually at least a portion of the genes within the family are tightly linked. Certainly, the ribosomal genes are an excellent example of clustered genes. In mapping studies, however, these clusters act as single Mendelian units rather than genes dispersed over 5–20 cM.

Other evidence for organization of the genome

Apart from these nearly tandemly repeated clusters of homologous loci, can we find other indications that genes are organized into larger units on the basis of function? In lower eukaryotes there exist numerous cases in which polycistronic messages are transcribed. Current evidence in multicellular plants suggests that this type of organization does not exist or is very rare. Table 2.1 presents the distribution of several types of functionally related genes in the garden pea. Five instances of possible clustering are evident. Two of these, the four loci on chromosome 7 coding plastid-specific enzymes, and the five loci on the same chromosome involved in carbohydrate metabolism (some overlap between the two groups exists), can be eliminated from further consideration because the respective loci are well distributed along the chromosome.

The large number of chlorophyll mutants mapped to chromosome 1 may be deceptive, for allelism tests have yet to be performed among all mutants and those few that are accurately mapped do not cluster in one region of the chromosome (Blixt, 1974). Two regions, one on chromosome 2 and one on chromosome 6, contain a gene or genes conferring resistance to several viruses (Provvidenti, 1987). The multiple virus resistances are inherited as a single Mendelian unit at each site. However, recombinants appear to have been produced in the wild, suggesting that these sites may consist of several clustered genes, each conditioning resistance to a separate virus.

Table 2.1. Distribution of functionally related genes on the garden pea linkage map.

Function	No. of genes on each chromosome that relate to function						
Chromosome:	1	2	3	4	5	6	7
Anthocyanin production	3	1	3	2	4	0	1
Male sterility	2	0	1	1	2	1	0
Chlorophyll synthesis	9	2	4	3	2	0	1
Nodule formation	5	1	0	1	1	2	3
Plastid-specific proteins	1	1	0	1	1	2	4
Flowering response	1	0	2	0	0	1	1
Virus resistance	0	3	1	0	0	3	0
Carbohydrate metabolism	2	2	1	1	2	1	5

The final possible cluster consists of genes involved in the nodulation response. LaRue and co-workers (Kneen & LaRue, 1988; Weeden *et al.*, 1991) have isolated 15 *sym* mutants that either fail to nodulate or have an abnormal nodulation response. Four of these mutants map to a small region on chromosome 1, very near the major cluster of leghaemoglobin genes (Weeden *et al.*, 1991). None of the *sym* mutants mapping to this region appears to be defective for leghaemoglobin (i.e. when nodules form they appear to function normally), suggesting that four non-complementing *sym* genes plus a cluster of several sequences homologous to pea cDNA for leghaemoglobin exist in a region of less than 20 cM in length. Whether this clustering has any functional significance has yet to be determined.

Genes related to chloroplast function

The nuclear genes coding plastid-specific proteins are a particularly interesting set of functionally related genes because there exist theoretical arguments for expecting these to show clustering. Several studies (Weeden, 1981; Cseke *et al.*, 1982; Weeden *et al.*, 1982; Scott & Timmis, 1984; Martin & Cerff, 1986; Shih *et al.*, 1986) have now demonstrated that many if not most of these genes originated in the genome of the prokaryotic endosymbiont that evolved into the chloroplast. At some time during the evolution of this organelle, the genes were transferred to the nuclear genome. The mechanism of transfer of these genes is still uncertain, but it is unlikely that single genes were neatly excised from the endosymbiont's genome and integrated into nuclear DNA. Rather, small or even large segments were probably released into the cytoplasm, perhaps when one of several 'protoplastids' in a primitive plant cell was broken or lysed, and dispersed into the nucleus, eventually to be incorporated into the nuclear DNA.

If such were the actual mechanism of transfer, one would expect flanking sequences adjacent to genes as those coding the plastid-specific glucosephosphate isomerase or glyceraldehyde-3-phosphate dehydrogenase also to consist of prokaryotic-derived sequences. The sequencing of flanking regions of the chlorophyll *a/b*-binding protein genes in tomato (Pichersky *et al.*, 1985) has revealed additional *cab* pseudogenes, but not other plastid-related sequences that would have been transferred simultaneously with the *cab* genes. However, other studies on DNA found both in the plastid genome and in the nucleus suggest that much of this DNA was transferred in segments of 2 kilobases or larger, and thus could have contained more than a single gene (Scott & Timmis, 1984).

Mapping of nuclear genes

The mapping of plant genomes has recently become one of the most rapidly expanding areas of research in plant genetics, proceeding at a rate unimaginable a few years ago. Even minor crops will soon have extensive linkage maps. As a counterpart to the human genome mapping project, plant geneticists will undoubtedly fully sequence the *Arabidopsis* genome, an immense project but technically feasible.

The dramatic explosion in genetic mapping capabilities has occurred as a result of improvements in techniques for handling and analysing DNA molecules. The most widely known and appreciated of these techniques is the use of cloned DNA segments to visualize restriction fragment length polymorphisms (RFLPs) among inbred lines, segregating progeny, or within populations (Beckmann & Soller, 1983). However, new techniques, such as random amplified polymorphic DNA markers (Williams *et al.*, 1990), or modifications of older techniques, such as use of hypervariable probes (Jeffreys *et al.*, 1985a) and two-dimensional gel electrophoresis (Uitterlinden *et al.*, 1989), continue to be introduced. The development of mapping technology is moving so rapidly (see Chapter 6) that specific protocols are usually modified within one or two years. For this reason the following discussion will refrain from presenting many details of procedures but focus instead on the types of applications and relative usefulness of the genetic markers now available.

Goals and practicalities of developing a linkage map

The chromosomal linkage map of an organism summarizes much of the genetic information available for that species, and can serve as a reference for the development and testing of additional genetic hypotheses. The development of conventional (markers every 10–20 cM) or saturated (markers every 1–2 cM) linkage maps is not an easy task because the length of a diploid plant genome is usually at least 1000–1500 cM. Thus, approximately 100 evenly spaced markers is a minimum for a conventional map and 1000 will be required for a saturated map. The first 25 markers mapped may cover a large fraction of the genome, but as more markers are mapped the probability that the next marker will be located in a region lacking other markers decreases rapidly. In order to saturate a map efficiently researchers will need to isolate particular regions of the genome and generate libraries from such fractions. Methods for performing such genomic fractionations are just beginning to appear, and all are technically rather difficult.

Although the generation of a complete linkage map remains a daunting task, partial maps can be readily constructed, and these can provide very interesting information on the organization of the genome. The hexaploid nature of cultivated wheat has been clearly demonstrated by molecular markers (Hart, 1979). The considerable number of duplicate isozyme loci in maize (Wendel *et al.*, 1986) strongly supported earlier conjectures that this taxon was an ancient tetraploid. Further evidence from RFLP studies has nicely confirmed this hypothesis (Helentjaris *et al.*, 1988). Molecular markers have provided evidence for conservation of linkage groups among genera in wheat and its relatives, in the tribe Vicieae (Leguminosae) and in the Solanaceae (Hart, 1979; Bonierbale *et al.*, 1988; Weeden *et al.*, 1988). The absence of linkage conservation has also been demonstrated, in this case between two genera within the Solanaceae, *Lycopersicon* and *Capsicum* (Tanksley *et al.*, 1988).

Types of markers: their advantages and limitations

For most current applications, markers should not affect the fitness or desirability of an individual and should not show pleiotropic or epistatic interactions with other markers. Although morphological markers are often the easiest to score, many fail to meet one or more of these latter requirements. At present, allozyme and DNA polymorphisms cannot be scored in the field but otherwise make very convenient 'tags' for traits or genes which are much more difficult to score.

The major disadvantage of markers is that the marker is not, by itself, the desired trait, and the linkage between the two can be broken by recombination or mutation. There is not much value in having a marker for an easily scored seedling character, nor is there much point in having a marker much over 10 cM away unless it can be used in combination with another marker to bracket the desirable locus (Tanksley, 1983). However, using techniques presently available, there is every reason to believe that polymorphic markers can be found at least every 5–10 cM in most taxa.

Single-gene markers

One of the most useful molecular markers for a monogenic trait has been the acid phosphatase, *Aps-1*, locus in tomato which was shown to be tightly linked to *Mi*, the gene conferring nematode resistance in tomato (Rick & Fobes, 1974). This marker has been widely used in hybrid seed production and tomato breeding programmes because of the simple assay conditions required for scoring the acid phosphatase phenotype, the high correlation between the inheritance of the marker and the resistance, and the difficulty of scoring directly for nematode resistance. The acid phosphatase assay can be performed on very small amounts of tissue, allowing plants to be non-destructively screened and the susceptible ones eliminated at an early age.

Numerous additional markers have been reported for other traits (see Table 2.2 for a partial list). Some of these are being used intensively while others are much less popular. For instance, *Adh-1* is not particularly practical as a marker for pea enation mosaic virus (PEMV) resistance because it is 5–6 cM from the resistance gene and the resistant phenotype is relatively easy to score in young plants. Why should a breeder experiment with a marker if its use will not significantly increase the efficiency or the accuracy

Table 2.2. Molecular markers reported for selected traits of importance to breeding programmes.

	Marker(s)	Reference
Monogenic character		
Nematode resistance (tomato)	*Aps-1*	Rick & Fobes, 1974
Male sterility (tomato)	*Prx-2*	Tanksley *et al.*, 1984
Self-incompatibility (tomato)	*Prx-1,* *Skdh*	Tanksley & Loaiza-Figueroa, 1985
Self-incompatibility (rye)	*Prx-7*	Wricke & Wehling, 1985
Self-incompatibility (apple)	*Got-2*	Manganaris & Alston, 1987
Bean yellow mosaic virus (pea)	*Pgm-p*	Weeden *et al.*, 1988
Pea enation mosaic virus	*Adh-1*	Weeden & Provvidenti, 1988
Rhizobium strain specificity	*Idh*	Kneen *et al.*, 1984
Cyanogenesis (clover)	RFLP	Hughes *et al.*, 1990
Fusarium resistance	RFLP	Sarfatti *et al.*, 1989
Fusarium resistance (tomato)		
Eyespot resistance (wheat)	RFLP	Chao *et al.*, 1990
Polygenic characters		
Soluble solids (tomato)	RFLPs	Osborn *et al.*, 1987 Tanksley & Hewitt, 1988
Hard-seededness (soybean)	RFLPs	Keim *et al.*, 1990
Cold tolerance (maize)	allozymes	Guse *et al.*, 1988
Maturity (maize)	allozymes	Guse *et al.*, 1988
Stigma exsertion (tomato)	allozymes	Tanksley *et al.*, 1982
Fertilization (maize)	allozymes	Wendel *et al.*, 1987
Cold tolerance (tomato)	*Pgi-1*	Vallejos & Tanksley, 1983
Water use efficiency	RFLPs	Martin *et al.*, 1989
Fruit mass (tomato)	RFLPs	Paterson *et al.*, 1988
Fruit pH (tomato)	RFLPs	Paterson *et al.*, 1988
Yield (maize)	*Acp-1*	Pollak *et al.*, 1984
Yield (maize)	allozymes	Frei *et al.*, 1986
Yield (maize)	allozymes	Stuber *et al.*, 1987

of the scoring? However, in New Zealand PEMV does not exist, nor can it be imported for screening breeding material. Breeders interested in producing disease-resistant material have had to send their material outside the country to be tested for PEMV resistance. The availability of the marker now permits these breeders to introgress the resistance gene efficiently into their lines without repeated screens at foreign locations.

Markers can also be used to explore more basic scientific questions. The genetic basis of self-incompatibility has been an intriguing problem because a very large number of alleles must be postulated at the self-incompatibility locus. Such high degrees of polymorphism are not characteristic of most other loci in these self-incompatible taxa, implying that self-incompatibility loci are a unique class of genes. Based on this observation and the fact that a self-incompatibility locus had yet to be mapped in any species, some investigators postulated that self-incompatibility was controlled by many loci rather than many alleles at a single locus (Mulcahy & Mulcahy, 1983). This hypothesis was at least partially rejected when self-incompatibility was mapped, using molecular markers, to a single locus in *Nicotiana* (Labroche *et al.*, 1983), tomato (Tanksley & Loaiza-Figueroa, 1985), rye (Wricke & Wehling, 1985) and apple (Manganaris & Alston, 1987).

Perhaps the most exciting application of a DNA marker is when it can serve as a starting-point for a chromosome walk to the gene of interest. The objective of isolating the gene for tobacco mosaic virus resistance or some equally important monogenic trait is one of the most challenging in molecular biology. Plant and animal genomes are so large that to walk even a few centiMorgans represents an enormous task, particularly because the presence of interspersed repetitive DNA sequences precludes going from one cloned segment of DNA to an adjacent piece by simply screening a lambda genomic library. With the development of pulsed-field gel electrophoresis equipment that can separate 1–9 megabase fragments of DNA (Lai *et al.*, 1989; van Daelan *et al.*, 1989), and yeast artificial chromosomes (YACs) that can be used to clone 250 kb pieces of DNA (Ward & Jen, 1990), the usefulness of tightly linked RFLP markers has grown significantly. It is now possible to identify, using a mapped probe, DNA fragments that are one or more megabases in size. Although the actual relationship between length in megabases and length in centiMorgans is uncertain and fluctuates even within a genome, current estimates for the conversion factor are about 1 megabase per centiMorgan. Thus a DNA marker identifying an RFLP 1–2 cM from a gene might also be able to identify a segment of DNA containing that gene.

Polygenic characters

Most traits of interest to breeders are influenced by more than one locus. In order to obtain the desired recombinant phenotypes from a cross, a breeder will often have to screen thousands of progeny, a process that can require the commitment of considerable time and resources. Theoretically, this process can be made much more efficient through the use of a set of molecular tags that can identify the various regions of the genome influencing the character. In practice, such an approach has not been as simple as originally envisioned (Ellis, 1986; Tanksley & Hewitt, 1988; J. Nienhuis, personal communication). However, there exist numerous examples where markers have played a vital role in the selection of breeding lines or represent the only practical method for isolating the desired recombinants. A few of these are discussed below and others are listed in Table 2.2.

The pyramiding in one cultivar of two or more genes for resistance to a certain pest is a goal of many breeding programmes because resistance based on one gene appears to be more likely to break down in the field than a resistance based on multiple genes. However, in order to combine two resistance genes, one must be able to distinguish plants containing both genes from plants containing either one or the other. This requirement is not always easily met. Apple breeders have available at least five dominant genes that bestow resistance to apple scab (*Venturia inaequalis*) (Dayton & Williams, 1968). Most of these genes produce a similar disease-free phenotype, and plants carrying the resistance genes at two loci cannot be distinguished from those with only one of the genes. The most reliable method for producing cultivars possessing more than one resistance gene is to have the genes tagged by molecular markers. Selection of those progeny displaying both markers can quickly generate a subpopulation of about a quarter the size of the original progeny. For apple breeders, reducing the numbers of plants that need to be grown in the orchard represents a distinct advantage.

A related problem involves the breeding for multiple disease resistance in cultivars. As described earlier, *Adh-1* is not an ideal marker for PEMV in garden peas. However, the marker is very useful if more than one virus resistance is segregating in the progeny. Screening a plant with more than one virus can be difficult and unreliable. By using *Adh-1* as a marker for PEMV, *Pgm-p* as a marker for bean yellow mosaic virus resistance and a direct screen for pea seed-borne mosaic virus, nearly all the regions of the pea genome containing virus resistance genes can be monitored simultaneously. Obviously this approach is not limited to virus resistance genes but is appropriate in any situation when the direct screening for one gene interferes with the analysis of one or more others.

The previous two examples describe how markers could be used to follow several monogenic traits simultaneously. Of much greater significance to plant genetics has been a revolution in attitude concerning the feasibility of using markers to investigate quantitative characters. As used here, quantitative characters include those traits controlled by several or multiple major genes as well as traits influenced by hundreds of loci. Formerly such complex traits were the domain of the quantitative geneticist. Sophisticated statistical models were used to aid in the interpretation of the data taken in the field or greenhouse or from the harvested material. Unfortunately, the results of such manipulations often did not lead to conclusions that could be used by the physiologist or breeder to simplify or focus further experiments or selection procedures.

Quantitative geneticists have known for many years that monogenic markers can be valuable tools for the analysis of quantitative characters (Breese & Mather, 1957, 1960; Spickett & Thoday, 1966; Law, 1967). Mathematical tools for the analysis of linkage between marker genes and quantitative trait loci (QTLs) were provided by a number of early investigators (Jayakar, 1970; Mather & Jinks, 1971; McMillan & Robertson, 1974). These tools have been refined for numerous applications, there now being available an excellent array of mathematical models and computer programs that can assist in the analysis of the tremendous amounts of data generated in RFLP studies (Paterson *et al.*, 1988; Jensen, 1989; Lander & Botstein, 1989; Knapp *et al.*, 1990; Lande & Thompson, 1990).

The availability of a relatively complete set of markers in maize, tomato and several other crops has encouraged geneticists working with these crops to attempt to identify and map genes controlling complex characters. A list of some of these characters is presented in Table 2.2. Some of these traits, such as soluble solids in tomato, have been important breeding objectives for several decades. Others, including hard-seededness in soybean, have been studied only recently. Investigations on characters such as hard-seededness were not seriously considered without the availability of molecular markers. The low probability of successfully elucidating the genetic base did not justify the major effort involved in data collection.

Several attempts have been made to examine the basis of yield (Pollak *et al.*, 1984; Frei *et al.*, 1986; Stuber *et al.*, 1987). Unfortunately, these studies have not been complete successes. The sections of the maize chromosomes affecting yield varied among the crosses analysed and presumably also varied with environmental conditions. Thus, for marking such complex characters as yield, the genetic analysis may have to be performed for each cross of interest and across many environments. At present, the high cost of scoring 100 or more RFLP markers in large populations makes it economically impossible for companies to perform thorough genetic analyses on every cross of interest (J. Nienhuis, personal communication).

Another even more serious limitation of RFLP technology in many crops is the dearth of polymorphic markers among the breeding lines. Despite the large number of DNA probes that can be generated from libraries, the identification of polymorphism can still be difficult. Certain taxa such as *Lycopersicon esculentum* (Rick, 1982) and *Triticum aestivum* (Chao *et al.*, 1990) contain very low degrees of genetic diversity. Hence, RFLPs are not abundant, and neither cDNA nor genomic libraries will reveal many polymorphic DNA fragments. To increase the amount of diversity present in such crops as tomato, soybean, cucumber and lentil researchers have resorted to interspecific crosses. In the case of hexaploid wheat, the B genome appears to possess more RFLPs than either the A or D genomes (Chao *et al.*, 1989). By concentrating on the B genome, markers may be found for all three genomes because of the high synteny displayed among the genomes.

Once an appropriate cross has been identified, one is still faced with the tedious process of screening the parents with different restriction enzyme/probe combinations in order to determine which combinations reveal polymorphisms. To maximize efficiency and minimize cost one needs to select a set of four or five endonucleases that will have a good chance of revealing RFLPs, a sometimes disconcerting proposition given the nearly 100 restriction enzymes to choose from. Initially, the choice of endonucleases included in the screen was based on ease of use, cost and availability. Endonucleases with six base recognition sequences appear to be preferable because these give DNA fragments in a convenient size range (500–15,000 bp) and generate more observable polymorphisms (Landry *et al.*, 1987a).

Table 2.3. Common restriction endonucleases, their recognition sequences, and their ability to reveal RFLPs.[a]

Endonuclease	Recognition sequence
Endonucleases often generating RFLPs	
*Bgl*II	AGATCT
*Dra*I	TTTAAA
*Eco*RI	GAATTC
*Eco*RV	GATATC
*Hind*III	AAGCTT
*Xba*I	TCTAGA
Endonucleases less likely to generate RFLPs	
*Bam*HI	GGATCC
*Pst*I	CTGCAG
*Pvu*I	CGATCG
*Sac*I	GAGCTC
*Sph*I	GCATGC
*Xho*I	CTCGAG

[a]Taxa studied: *Pisum, Lens, Cicer, Malus, Vitis.*

The considerable amount of empirical data generated in recent years indicates that certain endonucleases are more likely to detect RFLPs (Table 2.3). Those enzymes that can be included in the 'often polymorphic' group generally have a recognition sequence high in AT residues. These data confirm similar results in wheat (Chao *et al.*, 1990) and suggest that, in general, regions of the genome rich in AT residues are changing more rapidly than the remainder of the genome. However, Landry *et al.* (1987a) also compared restriction enzyme sites and numbers of RFLPs detected, concluding that, in lettuce, there was no difference between regions rich in GC and regions rich in AT. In this latter work the two endonucleases used that recognized sites high in GC were the four-base recognition enzymes

*Msp*I and *Pal*I. Since enzymes with four-base recognition sites were shown by these same authors to reveal lower degrees of polymorphism, their conclusion may be premature.

If regions high in AT are more likely to reveal RFLPs, at least in some species, other data may help clarify the source of the variation. A comparison of exon and intron sequences has revealed that introns are about 20% higher in AT content than exons (Montero *et al.*, 1990). Introns are known to display greater sequence variability than exons, and this difference may explain why endoncleases with recognition sites high in AT are more successful at identifying RFLPs. Further support for this hypothesis comes from the relative success of different types of libraries at generating probes that define RFLPs. Our own experience and results from other laboratories (Landry *et al.*, 1987a; Havey & Muehlbauer, 1989) indicate that cDNA clones will show RFLPs more frequently than genomic clones. The major difference between the two types of clones is that cDNA will usually hybridize to a larger genomic sequence than the size of the insert because of the introns intercalated within the coding sequence. The greater ability of cDNA to generate RFLPs is probably a direct consequence of these sequences spanning more introns than genomic clones of similar size or even significantly larger.

Limitations of RFLP analysis

Despite the dramatic success RFLP mapping has had in certain crops, such as maize (Helentjaris *et al.*, 1986), lettuce (Landry *et al.*, 1987b), rice (McCouch *et al.*, 1988) and crucifers (Chang *et al.*, 1988; Slocum *et al.*, 1990), other crops have been much more recalcitrant to this approach. As was mentioned above, many of the problem crops possess a low degree of genetic diversity and the classical RFLP approach does not reveal enough polymorphism to be particularly useful. In order to generate DNA clones that expose more variability, geneticists in these taxa are focusing on probes that hybridize to hypervariable regions of the genome. These regions include minisatellite DNA such as the sequences homologous to M13 first reported in humans (Jeffreys *et al.*, 1985a, b), but also found in other animals and certain plants (Rogstad *et al.*, 1988). M13 is a common and relatively inexpensive vector that can be purchased directly from many molecular biology firms and is therefore likely to be a very popular probe. Using such sequences, DNA 'fingerprints' containing many genomic fragments can be generated with common restriction enzymes. Often several of these fragments will display polymorphism. This approach has been very successful in several animal species (Love *et al.*, 1990). It is anticipated that the approach will work equally well in plants, since plant and animal genomes are very similar in structure. Data on plants, however, are only just beginning to appear in the literature (Nybom *et al.*, 1990).

Other approaches have been used to increase the frequency and diversity of DNA polymorphisms. Denaturing gel electrophoresis has been successfully applied in maize (Riedel *et al.*, 1990). In this technique DNA polymorphisms are indicated by changes in the melting-point (as determined in a chemical denaturant rather than temperature) of genomic DNA fragments. In maize, polymorphisms were detected at a frequency comparable to the RFLP techniques. Two-dimensional gel electrophoresis has been used to separate hundreds of DNA fragments on a single gel (Uitterlinden *et al.*, 1989; Yi *et al.*, 1990). Theoretically this technique could be used in combination with a microsatellite DNA probe to reveal the positions of hundreds of polymorphic fragments in one plant. Comparisons of 50 different autoradiographs generated from the two-dimensional gels of DNA from 50 different F_2 plants would permit the segregation pattern of all the polymorphic fragments to be determined simultaneously. However, significant practical problems, particularly in the maintenance of absolute reproducibility between gel runs, remain to be overcome before this approach can be widely used.

The major obstacles to general use of RFLP analysis or the markers identified by this approach continue to be the expense and technical difficulties involved. Although certain aspects of the procedures have been greatly simplified (Feinberg & Vogelstein, 1983; Reed & Mann, 1985), it still requires several days of skilled technician time to obtain information about a marker, and the cost of materials is by no means trivial. In addition, autoradiography remains the most reliable method for detecting single-copy sequences, although several very sensitive non-radioactive visualization systems have recently been marketed. At present, the expense and technical expertise necessary to analyse for RFLP markers places the method out of reach of most small seed companies and independent breeders and many people working on minor crops.

The random amplified polymorphic DNA technique

A major breakthrough with regard to DNA markers available to smaller-scale operations has been the development of the random amplified polymorphic DNA (RAPD) technique (Williams *et al.*, 1990). This method is based on the use of the polymerase chain reaction (PCR) to amplify specific sequences to such an extent that they can be visualized by ethidium bromide without the need for blotting or hybridization. The technique involves the use of a single primer approximately ten bases in length. This primer will hybridize to homologous genomic DNA sequences after the genomic DNA has been made single-stranded by heating at 92°C for 1 minute (see Fig. 2.1). A new DNA strand starting at the primer can be synthesized using the genomic DNA as a template. The enzyme *Taq* polymerase is used for this step because it is thermostable and can survive the 92°C denaturation step in the cycle.

The primer is short and will hybridize at many regions of the genome. Most of these regions will generate only one copy of the DNA template for every cycle through which the process is taken. However, if a second sequence homologous to the primer is found in an inverted orientation on the opposite strand within 2000 bases downstream, both strands will be replicated in this region (Fig. 2.1). For each cycle the number of copies of this region doubles, and after 20 cycles there should be over a million copies of this sequence for every original DNA molecule that contained the region. With a primer of ten bases, this inverted arrangement of sequences is expected to occur one to ten times in an angiosperm genome, depending on the size of the genome. Thus, while many portions of the genome will be replicated 40 times during the 40 cycles, only one to ten segments will be amplified 2^{40} times, generating a sufficient number of copies to be seen after electrophoresis and ethidium bromide staining.

The equipment and supplies necessary for this approach are inexpensive (about US$4000 and dropping) relative to RFLP mapping, and the expertise required is limited to the extraction of DNA and being scrupulously clean regarding the mixing of the components of the PCR reactions (any extraneous DNA may also be amplified). The data generated so far suggest a very high degree of polymorphism in the segments being amplified (Williams *et al.*, 1990; G.M. Timmerman & N.F. Weeden, unpublished results). It appears that these RAPD markers will show a frequency of variation comparable to that displayed by minisatellite DNA. This high degree of detected polymorphism implies that the selection of parents with diverse genetic backgrounds is much less critical when using RAPD markers than when working with RFLPs. Sufficient polymorphism may be present even between lines within a breeding programme. If so, the RAPD technique has two very favourable attributes with respect to applications to plant breeding programmes: (i) it does not require Southern blotting or labelled probes and thus is much more convenient and economical than most methods; and (ii) it reveals sufficient polymorphism to permit breeders and other geneticists to select crosses for analysis on the basis of commercially or scientifically interesting differences between the parents rather than on maximization of genetic differences.

Summary

Recent progress in plant molecular biology has opened doors long closed to plant breeders. The obstacles involved with the immense size and apparent lack of organization of the eukaryotic genome have been at least partially circumvented by the use of numerous polymorphic DNA markers. Virtually

RAPD Technique

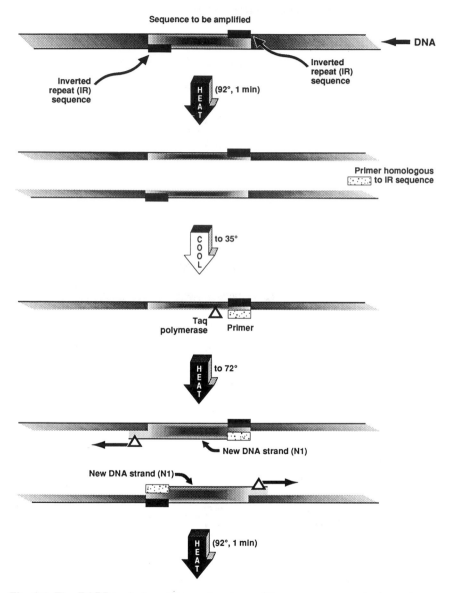

Fig. 2.1. The RAPD technique takes advantage of the rare occurrence when a ten-base sequence is present on opposite strands of the DNA molecule in an inverted orientation, the two sequences being separated by about 200–2000 base pairs as shown at top of the left hand side of the figure. After heating at 92°C to separate DNA strands, the 'primer' (which has a sequence complementary to the inverted repeat) is allowed to anneal at 35°C to the genomic DNA. The *Taq* polymerase will use the

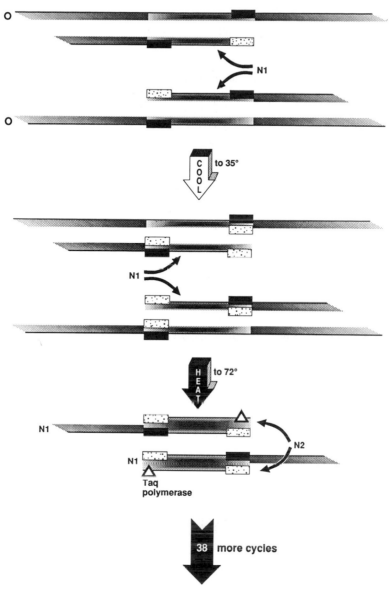

3′ end of the primer to initiate the synthesis of new DNA strands (N1 in figure) complementary to both of the original strands of the DNA (bottom of the left side of figure). Another 92°C heat step separates the newly synthesized N1 strands from the original DNA (top of right side of figure). Because there was an inverted sequence homologous to the primer, both of the N1 strands can bind the primer and be replicated up to the position of the original primer, forming a relatively short (200–2000 bp) N2 fragment (bottom of right side of figure). The number of N2 fragments will double for every additional time the cycle is repeated.

every heritable character, whether conditioned by one or hundreds of loci, can be dissected into its major genetic components and these loci placed on a chromosomal map through the use of molecular markers. However, the practicality of the molecular approach remains to be firmly established. The costs and technical requirements for RFLP analysis place this approach beyond the resources of many smaller companies and private breeders. The selection for certain quantitative traits by indirect means also appears to have significant problems in terms of costs versus benefits. Basic research programmes should be able to absorb the costs on the basis of the probable long-term pay-offs in understanding the genetic basis of complex characters and the availability of DNA markers near critical genes. For the strictly applied programmes, new techniques such as those based on amplification of specific sequences may eventually provide a large number of inexpensive and easily scored polymorphic markers.

Acknowledgements

Special thanks are extended to C. E. Vallejos and P. Gepts for sharing unpublished results on RFLP markers for *Phaseolus* and to J. Nienhuis for comments on the utility of RFLP markers and genetic maps in breeding programmes.

References

Appels, R. & Honeycutt, R. L. (1986) rDNA: evolution over a billion years. In: Dutta, S. K. (ed.), *DNA Systematics,* vol. II, *Plants.* CRC Press, Boca Raton, Florida, pp. 81–136.

Beadle, G. W. & Tatum, E. L. (1941) Genetic control of biochemical reactions in *Neurospora. Proceedings of the National Academy of Sciences USA* 27, 499–506.

Beckmann, J. S. & Soller, M. (1983) Restriction fragment length polymorphisms in genetic improvement: methodologies, mapping and costs. *Theoretical and Applied Genetics* 67, 35–43.

Bernardi, G., Olofsson, B., Filipski, J., Zerial, M., Salinas, J., Cuny, G., Meunier-Rotival, M. & Rodier, F. (1985) The mosaic genome of warm-blooded vertebrates. *Science* 228, 953–8.

Bernatzky, R. & Tanksley, S. D. (1986) Toward a saturated linkage map in tomato based on isozymes and random cDNA sequences. *Genetics* 112, 887–98.

Blackburn, E. H., Greider, C. W., Henderson, E., Lee, M. S., Shampay, J. & Shippen-Lentz, D. (1989) Recognition and elongation of telomers by telomerase. *Genome* 31, 553–60.

Blixt, S. (1974) The pea. In: King, R. C. (ed.), *Handbook of Genetics,* vol. 2. Plenum Press, New York, pp. 181–221.

Bonierbale, M. W., Plaisted, R. L. & Tanksley, S. D. (1988) RFLP maps based on a common set of clones reveal modes of chromosome evolution in potato and tomato. *Genetics* 120, 1095–103.

Bournival, B. L., Scott, J. W. & Vallejos, C. E. (1989) An isozyme marker for resistance to race 3 of *Fusarium oxysporum* f. sp. *lycopersici* in tomato. *Theoretical and Applied Genetics* 78, 489–94.

Breese, E. L. & Mather, K. (1957) The organization of polygenic activity within a chromosome in *Drosophila*. I. Hair characters. *Heredity* 11, 373–95.

Breese, E. L. & Mather, K. (1960) The organization of polygenic activity within a chromosome in *Drosophila*. II. Viability. *Heredity* 14, 375–400.

Casey, R., Domoney, C. & Ellis, N. (1986) Legume storage proteins and their genes. *Oxford Surveys in Plant Molecular and Cellular Biology* 3, 1–95.

Chang, C., Bowman, J. L., Dejohn, A. W., Lander, E. S. & Meyerowitz, E. M. (1988) Restriction fragment length polymorphism linkage map for *Arabidopsis*. *Proceedings of the National Academy of Sciences USA* 85, 6856–60.

Chao, S., Sharp, P. J., Worland, A. J., Warham, E. J., Koeber, R. M. D. & Gale, M. D. (1990) RFLP-based genetic maps of wheat homoeologous group 7 chromosomes. *Theoretical and Applied Genetics* 78, 495–504.

Choumanc, W. & Heizmann, P. (1988) Structure and variability of nuclear ribosomal genes in the genus *Helianthus*. *Theoretical and Applied Genetics* 76, 481–9.

Cordesse, F., Second, G. & Delseny, M. (1990) Ribosomal gene spacer length variability in cultivated and wild rice. *Theoretical and Applied Genetics* 79, 81–8.

Cseke, Cs., Weeden, N. F., Buchanan, B. B. & Uyeda, K. (1982) A special fructose bisphosphate functions as a cytoplasmic regulatory metabolite in green leaves. *Proceedings of the National Academy of Sciences USA* 79, 4322–6.

Davidson, E. H., Hough, B. R., Amensen, C. S. & Britten, R. J. (1975) Comparative aspects of DNA organization in Metazoa. *Chromosoma* 51, 253–9.

Dayton, D. F. & Williams, E. B. (1968) Independent genes in *Malus* for resistance to *Venturia inaequalis*. *Journal of the American Society for Horticultural Science* 92, 89–94.

Dobrowolski, B., Glund, K. & Metzlaff, M. (1989) Cloning of tomato nuclear ribosomal DNA: rDNA organization in leaves and suspension cultures. *Plant Science* 60, 199–205.

Ellis, T. H. N. (1986) Restriction fragment length polymorphism markers in relation to quantitative characters. *Theoretical and Applied Genetics* 72, 1–2.

Evans, I. J., James, A. M. and Barnes, S. R. (1983) Organization and evolution of repeated DNA sequences in closely related plant genomes. *Journal of Molecular Biology* 170, 803–26.

Fabijanski, S., Fedak, G., Armstrong, K. & Altosaar, I. (1990) A repeated sequence probe for the C genome in *Avena* (oats). *Theoretical and Applied Genetics* 79, 1–7.

Feinberg, A. P. & Vogelstein, B. (1983) A technique for radiolabeling DNA restriction fragments to a high specific activity. *Analytical Biochemistry* 132, 6–13.

Flavell, R. B. (1982) Chromosomal DNA sequences and their organization. In: Parthier, B. & Boulter, D. (eds), *Encyclopaedia of Plant Physiology. New Series*, vol. 14B. Springer-Verlag, Berlin, pp. 46–74.

Flavell, R. B., Rimpau, J. & Smith, D. B. (1977) Repeated sequence DNA relationships in four cereal genomes. *Chromosoma* 63, 205–22.

Frei, O. M., Stuber, C. W. & Goodman, M. M. (1986) Yield manipulation from selection on allozyme genotypes in a composite of elite conr lines. *Crop Science* 26, 917–21.

Gebhardt, C., Ritter, E., Debener, T., Schachtschabel, U., Walkmeier, B., Uhrig, H. & Salamini, F. (1989) RFLP analysis and linkage mapping in *Solanum tuberosum. Theoretical and Applied Genetics* 78, 65–75.

Greider, C. W. & Blackburn, E. H. (1989) A telomeric sequence in the RNA of *Tetrahymena* telomerase required for telomere repeat synthesis. *Nature* 337, 331–7.

Guse, R. A., Coors, J. G., Drolsom, P. N. & Tracy, W. F. (1988) Isozyme marker loci associated with cold tolerance and maturity in maize. *Theoretical and Applied Genetics* 76, 398–404.

Hart, G. E. (1979) Genetical and chromosomal relationships among the wheats and their relatives. *Stadler Symposia* 11, 9–29.

Hastie, N. D., Dempster, M., Dunlop, M. G., Thompson, A. M., Green, D. K. & Allshire, R. C. (1990) Telomere reduction in human colorectal carcinoma and with ageing. *Nature* 346, 866–8.

Havey, M. J. & Muehlbauer, F. J. (1989) Linkages between restriction fragment length, isozyme, and morphological markers in lentil. *Theoretical and Applied Genetics* 77, 395–401.

Helentjaris, T., Slocum, M., Wright, S., Schaefer, A. & Nienhuis, J. (1986) Construction of genetic linkage maps in maize and tomato using restriction fragment length polymorphisms. *Theoretical and Applied Genetics* 72, 761–9.

Helentjaris, T., Weber, D. F. & Wright, S. (1988) Identification of the genomic locations of duplicate nucleotide sequences in maize by analysis of restriction fragment length polymorphisms. *Genetics* 118, 353–63.

Hughes, M. A., Sharif, A. L., Dunn, M. A., Oxtoby, E. & Pancoro, A. (1990) Restriction fragment length polymorphism segregation analysis of the *Li* locus in *Trifolium repens* L. *Plant Molecular Biology* 14, 407–14.

Jacob, F. & Monod, J. (1961) Genetic regulatory mechanisms in the synthesis of proteins. *Journal of Molecular Biology* 3, 318–56.

Jayakar, S. D. (1970) On the detection and estimation of linkage between a locus influencing a quantitative character and a marker locus. *Biometrics* 26, 451–64.

Jeffreys, A. J. & Flavell, R. A. (1977) The rabbit β-globin gene contains a large insert in the coding sequence. *Cell* 12, 1097–108.

Jeffreys, A. J., Wilson, A. & Thein, S. L. (1985a) Hypervariable 'minisatellite' regions in human DNA. *Nature* 314, 67–73.

Jeffreys, A. J., Wilson, V. & Thein, S. L. (1985b) Individual-specific 'fingerprints' of human DNA. *Nature* 316, 76–9.

Jensen, J. (1989) Estimation of recombination parameters between a quantitative trait locus (QTL) and two marker gene loci. *Theoretical and Applied Genetics* 78, 613–18.

Johannsen, W. (1909) *Elemente der exakten Erblichkeitslehrer.* Fischer, Jena.

Jones, R. N. & Rees, H. (1969) Nuclear DNA variation in *Allium. Heredity* 23, 591–605.

Kavanaugh, T. A. & Timmis, J. N. (1988) Structure of melon rDNA and nucleotide sequence of the 17–25S spacer region. *Theoretical and Applied Genetics* 76, 673–80.

Keim, P. & Shoemaker, R. C. (1988) Construction of a random recombinant DNA library that is primarily single-copy sequence. *Soybean Genetics Newsletter* 15, 147–8.

Keim, P., Diers, B. W. & Shocmaker, R. C. (1990) Genetic analysis of soybean hard seededness with molecular markers. *Theoretical and Applied Genetics* 79, 465–9.

Knapp, S. J., Bridges, W. C. Jr & Birkes, D. (1990) Mapping and quantitative trait loci using molecular marker linkage maps. *Theoretical and Applied Genetics* 79, 583–92.

Kneen, B. E. & LaRue, T. A. (1988) Induced symbiosis mutants of pea (*Pisum sativum*) and sweetclover (*Melilotus alba annua*). *Plant Science* 58, 177–82.

Kneen, B., LaRue, T. A. & Weeden, N. F. (1984) Gene reported to affect symbiotic nitrogen fixation by peas. *Pisum Newsletter* 16, 31–4.

Koukalova, B., Reich, J., Matyasek, R., Kuhrova, V. & Bezdek, M. (1989) A BamHI family of highly repeated DNA sequences of *Nicotiana tabacum*. *Theoretical and Applied Genetics* 78, 77 80.

Labroche, Ph., Poirier-Hamon, S. & Pernes, J. (1983) Inheritance of leaf peroxidase isoenzymes in *Nicotiana alata* and linkage with the S-incompatibility locus. *Theoretical and Applied Genetics* 65, 163–70.

Lai, E., Birren, B. W., Clark, S. M., Simon, M. I. & Hood, L. (1989) Pulsed field gel electrophoresis. *BioTechniques* 7, 34–42.

Lande, R. & Thompson, R. (1990) Efficiency of marker-assisted selection in the improvement of quantitative traits. *Genetics* 124, 743–56.

Lander, E. S. & Botstein, D. (1989) Mapping Mendelian factors underlying quantitative traits using RFLP linkage maps. *Genetics* 121, 185–99.

Landry, B. S., Kesseli, R., Leung, H. & Michelmore, R. W. (1987a) Comparison of restriction endonucleases and sources of probes for their efficiency in detecting restriction fragment length polymorphisms in lettuce (*Lactuca sativa* L.). *Theoretical and Applied Genetics* 74, 646–53.

Landry, B. S., Kesseli, R. V., Farrara, B. & Michelmore, R. W. (1987b) A genetic map of lettuce (*Lactuca sativa* L.) with restriction fragment length polymorphisms, isozyme, disease resistance and morphological markers. *Genetics* 116, 331–7.

Law, C. N. (1967) The location of factors controlling a number of quantitative characters in wheat. *Genetics* 56, 445–61.

Lima-de-Faria, A. (1983) *Molecular Evolution and Organization of the Chromosome*. Elsevier, Amsterdam.

Love, J. M., Knight, A. M., McAleer, M. A. & Todd, J. A. (1990) Towards construction of a high resolution map of the mouse genome using PCR-analysed microsatellites. *Nucleic Acids Research* 18, 4123–30.

McCouch, S. R., Kochert, G., Yu, Z. H., Yang, Z. Y., Khush, G. S., Coffman, W. R. & Tanksley, S. D. (1988) Molecular mapping of rice chromosomes. *Theoretical and Applied Genetics* 76, 815–29.

McMillan, I. & Robertson, A. (1974) The power of methods for detection of major genes affecting quantitative characters. *Heredity* 32, 349–56.

Manganaris, A. G. & Alston, F. H. (1987) Inheritance and linkage relationships of glutamate oxaloacetate transaminase isoenzymes in apple. 1. The gene GOT-1, a marker for the S incompatibility locus. *Theoretical and Applied Genetics* 74, 154–61.

Martin, B., Nienhuis, J., King, G. & Schaefer, A. (1989) Restriction fragment length polymorphisms associated with water use efficiency in tomato. *Science* 243, 1725–7.

Martin, W. & Cerff, R. (1986) Procaryotic features of a nucleus-encoded enzyme. *European Journal of Biochemistry* 159, 323–31.

Mascia, P. N., Rubenstein, I., Phillips, R. L., Wang, A. S. & Xiang, L. Z. (1981) Localization of the 5S rRNA genes and evidence for diversity in the 5S rDNA region of maize. *Gene* 15, 7–20.

Mather, K. & Jinks, J. L. (1971) *Biometrical Genetics*. Cornell University Press, Ithaca, New York.

Montero, L. M., Salinas, J., Matassi, G. & Bernardi, G. (1990) Gene distribution and isochore organization in the nuclear genome of plants. *Nucleic Acids Research* 18, 1859–67.

Morgan, T. H. (1926) *The Theory of the Gene*. Yale University Press, New Haven, Connecticut.

Mulcahy, D. L. & Mulcahy, G. B. (1983) Gametophytic self-incompatibility reexamined. *Science* 220, 1247–51.

Murray, M. G., Cuellar, R. E. & Thompson, W. F. (1978) DNA sequence organization in the pea genome. *Biochemistry* 17, 5781–90.

Nakamura, Y., Leppert, M., O'Connell, P., Wolff, R., Holm, T., Culver, M., Martin, C., Fujimoto, E., Hoff, M., Kumlin, E. & White, R. (1987) Variable number of tandem repeat (VNTR) markers for human gene mapping. *Science* 235, 1616–22.

Nybom, H., Rogstad, S. H. & Schaal, B. A. (1990) Genetic variation detected by use of the M13 'DNA fingerprint' probe in *Malus, Prunus,* and *Rubus* (Rosaceae). *Theoretical and Applied Genetics* 79, 153–6.

Osborn, T. C., Alexander, D. C. & Fobes, J. F. (1987) Identification of restriction fragment length polymorphisms linked to genes controlling soluble solids content of tomato fruit. *Theoretical and Applied Genetics* 73, 350–6.

Paterson, A. H., Lander, E. S., Hewitt, J. D., Peterson, S., Lincoln, S. E. & Tanksley, S. D. (1988) Resolution of quantitative traits into Mendelian factors by using a complete linkage map of restriction fragment length polymorphisms. *Nature* 335, 721–6.

Pichersky, E., Bernatzky, R., Tanksley, S. D., Breidenbach, R. B., Kausch, R. B. & Cashmore, A. P. (1985) Molecular characterization and genetic mapping of two clusters of genes encoding chlorophyll *a/b*-binding proteins in *Lycopersicon esculentum*. *Gene* 40, 247–58.

Polans, N. O., Weeden, N. F. & Thompson, W. F. (1986) The distribution, inheritance and linkage relationships of ribosomal DNA spacer length variants in pea. *Theoretical and Applied Genetics* 72, 289–95.

Pollak, L. J., Gardner, C. O. & Parkhurst, A. M. (1984) Relationships between enzyme marker loci and morphological traits in two mass selected maize populations. *Crop Science* 24, 1174–9.

Provvidenti, R. (1987) List of genes in *Pisum sativum* for resistance to viruses. *Pisum Newsletter* 19, 48–9.

Reed, K. C. & Mann, D. A. (1985) Rapid transfer of DNA from agarose gels to nylon membranes. *Nucleic Acids Research* 13, 7207–21.

Rees, H., Cameron, F. M., Hazarika, M. H. & Jones, G. H. (1966) Nuclear variation between diploid angiosperms. *Nature* 211, 828–30.

Rick, C. M. (1971) Some cytogenetic features of the genome in diploid plant species. *Stadler Symposia* 1/2, 153–71.

Rick, C. M. (1982) Tomato (*Lycopersicon*). In: Tanksley, S. D. & Orton, T. J. (eds), *Isozymes in Plant Genetics and Breeding,* Part B. Elsevier, Amsterdam, pp. 147–65.

Rick, C. M. & Fobes, J. F. (1974) Association of an allozyme with nematode resistance. *Report of the Tomato Genetics Cooperative* 24, 25.

Riedel, G. E., Swangerg, S. L., Durands, K. D., Marquette, K., LaPan, P., Bledsoe, P., Kennedy, A. & Lin, B.-Y. (1990) Denaturing gradient gel electrophoresis identifies genomic DNA polymorphism with high frequency in maize. *Theoretical and Applied Genetics* 80, 1–10.

Rogers, S. O. & Bendich, A. J. (1987) Ribosomal RNA in plants: variability in copy number and in the intergenic spacer. *Plant Molecular Biology* 9, 509–20.

Rogstad, S. H., Patton, J. C. II and Schaal, B. A. (1988) M13 repeat probe detects DNA minisatellite-like sequences in gymnosperms and angiosperms. *Proceedings of the National Academy of Sciences USA* 85, 9176–8.

Saghai-Maroof, M. A., Soliman, K. M., Jorgensen, R. A. & Allard, R. W. (1984) Ribosomal DNA spacer-length polymorphism in barley: Mendelian inheritance, chromosomal location, and population dynamics. *Proceedings of the National Academy of Sciences USA* 81, 8014–18.

Sakowicz, T., Galazka, G., Konarzewska, A., Kwinkowski, M. & Klysik, J. (1986) An unusually high number of direct repeats detected by sequence analysis of the dispersed EcoRI-family fragments in *Lupinus luteus* L. *Planta* 168, 207–13.

Salinas, J., Matassi, G., Montero, L. M. & Bernardi, G. (1988) Compositional compartmentalization and compositional patterns in the nuclear genomes of plants. *Nucleic Acids Research* 16, 4269–85.

Sarfatti, M., Katan, J., Fluhr, R. & Zamir, D. (1989) An RFLP marker in tomato linked to *Fusarium oxysporum* resistance gene *I2*. *Theoretical and Applied Genetics* 78, 755–9.

Scott, N. S. & Timmis, J. N. (1984) Homologies between nuclear and plastid DNA in spinach. *Theoretical and Applied Genetics* 67, 279–88.

Shih, M.-C., Lazar, G. & Goodman, H. M. (1986) Evidence in favor of the symbiotic origin of chloroplasts: primary structure and evolution of tobacco glyceraldehyde-3-phosphate dehydrogenases. *Proceedings of the National Academy of Sciences USA* 79, 7624–8.

Slocum, M. K., Figdore, S. S., Kennard, W. C., Suzuki, J. Y. & Osborn, T. C. (1990) Linkage arrangement of restriction fragment length polymorphism loci in *Brassica oleracea*. *Theoretical and Applied Genetics* 80, 57–64.

Spickett, S. G. & Thoday, J. M. (1966) Regular responses to selection. 3. Interaction between located polygenes. *Genetics Research* 7, 96–121.

Stuber, C. W., Edwards, M. D. & Wendel, J. F. (1987) Molecular-marker-facilitated investigations of quantitative-trait loci in maize. II. Factors influencing yield and its component traits. *Crop Science* 27, 639–48.

Sutton, W. S. (1903) The chromosomes in heredity. *Biological Bulletin* 4, 231–51.

Tanksley, S. D. (1983) Molecular markers in plant breeding. *Plant Molecular Biology Reporter* 1, 3–8.

Tanksley, S. D. & Hewitt, J. (1988) Use of molecular markers in breeding for soluble solids content in tomato – a re-examination. *Theoretical and Applied Genetics* 75, 811–23.

Tanksley, S. D. & Loaiza-Figueroa, F. (1985) Gametophytic self-incompatibility is controlled by a single major locus on chromosome 1 in *Lycopersicon peruvianum*. *Proceedings of the National Academy of Sciences USA* 82, 5093–6.

Tanksley, S. D., Medina, H. & Rick, C. M. (1982) Use of naturally-occurring enzyme variation to detect and map genes controlling quantitative traits in an interspecific backcross of tomato. *Heredity* 49, 11–25.

Tanksley, S. D., Rick, C. M. & Vallejos, C. E. (1984) Tight linkage between a nuclear male-sterile locus and an enzyme marker in tomato. *Theoretical and Applied Genetics* 68, 109–13.

Tanksley, S. D., Bernatzky, R. Lapitan, N. L. & Prince, J. P. (1988) Conservation of gene repertoir but not gene order in pepper and tomato. *Proceedings of the National Academy of Sciences USA,* 85, 6419–23.

Uitterlinden, A. G., Slagboom, P. E., Knook, D. L. & Vijg, J. (1989) Two-dimensional DNA fingerprinting of human individuals. *Proceedings of the National Academy of Sciences USA* 86, 2742–6.

Vallejos, C. E. & Tanksley, S. D. (1983) Segregation of isozyme markers and cold tolerance in an interspecific backcross of tomato. *Theoretical and Applied Genetics* 66, 241–7.

Vallejos, C. E., Tanksley, S. D. & Bernatzky, R. (1986) Localization in the tomato genome of DNA restriction fragments containing sequences homologous to the rRNA (45S), the major chlorophyll *a/b* binding polypeptide and the ribulose bisphosphate carboxylase genes. *Genetics* 112, 93–105.

van Daelen, R. A. J., Jonkers, J. J. & Zabel, P. (1989) Preparation of megabase-sized tomato DNA and separation of large restriction fragments by field inversion gel electrophoresis (FIGE). *Plant Molecular Biology* 12, 341–52.

Ward, E. R. & Jen, G. C. (1990) Isolation of single-copy-sequence clones from a yeast artificial chromosome library of randomly-sheared *Arabidopsis thaliana* DNA. *Plant Molecular Biology* 14, 561–8.

Watson, J. D. & Crick, F. H. C (1953) A structure for desoxyribose nucleic acids. *Nature* 171, 737–8.

Weeden, N. F. (1981) Genetic and biochemical implications of the endosymbiotic origin of the chloroplast. *Journal of Molecular Evolution* 17, 133–9.

Weeden, N. F. (1990). A leghemoglobin cluster is located near D on chromosome 1. *Pisum Newsletter* 22, 69–70.

Weeden, N. F. & Provvidenti, R. (1988) A marker locus, *Adh-1*, for resistance to pea enation mosaic virus in *Pisum sativum*. *Journal of Heredity* 79, 128–31.

Weeden, N. F. & Wolko, B. (1990) Linkage map for the garden pea (*Pisum sativum*) based on molecular markers. In: O'Brien, S. J. (ed.), *Genetic Maps,* 5th edn. Cold Spring Harbor Laboratory Press, Cold Spring Harbor, New York, pp. 6.106–6.112.

Weeden, N. F., Higgins, R. C. & Gottlieb, L. D. (1982) Immunological similarity between a cyanobacterial enzyme and a nuclear DNA-encoded plastic-specific isozyme from spinach. *Proceedings of the National Academy of Sciences USA* 79, 5953–5.

Weeden, N. F., Zamir, D. & Tadmor, Y. (1988) Applications of isozyme analysis in pulse crops. In: Summerfield, R. J. (ed.), *World Crops: Cool Season Food Legumes.* Kluwer Academic Publishers, Dordrecht, pp. 979–87.

Weeden, N. F., Kneen, B. E. & LaRue, T. A. (1991) Genetic analysis of *sym* genes and other nodule-related genes in *Pisum sativum*. In: Gresshoff, P. (ed.), *Nitrogen Fixation: Achievements and Objectives*. Chapman and Hall, New York.

Wendel, J. F., Stuber, C. W., Edwards, M. D. & Goodman, M. M. (1986) Duplicated chromosomal segments in maize (*Zea mays* L.): further evidence from hexokinase isozymes. *Theoretical and Applied Genetics* 72, 178–85.

Wendel, J. F., Edwards, M. D. & Stuber, C. W. (1987) Evidence for multilocus genetic control of preferential fertilisation in maize. *Heredity* 58, 297–301.

Williams, J. G. K., Kubelik, A. R., Livak, K. J., Rafalski, J. A. & Tingey, S. V. (1990) DNA polymorphisms amplified by arbitrary primers are useful as genetic markers. *Nucleic Acids Research* 18, 6531–5.

Wricke, G. & Wehling, P. (1985) Linkage between an incompatibility locus and a peroxidase isozyme locus (Prx7) in rye. *Theoretical and Applied Genetics* 71, 289–91.

Yi, M., Au, L., Ichikawa, N. & Ts'o, P. O. P. (1990) Enhanced resolution electrophoresis and double-labeling. *Proceedings of the National Academy of Sciences USA* 87, 3919–23.

Zakian, V. A. (1989) Structure and function of telomeres. *Annual Review of Genetics* 23, 579–604.

Zamir, D. & Tanksley, S. D. (1988) Tomato genome is comprised largely of fast-evolving, low copy-number sequences. *Molecular and General Genetics* 213, 254–61.

Zhao, X., Wu, T., Xie, Y. & Wu, (1989) Genome-specific repetitive sequences in the genus *Oryza*. *Theoretical and Applied Genetics* 78, 201–9.

Zimmer, E. A., Jupe, E. R. & Walbot, V. (1988) Ribosomal gene structure, variation and inheritance in maize and its ancestors. *Genetics* 120, 1125–36.

Chapter 3
Gene Transfer to Plants Using *Agrobacterium*

J. E. Grant, E. M. Dommisse, M. C. Christey &
A. J. Conner

DSIR Crop Research, Private Bag, Christchurch,
New Zealand

Introduction

Agrobacterium is a soil-dwelling bacterium that infects wound sites on a wide range of plant species and induces the development of crown gall tumours or hairy roots. These growth responses result from a natural genetic engineering event, in which a specific region of DNA from a Ti (tumour-inducing) or Ri (root-inducing) plasmid is transferred from the *Agrobacterium* to a plant cell. This T-DNA (transferred DNA) is integrated and expressed in the nuclear genome of the plant cells. It encodes enzymes responsible for the biosynthesis of phytohormones and/or proteins affecting the sensitivity of plant cells to phytohormones. The expression of these genes results in the development of tumours or hairy roots. The T-DNA also encodes genes specifying enzymes involved in the production and secretion of opines (derivatives of amino acids) which the *Agrobacterium* utilizes as a food source. *Agrobacterium* itself does not appear to express genes on the T-DNA.

The importance of *Agrobacterium* for plant genetic engineering is its natural ability to transfer a segment of DNA into plant cells. Over the last ten years plant scientists have capitalized on molecular biology technology to manipulate the T-DNA of *Agrobacterium* for the development of gene vectors to produce transgenic plants. To use *Agrobacterium* for plant transformation a number of experimental steps must be optimized in both the bacteria and the plant material. These include: the identification of an *Agrobacterium* strain which 'infects' the appropriate plant genotype; the design and construction of modified T-DNA to allow gene expression in plant cells; the transfer to and maintenance of the modified T-DNA in a specified *Agrobacterium* strain; the frequency of T-DNA transfer events to plant cells being high enough to be detected; and the selection and regeneration of transformed plant cells.

The perspective of this chapter is that *Agrobacterium* is a tool for plant breeding and improvement, via the transfer of desirable genes into crop plants. We present an overview of the basis for the use of *Agrobacterium*, summarize the methods involved, and discuss the limitations and current status of the system with respect to horticultural and agricultural applications.

Basis for crown gall and hairy root formation

Species, plasmids and opines

Strains of *Agrobacterium* are generally classified into three biotypes which can be considered as separate species: *A. tumefaciens, A. rubi* and *A. rhizogenes* (Kerr & Brisbane, 1983). The characteristics of *Agrobacterium*-induced neoplastic growths are determined by the Ti or Ri plasmid content of an *Agrobacterium* strain. Saprophytic strains without Ti or Ri plasmids are also common and are avirulent with respect to crown gall and hairy root induction. A tumorigenic strain is not necessarily designated *A. tumefaciens*, or a hairy root strain *A. rhizogenes*, since transconjugants can be selected in which any strain can harbour a Ti or Ri plasmid. For this reason it is preferable to designate *Agrobacterium* strains with their plasmid complement.

The Ti and Ri plasmids are named according to the *Agrobacterium* strain from which they were originally isolated. They are large plasmids and range in size from 140 to 235 kilobase pairs (Melchers & Hooykaas, 1987). The T-DNA region which is transferred to plant cells ranges in size from 14 to 42 kb, and is bordered by conserved 25 base pair sequences (see below). Any sequences between these borders may be integrated into plant nuclear DNA. There are two other important regions on Ti and Ri plasmids. One is the virulence region which encodes the genes responsible for the excision, transfer and integration of T-DNA from *Agrobacterium* into the genome of plant cells (see below). The other region encodes genes responsible for the catabolism of opines, thus allowing *Agrobacterium* to utilize the opines secreted in tumours and hairy roots.

Opines are secreted from transformed plant cells into the intercellular regions of a tumour or rhizosphere of hairy roots where *Agrobacterium* lives (Petit & Tempé, 1985). These compounds cannot be metabolized by the plant cells but serve as a carbon and nitrogen source for the bacteria. The opine synthase genes carried by each Ti and Ri plasmid determine the opine type produced by the tumour or hairy root cells. A strain carrying specific opine synthase genes on its T-DNA also carries genes for the catabolism of these specific opines elsewhere on its Ti or Ri plasmid. For this reason,

Table 3.1. Opine types classified on a biological basis.

Opine type	Representative opines present in neoplastic growths	*Agrobacterium* strain examples
Ti plasmids		
Octopine	Octopine, octopinic acid, lysopine, histopine, agropine, agropinic acid, mannopine, mannopinic acid	B6, ACH5
Nopaline	Nopaline, nopalinic acid, agrocinopine A	C58, T37
Agropine	Agropine, agropinic acid, mannopine, mannopinic acid, agrocinopine C	AT1, AT4
Succinamopine	Succinamopine, succinamopine lactam, succinopine lactam	Eu6, 181
Grapevine	Octopine, cucumopine	K305, K308
Ri plasmids		
Agropine	Agropine, agropinic acid, mannopine, mannopinic acid, agrocinopine A	A4, TR105
Mannopine	Mannopine, mannopinic acid, agrocinopine C, agropinic acid	TR7, 8196
Cucumopine	Cucumopine	2655, 2657

Derived from Petit & Tempé, 1985; Davioud *et al.,* 1988.

Agrobacterium strains and their Ti and Ri plasmids are often classified on the basis of opine types (see Table 3.1).

Nopaline and succinoamopine Ti plasmids and mannopine Ri plasmids have one continuous T-DNA flanked by two border sequences (Melchers & Hooykaas, 1987). In contrast, octopine Ti plasmids and agropine Ri plasmids have non-continuous T-DNAs, known as the TL-DNA and TR-DNA, which are each bound by two border sequences.

Tumorigenic and rhizogenic genes

In vitro cultures of tumours and hairy roots continue to grow and proliferate after excision from the site of inoculation. This growth occurs without exogenously supplied hormones and is due to the expression of stably integrated T-DNA genes. These genes are either responsible for phytohormone production (the *onc* (oncogenicity) genes of Ti plasmids) or increased sensitivity to auxins (the *rol* (root locus) genes of Ri plasmids). Tumour tissues continue to proliferate as a disorganized callus culture, whereas hairy roots proliferate into a highly branched plagiotropic root system (Tepfer, 1989).

Homologous DNA sequences have been found in nopaline T-DNA and octopine TL-DNA (Engler *et al.*, 1981). These regions of sequence similarity contain the tumour-producing phytohormone genes encoding enzymes in the pathways for the production of isopentenyl-AMP (a cytokinin) and indole-acetic acid (an auxin). A recently characterized *onc* gene (*6b*) has also been found to have growth-inducing properties (Hooykaas *et al.*, 1988). This gene induces tumours on *Nicotiana glauca*, *N. rustica*, *Kalanchoe tubiflora* and grapevine, and may act by reducing the inhibitory effects of high auxin concentrations, thereby maintaining cells in an undifferentiated state (Tinland *et al.*, 1990).

The T-DNA of Ri plasmids has a series of four *rol* genes known as *rol A, B, C* and *D*. The transfer to, and expression of, these genes in plant cells leads to the hairy root phenotype (Tepfer, 1989). The exact cellular effects of the *rol* genes are still not clearly understood, however, they are known to enhance the sensitivity of plant cells to endogenous auxins (Shen *et al.*, 1988). In agropine Ri plasmids, the TR-DNA contains auxin biosynthetic genes which may play a minor role in the development of hairy roots (Gelvin, 1990).

Host range of *Agrobacterium*

A wide range of plants is susceptible to tumour and hairy root formation induced by *Agrobacterium*. The hosts for *Agrobacterium* are mostly dicotyledonous plants, but also include a few gymnosperms (De Cleene & De Ley, 1976, 1981) and several monocotyledonous plants (Suseelan *et al.*, 1987; Dommisse & Conner, 1989; Dommisse *et al.*, 1990). The strain specificity for plant genotypes and tissue types has been well documented. Examples include peas (Hobbs *et al.*, 1989), soybeans (Byrne *et al.*, 1987), pine (Morris *et al.*, 1989) and *Brassica* (Charest *et al.*, 1989).

The nature of the interaction between bacteria and plant cells is still unclear. It is important to screen a series of *Agrobacterium* strains on a range of crop genotypes for virulent strain–genotype combinations. For example, to obtain transgenic soybean, Hinchee *et al.* (1988) first surveyed 100 cultivars for their *in vitro* response to *A. tumefaciens*-mediated transformation and then used the three most responsive in subsequent experiments. Furthermore, the age, type and physiological state of the plant tissue prior to inoculation can also be important in determining strain–genotype compatibilities (Hernalsteens *et al.*, 1984).

There are many hypotheses to explain the inability of *Agrobacterium* to induce tumours or hairy roots in most monocotyledonous species (and other non-susceptible plants). These include: lack of binding of *Agrobacterium* to plant cell walls (Rao *et al.*, 1982), reduced activity of T-DNA promoters (Graves & Goldman, 1986), inhibition of *vir* gene induction (Sahi *et al.*, 1990) and 'abnormal' auxin–cytokinin balance in monocotyledonous cells

(Schafer *et al.*, 1987). The lack of a pronounced wound response is thought to be an important factor, since only plants and tissues which proliferate large populations of wound-adjacent cells are considered competent for efficient transformation (Potrykus, 1990).

Molecular events involved in T-DNA transfer

Agrobacterium chromosomal genes

The first step toward gene transfer by *Agrobacterium* is the attachment of the bacteria to host plant cells at wound sites. The nature of the plant cell receptor to which *Agrobacterium* binds is unknown, but the availability of a receptor for attachment is considered to be one factor important in determining the host range of *Agrobacterium* (Matthysse, 1986). The plant cell is believed to be a passive partner in the attachment process. *Agrobacterium* plays a more active role, with several constitutively expressed chromosomal virulence genes being important. These genes (*chvA*, *chvB*, *pscA*, *att*) affect the biosynthesis and secretion of polysaccharides such as beta-1,2-glucan and succinoglycan. Mutations in any of these chromosomal loci result in defective plant cell attachment and consequent lack of virulence (Garfinkel & Nester, 1980; Matthysse, 1987; Thomashow *et al.*, 1987).

The T-DNA border sequences

The T-DNA regions of *Agrobacterium* Ti and Ri plasmids are flanked by border sequences that define the segment of DNA to be transferred to recipient plant cells. These border sequences are 25 bp imperfect direct repeats and are the only elements required in *cis* orientation for mobilization of the DNA into the plant cell (Zambryski, 1988). Any DNA sequences placed between these borders can be transferred into plant cells.

Deletion of the first 6 bp or the last 10 bp of the 25 bp sequences blocks T-DNA transfer (Wang *et al.*, 1987). The right border is a key element as T-DNA transfer is initiated at this site (see below). Consequently the correct orientation is important; inversion results in a substantial reduction in efficiency of T-DNA transfer. Neighbouring the right border (just outside the T-DNA) is a 24 bp enhancer element which contributes to the efficiency of T-DNA transfer (Peralta *et al.*, 1986). This element, known as overdrive (*ode*) is located 13–14 bp from the right border, although it is active up to 7 kb away, and can function in either orientation.

The virulence region of the Ti and Ri plasmids

The virulence region of Ti and Ri plasmids occurs outside the T-DNA and is a segment of approximately 40 kb consisting of six distinct operons (*virA*, *virB, virG, virC, virD, virE,* reading clockwise towards the T-DNA) encoding *trans*-acting factors essential for T-DNA transfer. There are varying numbers of open reading frames within each of these operons, which are strongly, co-ordinately induced by phenolic compounds leached from wound sites on plants (Stachel *et al.,* 1985; Bolton *et al.,* 1986). The ability of plants to produce these signal molecules may be an important factor contributing to the host range of *Agrobacterium.*

The *virA* and *virG* loci are required for induction of the remainder of the *vir* operons by external signals. The primary signal molecules present in tobacco exudates from wounded cells were purified and identified as acetosyringone and alpha-hydroxyacetosyringone (Stachel *et al.,* 1985), although other phenolic compounds were capable of partially or fully inducing activity. Some opines enhance the induction of *vir* operons by phenolic compounds (Veluthambi *et al.,* 1989).

The virA and virG proteins form part of a two-component positive regulatory system (i.e. sensor and activator). The virA protein appears to be a transmembrane protein, and therefore is in the correct position to act as the primary signal receptor (Hooykaas, 1989). Once the virA protein binds the phenolic inducer, the resulting complex is thought to act as a kinase phosphorylating the virG protein. This converts the virG protein to a form capable of binding to specific DNA consensus sequences of various *vir* gene promoters. In this manner the activated virG protein increases transcription of its own gene and induces the transcription of the *virB, virC, virD* and *virE* operons (Zambryski, 1988).

Two proteins encoded by the *virD* operon, virD1 and virD2, have an endonuclease activity capable of generating single-stranded site-specific nicks between the third and fourth base pairs on the bottom strand of the T-DNA border repeats (Hooykaas, 1989). The virD2 protein attaches to the 5′ terminus of the nicked right border T-DNA and a replicative process synthesizes a single-stranded DNA molecule from the bottom strand of the T-DNA. The single-stranded nick at the left border repeat permits release of a single-stranded DNA molecule known as the T-strand (Zambryski, 1988). The overdrive enhancer element (*ode*) stimulates T-strand formation by interacting with the virD2 protein and one or both proteins encoded by the *virC* operon. A non-specific single-stranded DNA-binding protein, the virE2 protein, is thought to bind to and protect the T-strands. This T-strand–protein complex is believed to be an intermediary for T-DNA transfer to plant cells which must be exported by the *Agrobacterium.*

The *virB* operon encodes at least 11 proteins thought to form membrane-associated structures that may form a channel(s) through which the T-strand–protein complex is exported (Hooykaas, 1989).

Other accessory virulence genes may be present in specific Ti or Ri plasmids and may help to determine host range for specific plant species (Hooykaas, 1989). For example, in octopine strains there is a *virF* locus encoding a 22.4 kDa protein necessary for tumour formation on *Nicotiana glauca*. Nopaline strains lack the *virF* locus, but instead contain a *tzs* gene encoding biosynthesis of the cytokinin trans-zeatin in *Agrobacterium*. The exact role that these genes play in the T-DNA transfer process is not well understood.

Vector systems for gene transfer to plant cells

Disarmed *Agrobacterium* strains

The phytohormone biosynthetic genes encoded on the T-DNA of wild-type Ti plasmids interfere with the protocols for the regeneration of plant cells. This led to the development of 'disarmed' Ti plasmids, in which the phytohormone biosynthetic genes are deleted from the T-DNA. As a result normal morphogenesis of transformed plant cells into complete plants is then possible in tissue culture (Fraley *et al.*, 1986). The phytohormone biosynthetic genes can be completely removed as first described by Zambryski *et al.* (1983), partially removed (e.g. Barton *et al.*, 1983) or arranged in such a way that the T-DNA border sequences do not allow the phytohormone biosynthetic genes to be inserted into the plant genome.

Alternatively, the T-DNA may be completely removed from the resident Ti plasmid. The utility of this approach was first demonstrated by Hoekema *et al.* (1983), who constructed pAL4404, a modified pTiACH5 which carried no T-DNA but retained the *vir* region responsible for T-DNA transfer. Such plasmids are used in conjunction with binary vectors (see below).

Co-integrate vectors

Co-integrate vectors involve the integration of foreign DNA into the resident Ti or Ri plasmid. The foreign DNA is firstly inserted into a vector that cannot replicate in *Agrobacterium* cells but can recombine with the Ti/Ri plasmid through a single or double crossover event at a homologous site. This produces a co-integrate between the two plasmids. There are two main methods for their construction.

1. Zambryski *et al.* (1983) deleted a large region of T-DNA from the nopaline Ti plasmid, pTiC58. This completely removed the oncogenic functions of the T-DNA except for the border sequences and the nopaline synthase gene, which was retained as a T-DNA specific marker. In place of the T-DNA a bacterial plasmid pBR322 was inserted. The resulting plasmid pGV3850 is an active co-integrate vector for transfer of T-DNA to plant cells. Because the pBR322 sequence lies between the

T-DNA borders, any foreign gene inserted into a plasmid containing homologous sequences can be introduced into the *Agrobacterium* by a single crossover event.
2. Fraley *et al.* (1985) modified the octopine Ti plasmid pTiT37. They removed 70% of the TL-DNA and the entire TR-DNA, leaving the left border and 3 kb of the TL-DNA containing a sequence allowing the introduction of genes via a single crossover event. The plasmid to be co-integrated into the modified Ti plasmid must have the homologous sequence, as well as a T-DNA right border sequence. This is known as the SEV (split-end vector) system as the border sequences are initially present on separate plasmids. If designed correctly, the SEV system eliminates the presence of bacterial genes between the border sequences, a feature in the Zambryski system.

The main advantage of the co-integrate vectors is their high stability in *Agrobacterium*. However, two disadvantages are the detailed knowledge required of the Ti plasmid before it can be manipulated, and the relatively low rates of co-integrate formation (about 10^{-5}).

Binary vectors

Binary vectors involve foreign genes being located on an additional plasmid capable of autonomous replication in *Agrobacterium* cells (Fraley *et al.*, 1986; Klee *et al.*, 1987). These vectors do not require a region of homology with the resident Ti or Ri plasmid. The binary vectors need to contain at least one border sequence, although they usually contain both the left and right borders. The main advantages of the binary vectors are their independence of specific Ti and Ri plasmids and the high frequency of introduction into *Agrobacterium* (about 10^{-1}; Klee *et al.*, 1987). Therefore, binary vectors are currently used to the virtual exclusion of co-integrate vectors.

Binary vectors are especially useful in conjunction with disarmed *Agrobacterium* strains harbouring plasmids carrying an intact *vir* region and no T-DNA, such as LBA4404 (Hoekema *et al.*, 1983) and EHA101 (Hood *et al.*, 1986). However, binary vectors can also be used in *Agrobacterium* strains that have not been disarmed (e.g. Fillatti *et al.*, 1987b), allowing genotype-specific strains of *Agrobacterium* to be more readily used for transformation. In this situation transformed cells can be of three types: those that have the binary vector T-DNA plus the native DNA from the Ti or Ri plasmid; those cells that have only the binary vector DNA; and those cells that have only the native T-DNA.

Components of the modified T-DNA

The genes used in transformation experiments are often of non-plant origin.

Their regulatory signals must be appropriately modified to allow expression in plant cells. This requires selection of suitable promoter, termination and translation initiation regions to achieve the desired magnitude and tissue specificity of gene expression (Fraley *et al.,* 1986). These genes are chimeric in that the various components are often derived from diverse sources.

A selectable marker gene is an essential component of the modified T-DNA. It allows the preferential growth of the rare transformed cells by conferring the ability to grow in tissue culture on media containing normally toxic concentrations of a particular selective agent. Features of a good selectable system for plant transformation include: ease of selection, absence or low frequency of escapes, non-toxicity to resistant plant tissue, and inheritance as a dominant trait. The most commonly used selectable marker gene has been the neomycin phosphotransferase (NPTII) gene, which confers resistance to kanamycin and related antibiotics (Fraley *et al.,* 1986). The NPTII gene was derived from bacterial transposon Tn5, and acts by inactivating kanamycin via a phosphorylation reaction, thereby permitting transformed cells to grow and differentiate. A range of other useful selectable marker systems involving antibiotic and herbicide resistance genes has also been developed. These include resistance to hygromycin (Van de Elzen *et al.,* 1985), bleomycin (Hille *et al.,* 1986), streptomycin (Jones *et al.,* 1987b), gentamicin (Hayford *et al.,* 1988), phosphinothricin (De Block *et al.,* 1987), bromoxynil (Stalker *et al.,* 1988) and sulphonamide (Guerineau *et al.,* 1990).

Another useful component of the modified T-DNA is a reporter gene whose expression can be easily monitored upon transfer to plant cells. Features of a good reporter gene include: no endogenous expression in plant cells, no expression in *Agrobacterium*, and an easily assayable product. The most common and versatile reporter gene is the beta-glucuronidase (GUS) gene (*uidA*) from *Escherichia coli*. Activity of the encoded enzyme cleaves a wide range of beta-glucuronide substrates, which can be conveniently measured using fluorometric, spectrophotometric or histochemical assays (Jefferson *et al.,* 1987). Janssen & Gardner (1990) developed a version of the GUS gene which lacks a bacterial ribosome binding site and shows negligible expression in *Agrobacterium*. This gene is especially useful when developing and optimizing *Agrobacterium* transformation protocols for specific plant genotypes or species.

Selection and regeneration of transformed plant cells

The most widely used method for *A. tumefaciens* is the co-cultivation of disarmed strains carrying a modified T-DNA with explants of plant tissue. For *A. rhizogenes*-mediated transformation the usual approach is to co-cultivate with strains harbouring binary vectors and select transformed hairy roots from which complete plants are regenerated in culture. Other approaches not involving the use of tissue culture include the co-cultivation of *Agrobacterium* with imbibed seeds, e.g. *Arabidopsis thaliana* (Feldmann & Marks, 1987), and the injection of *Agrobacterium* into germinating seeds, e.g. soybean (Chee *et al.*, 1989). When transformation occurs using the latter methods the resulting plants are expected to be chimeric for sectors of untransformed tissues and sectors of transformed tissue that may have arisen from one or more independent transformation events. Progeny of the next generation can be screened for totally transgenic plants resulting from individual transformation events.

Co-cultivation with *Agrobacterium*

The initial studies on transfer of foreign genes to plants involved the co-cultivation of plant protoplasts with *Agrobacterium* (Fraley *et al.*, 1983; Herrera-Estrella *et al.*, 1983). A major technical advance was the demonstration that transgenic plants could be regenerated from leaf discs following co-cultivation with *Agrobacterium* (Horsch *et al.*, 1985). Subsequently transgenic plants have been produced from many families using this approach, or modifications of it (see Table 3.2). Virtually every plant explant source has been co-cultivated with *Agrobacterium* and transgenic plants obtained. Such explants include cotyledons, leaves, thin cell layers, cotyledonary petioles, peduncles, hypocotyls, stems, microspores and proembryos.

The explant co-cultivation method involves dipping explants into a culture of modified *Agrobacterium*, blotting on sterile filter paper, and culturing on callusing or regeneration media. After 24–72 h incubation, explants are transferred to similar medium containing an antibiotic (e.g. cefotaxime or carbenicillin) to suppress *Agrobacterium* growth. Selection for transgenic cells, by inclusion of the selective agent in the culture medium, is usually initiated at this time. In some instances it is preferable to delay selection until 6–8 days after co-cultivation, e.g. potatoes (Conner *et al.*, 1991). The concentrations of the selective agent used vary widely depending on the sensitivity of the plant species and/or explant source. In *Brassica napus* 15 mg/l of kanamycin was used (Moloney *et al.*, 1989), whereas 300 mg/l was necessary in petunias, tobacco and tomato (Horsch *et al.*, 1985). Cell colonies and/or shoots growing on the selection medium are transferred to fresh medium as required for the development of complete plants. The selective agent is usually maintained in all culture media throughout plant development. In some instances it may inhibit shoot regeneration or root initiation and it may become necessary to omit or reduce the concentration of the selective agent, e.g. potatoes (Conner *et al.*, 1991).

The methods for *A. rhizogenes* either involve co-cultivation (as above) or wounding axenic plant tissue and smearing actively growing bacterial cells

Table 3.2. Examples of transgenic crop plants, containing foreign genes, produced via *Agrobacterium*-mediated transformation.

Vegetables	Arable crops	Ornamental and medicinal	Fruit and trees	Pasture crops
Potato (1)	Sunflower (13)	*Petunia* (21)	Pepino (26)	Alfalfa/lucerne (36)
Tomato (2)	Sugarbeet (14)	Tobacco (21)	Tamarillo (27)	Lotus (37)
Lettuce (3)	Oilseed rape (15)	*Kalanchoe* (22)	Apple (28)	*Stylosanthes* (38)
Celery (4)	Soybean (16)	Chrysanthemum (23)	Kiwifruit (29)	White clover (39)
Cucumber (5)	Cotton (17)	Geranium (24)	Walnut (30)	
Carrot (6)	Flax/linseed (18)	*Lisianthus* (25)	*Populus* (31)	
Cauliflower (7)	Kale (19)		Strawberry (32)	
Broccoli (8)	Mustard (20)		*Citrus* (33)	
Cabbage (9)			Muskmelon (34)	
Asparagus (10)			*Azadirachta* (35)	
Eggplant (11)				
Pea (12)				

1 De Block, 1988
2 McCormick *et al.*, 1986
3 Michelmore *et al.*, 1987
4 Catlin *et al.*, 1988
5 Trulson *et al.*, 1986
6 Scott & Draper, 1987
7 De Block *et al.*, 1988
8 Christey & Earle, 1989
9 Shahin & Yashar, 1986
10 Conner *et al.*, 1988
11 Guri & Sink, 1988
12 Puonti-Kaerlas *et al.*, 1990
13 Everett *et al.*, 1987

14 Gasser & Fraley, 1989
15 Fry *et al.*, 1987
16 Hinchee *et al.*, 1988
17 Firoozabady *et al.*, 1987
18 Basiran *et al.*, 1987
19 Christey & Sinclair, 1990
20 Mathews *et al.*, 1990
21 Horsch *et al.*, 1985
22 Jia *et al.*, 1989
23 Ledger *et al.*, 1991
24 Butcher *et al.*, 1990
25 Deroles *et al.*, 1990
26 Atkinson & Gardner, 1991

27 Atkinson *et al.*, 1990
28 James *et al.*, 1989
29 Janssen, personal communication
30 McGranahan *et al.*, 1988
31 Fillatti *et al.*, 1987b
32 Nehra *et al.*, 1990
33 Hidaka *et al.*, 1990
34 Fang & Grumet, 1990
35 Naina *et al.*, 1989
36 Shahin *et al.*, 1986
37 Jensen *et al.*, 1986
38 Manners & Way, 1989
39 White & Greenwood, 1987

on to the exposed surface. When a small swelling has developed it is excised and placed on culture media with appropriate selective compounds for the development of hairy root cultures. Such hairy roots can be selected in a wide range of plant species, many of which can be regenerated into plants (Tepfer, 1989). Examples of transgenic plants produced in this manner include *Lotus* (Jensen *et al.*, 1986), kale (Christey & Sinclair, 1990), cucumber (Trulson *et al.*, 1986) and tobacco (Comai *et al.*, 1985). Plants regenerated from hairy roots often exhibit an unusual phenotype due to the expression of the *rol* loci. This Ri phenotype involves wrinkled leaves, shortened internodes, reduced apical dominance and plagiotropic roots etc. (Tepfer, 1989). In most transgenic plants regenerated from hairy roots, the *rol* loci and the binary vector T-DNA segregate independently of one another in the next generation. This allows the identification of phenotypically normal plants transformed with only the binary vector T-DNA (Manners & Way, 1989).

Enhancing transformation

The frequencies at which transgenic plants have been selected and re-generated following *Agrobacterium* transformation vary considerably between plant species. These range from 100% of explants producing trans-genic shoots in potato (De Block, 1988), to less than 1% in sunflower (Schrammeijer *et al.*, 1990) and asparagus (Conner *et al.*, 1988). Marked variations in transformation frequencies are also common among genotypes of the same crop (Conner *et al.*, 1991).

Identification of suitable *Agrobacterium* strain and plant genotype com-binations is important for optimizing transformation rates (see above). The choice of selectable marker system, components of the plant culture media and environment, and explant tissue to be co-cultivated are all important. Since plant tissues often decrease in their tissue culture aptitude following co-cultivation, the cells which interact with *Agrobacterium* must have a high potential for growth and regeneration. Moloney *et al.* (1989) were able to obtain higher transformation rates than previously reported in *Brassica napus* (55% vs. < 10%) solely by the use of intact petioles from cotyledons instead of other explants. In some instances the co-cultivation of explants with *Agrobacterium* results in a severe hypersensitive response, which dramatically reduces shoot regeneration. By using agarose instead of agar, Charest *et al.* (1988) reduced tissue necrosis in *Brassica napus*, thereby allowing shoot regeneration and production of transgenic plants from thin layers of epidermal cells.

An essential component of the *Agrobacterium* system is the induction of the *vir* operons (see above). A number of methods have been used to induce these operons, thereby enhancing transformation. For example, preincuba-tion of *A. tumefaciens* with 20 μM acetosyringone increased the trans-formation frequency of *Arabidopsis thaliana* from 2–3 to 64% of explants producing transgenic plants (Sheikholeslam & Weeks, 1987). In the monocotyledonous *Dioscorea bulbifera* (yam), *in vitro* tumour production was achieved only after induction of *A. tumefaciens* with wound exudates from potato tuber tissue (Schafer *et al.*, 1987).

Using a cell suspension feeder layer during the co-cultivation period has been found to have beneficial effects on the transformation frequency in some species. For example, in tomato the proportion of explants producing transformation events increased 15% following co-cultivation on media with tobacco cells (Fillatti *et al.*, 1987a). In lettuce, Michelmore *et al.* (1987) found the presence of a feeder layer, although not absolutely necessary, ensured repeatable high rates of transformation. The use of feeder cells, such as plated tobacco cell suspensions, may enhance transformation via two mechanisms: (i) inducing *Agrobacterium* *vir* genes (Veluthambi *et al.*, 1987); and (ii) promoting cell division and growth of individual plant cells (Horsch & Jones, 1980).

Analyses of transformed plant cells

Selection and growth of plant cells on selective media provide initial phenotypic evidence for transformation. However, spontaneous variants with increased resistance to many chemicals can be readily selected in plant tissue culture (Conner & Meredith, 1989). This includes resistance to kanamycin (Owens, 1981), the most commonly used selection agent for plant transformation. Therefore biochemical and molecular evidence is essential to confirm expression and integration of transferred genes. Studying segregation in subsequent generations to examine inheritance and genetic stability of the transferred gene(s) is also important.

Gene expression

Biochemical and molecular evidence for the expression of the inserted gene(s) can be obtained via northern analyses, enzyme assays and western blotting. Northern analyses can confirm the presence of the expected RNA transcript and that it is of the correct size. Western blotting allows detection of the anticipated translation products and their quantitation, whereas enzyme assays (when possible) confirm the presence of a functionally active gene-specified protein.

The presence of a selectable marker allows selection of transgenic cells but does not guarantee 100% co-transmission and/or expression of the other genes on the same section of T-DNA. Independently selected transgenic plants often show varying degrees of gene expression (e.g. Hoekema *et al.*, 1989). Several factors other than the absolute effects of the regulatory sequences can influence the magnitude of gene expression. These include copy number, position effects resulting from the site of insertion and methylation of the transferred genes (Horsch *et al.*, 1985; Jones *et al.*, 1987a).

Gene integration

The integration of the foreign gene(s) into the plant nuclear genome can be determined via Southern analyses and the use of the polymerase chain reaction (PCR). Southern analyses allow the number of copies and nature of the integration of specific genes or DNA regions to be determined. PCR is a new and powerful technique for confirming DNA insertion in transgenic plants (Lassner *et al.*, 1989). Primers can be designed which simultaneously amplify specific genes or DNA regions on the T-DNA that are expected to be integrated into the genome of plants. Advantages include the rapid manner in which large collections of transgenic plants can be analysed and the very small amount of plant tissue required.

Insertion of modified T-DNA into the genome of the recipient plant generally occurs in a random manner. This has been determined from mapping T-DNA inserts in ten independent tomato lines (Chyi *et al.*, 1986) and nine independent petunia lines (Wallroth *et al.*, 1986). Furthermore, there is no control over the number of integration events, or whether the modified T-DNA is transferred in a complete, truncated or rearranged manner. In studies of large collections of transgenic plants (Spielmann & Simpson, 1986; Jones *et al.*, 1987a; Jorgensen *et al.*, 1987; Deroles & Gardner, 1988b) integration of intact T-DNA regions was common. Deviations usually involve deletions of the T-DNA ends, inverted or tandem repeats, and complex rearrangements. These studies also establish that the majority of events are single insertions, although up to 4–5 independent genome sites in the same cell may be involved.

Inheritance of T-DNA inserts

The primary transgenic plants are expected to be heterozygous for the T-DNA insertion events. Their progeny segregate for the T-DNA. In studies of large collections of kanamycin-resistant transgenic plants (Budar *et al.*, 1986; Deroles & Gardner, 1988a; Heberle-Bors *et al.*, 1988), the majority segregated for kanamycin resistance in ratios expected for a Mendelian inherited, single, dominant, nuclear locus. In each of these studies there was also a significant proportion of plants segregating for two independently inherited loci. The segregation ratios following selfing in some plants suggested that T-DNA insertion may have caused recessive lethal mutations by disrupting an important gene. Anomalous segregation patterns in undefined genetic ratios were also found in some transgenic plants and were often associated with high copy number of T-DNA (Jones *et al.*, 1987a; Jorgensen *et al.*, 1987; Deroles & Gardner, 1988b).

Transgenic plants homozygous for single T-DNA insertions continue to breed true and show high meiotic stability upon backcrossing to wild-type plants (Muller *et al.*, 1987). This genetic stability is essential for the use of *Agrobacterium*-mediated transformation in agriculture.

Agricultural perspective

Plants transformed using *Agrobacterium* have been obtained in a wide range of crops (Table 3.2). In virtually all instances kanamycin resistance has been used as a selectable marker gene. The information presented in Table 3.2 shows that *Agrobacterium*-mediated transformation has been highly successful. However, in many crops only one or very few transgenic plants have been produced (e.g. asparagus, broccoli, walnut). In other crops, considerable research is still required for the efficient production of transgenic plants in agriculturally useful genotypes. Sometimes *Agrobacterium*-mediated transformation of élite genotypes has only operated well in single or very few laboratories (e.g. lettuce, kale, cotton, clover). The methods are routine in solanaceous crops (e.g. potato, tomato, tobacco), with systems also being well developed in oilseed rape, soybean, alfalfa, and *Lotus*.

The main obstacle to efficient *Agrobacterium*-mediated transformation in many crops is combining selection of transformed cells and their subsequent regeneration. Although each of these components can be achieved independently in most crops, it is often difficult to obtain both *Agrobacterium* transformation and plant regeneration in the same cell. Genetic variation for sensitivity to *Agrobacterium* transformation (see above) and response in cell culture (Conner & Meredith, 1989) are well known within crop species. Selection and breeding for enhanced regeneration from tissue culture have proved successful, even after only one or two generations of selection (Conner & Meredith, 1989). In many crops it may be necessary to use a genotype specifically developed for transformation. It would then be necessary to incorporate newly introduced genes into élite genotypes or existing cultivars by conventional plant breeding techniques, although this would increase the time to cultivar release.

Despite these limitations considerable progress has already been made towards gene transfer of agriculturally important traits. This principally involves the development of plants with improved resistance to herbicides, insect pests and viral diseases (see other chapters in this book). Many transgenic plants expressing agriculturally useful traits have already reached field testing stage (Table 3.3).

For the successful integration of *Agrobacterium*-mediated transformation into plant breeding programmes, procedures must be developed for efficient, large-scale selection of transformed cells and rapid regeneration of transgenic plants. Many different independently transformed lines can be screened for those that retain all or most of their previous élite traits, but incorporate the desired genetic change. This is important for two main reasons.

1. The point of T-DNA integration can interrupt the functioning of existing genes in plant cells via insertional mutagenesis (e.g. Feldmann *et al.*, 1989) and result in position effects on gene expression (see above). Since these events differ among independently transformed lines, plants exhibiting such effects can be discarded.

2. Genetic changes (somaclonal variation) are known to occur during the cell culture and plant regeneration phase associated with plant transformation. This variation is unrelated to the specific modification sought, and is accentuated by plant cell culture over prolonged periods. Rapid regeneration of many independently transformed lines will help reduce the frequency of such changes and allow selection of transgenic lines that are phenotypically normal when grown in the field.

Table 3.3. Examples of transgenic crop plants resulting from *Agrobacterium*-mediated gene transfer to have reached field testing stage of development.

Transgenic phenotype	Crop plants
Herbicide resistance	
Phosphinothricin	Tobacco, tomato, potato, oilseed rape, sugarbeet, alfalfa, poplar
Glyphosate	Tobacco, tomato, oilseed rape, flax, sugarbeet, alfalfa, soybean, cotton
Bromoxynil	Tobacco, tomato, cotton
Sulphonylurea	Tobacco, tomato, potato, flax
Atrazine	Tobacco
Insect resistance	
Insecticidal BT proteins	Tobacco, tomato, potato, cotton
Proteinase inhibitors	Tobacco
Virus resistance	
Alfalfa mosaic virus	Tobacco, alfalfa
Tobacco mosaic virus	Tomato
Tomato mosaic virus	Tomato
Potato virus X	Potato
Potato virus Y	Potato
Potato leaf roll virus	Potato
Rhizomania virus	Sugarbeet
Cucumber mosaic virus	Cucumber
Other characters	
Heavy metal resistance	Tobacco
Firm fruit	Tomato
Thaumatin production	Potato
Modified seed storage proteins	Oilseed rape

Extracted from OECD (1990).

The number of independently selected transformed lines that constitute a large collection is unknown, and will no doubt differ for the specific crop and the nature of its breeding system(s). For seed-propagated crops it may be more efficient to produce fewer transformed lines and then eliminate undesirable somaclonal changes via conventional plant breeding. However, in clonal crops further breeding would be undesirable, since the genetic integrity of élite clones would be lost. For this reason larger numbers of primary transgenic plants may be important for clonal crops.

Conclusions

The natural gene-transferring ability of *Agrobacterium* can be exploited to genetically modify plants. The transfer of DNA from *Agrobacterium* to plant cells is a consequence of specific DNA–protein interactions. However, understanding of the precise mechanisms involved is incomplete. Although further research on the processes involved may enhance the efficiency of gene transfer by *Agrobacterium* and extend the use of such a system to a broader range of plant species, it is not crucial for the successful transfer of genes to many crop plants.

Agrobacterium-mediated transformation offers an exciting, proven approach for the genetic manipulation of crop plants. It is now technically possible to transfer genes across all taxonomic boundaries into plants – from other plants, animals and microbes – or even to introduce totally artificial genes. This offers considerable potential for genetic improvement of crop plants, especially for disease and pest resistance, and improved quality characteristics. In addition to applications in crop improvement, *Agrobacterium*-mediated transformation offers a powerful research tool for studying plant biology, especially the control mechanisms in gene expression and development.

Acknowledgements

We thank Drs Gail Timmerman, Richard Gardner and Nick Ashby for helpful comments on the manuscript.

References

Atkinson, R. G. & Gardner, R. C. (1991) Regeneration of transgenic pepino plants. *Plant Cell Reports*, in press.

Atkinson, R. G., Hutching, D. & Gardner, R. C. (1990) Transformation and regeneration of the pepino and the tamarillo. *New Zealand Genetical Society, 36th Annual Meeting*, Abstract 2.

Barton, K. A., Binns, A. N., Matzke, A. & Chilton, M. D. (1983) Regeneration of intact tobacco plants containing full length copies of genetically engineered T-DNA to R1 progeny. *Cell* 32, 1039–43.

Basiran, N., Armitage, P., Scott, R. J. & Draper, J. (1987) Genetic transformation of flax (*Linum usitatissimum*): regeneration of transformed shoots via a callus phase. *Plant Cell Reports* 6, 396–9.

Bolton, G. W., Nester, E. W. & Gordon, M. P. (1986) Plant phenolic compounds induce expression of the *Agrobacterium tumefaciens* loci needed for virulence. *Science* 232, 983–5.

Budar, F., Thia-Toong, L., van Montagu, M. & Hernalsteens, J.-P. (1986) *Agrobacterium*-mediated gene transfer results mainly in transgenic plants transmitting T-DNA as a single Mendelian factor. *Genetics* 114, 303–13.

Butcher, S. M., Deroles, S. C. & Ledger, S. E. (1990) Transformation of *Geranium*. *New Zealand Genetical Society, 36th Annual Meeting,* Abstract 95.

Byrne, M. C., McDonnell, R. E., Wright, M. S. & Carnes, M. G. (1987) Strain and cultivar specificity in the *Agrobacterium*–soybean interaction. *Plant Cell Tissue and Organ Culture* 8, 3–15.

Catlin, D., Ochoa, O., McCormick, S. & Quiros, C. F. (1988) Celery transformation by *Agrobacterium tumefaciens:* cytological and genetic analysis of transgenic plants. *Plant Cell Reports* 7, 100–3.

Charest, P. J., Holbrook, L. A., Gabard, J., Iyer, V. N. & Miki, B. L. (1988) *Agrobacterium*-mediated transformation of thin cell layer explants from *Brassica napus* L. *Theoretical and Applied Genetics* 75, 438–45.

Charest, P. J., Iyer, V. N. & Miki, B. L. (1989) Virulence of *Agrobacterium tumefaciens* strains with *Brassica napus* and *Brassica juncea*. *Plant Cell Reports* 8, 303–6.

Chee, P. P., Fober, K. A. & Slightom, J. L. (1989) Transformation of soybean (*Glycine max*) by infecting germinating seeds with *Agrobacterium tumefaciens*. *Plant Physiology* 91, 1212–18.

Christey, M. C. & Earle, E. D. (1989) Genetic manipulation of *Brassica oleracea* var. *italica* (broccoli) via protoplast fusion and transformation. *Australian Society of Plant Physiologists, 29th Annual Meeting,* Abstract 40.

Christey, M. C. & Sinclair, B. K. (1990) Selection of transformed hairy root lines in *Brassica oleracea, B. napus* and *B. campestris*. *Proceedings of the 6th Crucifer Genetics Workshop, Cornell University, Ithaca,* p.20.

Chyi, Y.-S., Jorgensen, R. A., Goldstein, D., Tanksley, S. D. & Loaiza-Figueroa, F. (1986) Locations and stability of *Agrobacterium*-mediated T-DNA insertions in the *Lycopersicon* genome. *Molecular and General Genetics* 204, 64–9.

Comai, L., Facciotti, D., Hiatt, W. R., Thompson, G., Rose, R. E. & Stalker, D. M. (1985) Expression in plants of a mutant *aroA* gene from *Salmonella typhimurium* confers tolerance to glyphosate. *Nature* 317, 741–4.

Conner, A. J. & Meredith, C. P. (1989) Genetic manipulation of plant cells. In: Marcus, A. (ed.), *The Biochemistry of Plants: A Comprehensive Treatise,* vol. 15, *Molecular Biology*. Academic Press, Orlando, pp. 653–88.

Conner, A. J., Williams, M. K., Deroles, S. C. & Gardner, R. C. (1988) *Agrobacterium*-mediated transformation of asparagus. In: McWhirter, K. S., Downes, R. W. & Reid, B. J. (eds), *Ninth Australian Plant Breeding Conference, Proceedings*. Agricultural Research Institute, Wagga Wagga, pp. 131–2.

Conner, A. J., Williams, M. K., Gardner, R. C., Deroles, S. C., Shaw, M. L. & Lancaster, J. E. (1991) *Agrobacterium*-mediated transformation of New Zealand potato cultivars. *New Zealand Journal of Crop and Horticultural Sciences,* in press.

Davioud, E., Petit, A., Tate, M. E., Ryder, M. H. & Tempe, J. (1988) Cucumopine – a new T-DNA-encoded opine in hairy root and crown gall. *Phytochemistry* 27, 2429–33.

De Block, M. (1988) Genotype-independent leaf disc transformation of potato (*Solanum tuberosum*) using *Agrobacterium tumefaciens*. *Theoretical and Applied Genetics* 76, 767–74.

De Block, M., Botterman, J., Vandewiele, M., Dockx, J., Thoen, C., Gossele, V., Roa Movva, N., Thompson, C., van Montagu, M. & Leemans, J. (1987) Engineering herbicide resistance in plants by expression of a detoxifying enzyme. *EMBO Journal* 6(9), 2513–18.

De Block, M., De Brouwer, D. & Tenning, P. (1988) Transformation of *Brassica napus* and *Brassica oleracea* using *Agrobacterium tumefaciens* and the expression of the *bar* and *neo* genes in the transgenic plants. *Plant Physiology* 91, 694–701.

De Cleene, M. & De Ley, J. (1976) The host range of crown gall. *Botanical Review* 42, 389–466.

De Cleene, M. & De Ley, J. (1981) The host range of infectious hairy root. *Botanical Review* 47, 147–94.

Deroles, S. C. & Gardner, R. C. (1988a) Expression and inheritance of kanamycin resistance in a large number of transgenic petunias generated by *Agrobacterium*-mediated transformation. *Plant Molecular Biology* 11, 355–64.

Deroles, S. C. & Gardner, R. C. (1988b) Analysis of T-DNA structure in a large number of transgenic petunias generated by *Agrobacterium*-mediated transformation. *Plant Molecular Biology* 11, 365–77.

Deroles, S., Ledger, S., Markham, K., Given, N. & Davies, K. (1990) Changing the colour of *Lisianthus*. *New Zealand Genetical Society, 36th Annual Meeting*, Abstract 97.

Dommisse, E. M. & Conner, A. J. (1989) Monocotyledonous plants as hosts for *Agrobacterium*. *Australian Society of Plant Physiologists, 29th Annual Meeting*, Abstract 36.

Dommisse, E. M., Leung, D. W. M., Shaw, M. L. & Conner, A. J. (1990) Onion is a monocotyledonous host for *Agrobacterium*. *Plant Science* 69, 249–57.

Engler, G., Depicker, A., Maenhaut, R., Villarroel-Mandiola, R., van Montagu, M. & Schell, J. (1981) Physical mapping of DNA base sequence homologies between an octopine and a nopaline Ti plasmid of *Agrobacterium tumefaciens*. *Journal of Molecular Biology* 152, 183–208.

Everett, N. P., Robinson, K. E. P. & Mascarenhas, D. (1987) Genetic engineering of sunflower (*Helianthus annuus*). *Bio/Technology* 5, 1201–4.

Fang, G. & Grumet, R. (1990). *Agrobacterium tumefaciens* mediated transformation and regeneration of musk melon plants. *Plant Cell Reports* 9, 160–4.

Feldmann, K. A. & Marks, M. D. (1987) *Agrobacterium*-mediated transformation of germinating seeds of *Arabidopsis thaliana*: a non-tissue culture approach. *Molecular and General Genetics* 208, 1–9.

Feldmann, K. A., Marks, M. D., Christianson, M. L. & Quatrano, R. S. (1989) A dwarf mutant of *Arabidopsis* generated by T-DNA insertion mutagenesis. *Science* 243, 1351–4.

Fillatti, J. J., Kiser, J., Rose, R. & Comai, L. (1987a) Efficient transfer of a glyphosate tolerance gene into tomato using a binary *Agrobacterium tumefaciens* vector. *Bio/Technology* 5, 726–30.

Fillatti, J. J., Sellmer, J., McCown, J. B., Haissig, B. & Comai, L. (1987b) *Agrobacterium*-mediated transformation and regeneration of *Populus*. *Molecular and General Genetics* 206, 192–201.

Firoozabady, E., DeBoer, D. L., Merlo, D. J., Halk, E. L., Amerson, L. N., Rashka, K. E. & Murray, E. E. (1987) Transformation of cotton (*Gossypium hirsutum* L.) by *Agrobacterium tumefaciens* and regeneration of transgenic plants. *Plant Molecular Biology* 10, 105–16.

Fraley, R. T., Rogers, S. G., Horsch, R. B., Sanders, P. R., Flick, J. S., Adams, S. P., Bittner, M. L., Brand, L. A., Fink, C. L., Fry, J. S., Galluppi, G. R., Goldberg, S. B., Hoffmann, N. L. & Woo, S. C. (1983) Expression of bacterial genes in plant cells. *Proceedings of the National Academy of Sciences USA* 80, 4803–7.

Fraley, R. T., Rogers, S. G., Horsch, R. B., Eichholtz, D. A., Flick, J. S., Fink, C. L., Hoffman, N. L. & Sanders, P. R. (1985) The SEV system: a new disarmed Ti plasmid vector system for plant transformation. *Bio/Technology* 3, 629–35.

Fraley, R. T., Rogers, S. G. & Horsch, R. B. (1986) Genetic transformation in higher plants. *CRC Critical Reviews in Plant Sciences* 4(1), 1–46.

Fry, J., Barnason, A. & Horsch, R. B. (1987) Transformation of *Bassica napus* with *Agrobacterium tumefaciens* based vectors. *Plant Cell Reports* 6, 321–5.

Garfinkel, D. J. & Nester, E. W. (1980) *Agrobacterium tumefaciens* mutants affected in crown gall tumorigenesis and opine catabolism. *Journal of Bacteriology* 144, 732–43.

Gasser, C. S. & Fraley, R. T. (1989) Genetically engineering plants for crop improvement. *Science* 244, 1293–9.

Gelvin, S. B. (1990) Crown gall disease and hairy root disease. A sledgehammer and a tackhammer. *Plant Physiology* 92, 281–5.

Graves, A. & Goldman, S. (1986) The transformation of *Zea mays* seedlings with *Agrobacterium tumefaciens*. *Plant Molecular Biology* 7, 43–50.

Guerineau, F., Brooks, L., Meadows, J., Lucy, A., Robinson, C. & Mullineaux, P. (1990) Sulfonamide resistance gene for plant transformation. *Plant Molecular Biology* 15, 127–36.

Guri, A. & Sink, K. C. (1988) *Agrobacterium* transformation of eggplant. *Journal of Plant Physiology* 133, 52–5.

Hayford, M. B., Medford, J. I., Hoffman, N. L., Rogers, S. G. & Klee, H. J. (1988) Development of a plant transformation selection system based on expression of genes encoding gentamicin acetyltransferases. *Plant Physiology* 86, 1216–22.

Heberle-Bors, E., Charvat, B., Thompson, D., Schernthaner, J. P., Barta, A., Matzke, A. J. M. & Matzke, M. A. (1988) Genetic analysis of T-DNA insertions into the tobacco genome. *Plant Cell Reports* 7, 571–4.

Hernalsteens, J. P., Tiang-Toong, L., Schell, J. & van Montagu, M. (1984) An *Agrobacterium*-transformed cell culture from the monocot *Asparagus officinalis*. *EMBO Journal* 3, 3039–41.

Herrera-Estrella, L., De Block, M., Van Montagu, M. & Schell, J. (1983) Chimeric genes as dominant selectable markers in plant cells. *EMBO Journal* 2, 987–95.

Hidaka, T., Omura, M., Ugaki, M., Tomiyama, M., Kato, A., Ohshima, M. & Motoyoshi, F. (1990) *Agrobacterium*-mediated transformation and regeneration of *Citrus* spp. from suspension cells. *Japanese Journal of Breeding* 40, 199–207.

Hille, J., Verheggen, F., Roelvink, P., Franssen, H., van Kammen, A. & Zabel, P. (1986) Bleomycin resistance: a new dominant selectable marker for plant transformation. *Plant Molecular Biology* 7, 171–6.

Hinchee, M. A. W., Connor-Ward, D. V., Newell, C. A., McDonnell, R. E., Sato, S. J., Gasser, C. S., Fischhoff, D. A., Re, D. B., Fraley, R. T. & Horsch, R. B. (1988) Production of transgenic soybean plants using *Agrobacterium*-mediated DNA transfer. *Bio/Technology* 6, 915–22.

Hobbs, S. L. A., Jackson, J. A. & Mahon, J. D. (1989) Specificity of strain and genotype in the susceptibility of pea to *Agrobacterium tumefaciens. Plant Cell Reports* 8, 274–7.

Hoekema, A., Hirsch, P., Hooykaas, J. & Schilperoort, R. (1983) A binary plant vector strategy based on separation of *vir-* and T-region of the *Agrobacterium tumefaciens* Ti-plasmid. *Nature* 303, 179–80.

Hoekema, A., Huisman, M. J., Molendijk, L., Van de Elzen, P. J. M. & Cornelissen, B. J. C. (1989) The genetic engineering of two commercial potato cultivars for resistance to potato virus X. *Bio/Technology* 7, 273–8.

Hood, E. E., Helmer, G. L., Fraley, R. T. & Chilton, M.-D. (1986) The hypervirulence of *Agrobacterium tumefaciens* A281 is encoded in a region of pTiBo542 outside of T-DNA. *Journal of Bacteriology* 168, 1291–301.

Hooykaas, P. J. J. (1989) Transformation of plant cells via *Agrobacterium. Plant Molecular Biology* 13, 327–36.

Hooykaas, P. J. J., Den Dulk-Ras, H. & Schilperoort, R. A. (1988) The *Agrobacterium tumefaciens* T-DNA gene 6b is an *onc* gene. *Plant Molecular Biology* 11, 791–4.

Horsch, R. B. & Jones, G. E. (1980) A double filter paper technique for plating cultured plant cells. *In Vitro* 16, 103–8.

Horsch, R. B., Fry, J. E., Hoffmann, N. L., Eichholtz, D., Rodgers, S. G. & Fraley, R. T. (1985) A simple and general method for transferring genes into plants. *Science* 227, 1229–31.

James, D. J., Passay, A. J., Barbara, D. J. & Bevan, M. (1989) Genetic transformation of apple (*Malus pumila* Mill.) using a disarmed Ti-binary vector. *Plant Cell Reports* 7, 658–61.

Janssen, B.-J. & Gardner, R. C. (1990) Localised transient expression of GUS in leaf discs following cocultivation with *Agrobacterium. Plant Molecular Biology* 14, 61–72.

Jefferson, R. A., Kavanagh, T. A. & Bevan, M. W. (1987) GUS fusions: β-glucuronidase as a sensitive and versatile gene fusion marker in higher plants. *EMBO Journal* 6, 3901–7.

Jensen, J. S., Marcker, K. A., Otten, L. & Schell, J. (1986) Nodule specific expression of a chimeric soybean leghaemoglobin gene in transgenic *Lotus corniculatus. Nature* 321, 669–74.

Jia, S.-R., Yang, M.-Z., Ott, R. & Chua, N.-H. (1989) High frequency transformation of *Kalanchoe laciniata. Plant Cell Reports* 8(6), 336–40.

Jones, J. D. G., Gilbert, D. E., Grady, K. L. & Jorgensen, R. A. (1987a) T-DNA structure and gene expression in petunia plants transformed by *Agrobacterium tumefaciens* C58 derivatives. *Molecular and General Genetics* 207, 478–85.

Jones, J. D. G., Svab, Z., Harper, E. C., Hurwitz, C. D. & Maliga, P. (1987b) A dominant nuclear streptomycin resistance marker for plant cell transformation. *Molecular and General Genetics* 210, 86–91.

Jorgensen, R., Snyder, C. & Jones, J. D. G. (1987) T-DNA is organized predominantly in inverted repeat structures in plants transformed with *Agrobacterium tumefaciens* C58 derivatives. *Molecular and General Genetics* 207, 471–7.

Kerr, A. & Brisbane, P. G. (1983) *Agrobacterium.* In: Faley, P. C. & Persley, G. J. (eds), *Plant Bacterial Diseases: a Diagnostic Guide.* Academic Press, Sydney, pp. 27–43.

Klee, H., Horsch, R. & Rogers, S. (1987) *Agrobacterium*-mediated plant transformation and its further applications to plant biology. *Annual Review of Plant Physiology* 38, 467–86.

Lassner, M. W., Petersen, P. & Yoder, J. I. (1989) Simultaneous amplification of multiple DNA fragments by polymerase chain reaction in the analysis of transgenic plants and their progeny. *Plant Molecular Biology Reporter* 7, 116–28.

Ledger, S. E., Deroles, S. C. & Given, N. K. (1991) Regeneration and *Agrobacterium*-mediated transformation of *Chrysanthemum*. *Plant Cell Reports,* in press.

McCormick, S., Niedermeyer, J., Fry, J., Barnason, A., Horsch, R. & Fraley, R. (1986) Leaf disc transformation of cultivated tomato (*Lycopersicon esculentum*) using *Agrobacterium tumefaciens*. *Plant Cell Reports* 5, 81–4.

McGranahan, G. H., Leslie, C. A., Uratsu, S. L., Martin, L. A. & Dandekar, A. M. (1988) *Agrobacterium* mediated transformation of walnut somatic embryos and regeneration of transgenic plants. *Bio/Technology* 6, 800–4.

Manners, J. M. & Way, H. (1989) Efficient transformation with regeneration of the tropical pasture legume *Stylosanthes humilis* using *Agrobacterium rhizogenes* and a Ti plasmid-binary vector system. *Plant Cell Reports* 8, 341–5.

Mathews, H., Bharathan, N., Litz, R. E., Narayanan, K. R., Rao, P. S. & Bhatia, C. R. (1990) Transgenic plants of mustard *Brassica juncea* (L.) Czern and Coss. *Plant Science* 72, 245–52.

Matthysse, A. G. (1986) Initial interactions of *Agrobacterium tumefaciens* with plant host cells. *CRC Critical Reviews in Microbiology* 13(3), 281–307.

Matthysse, A. G. (1987) Characterization of nonattaching mutants of *Agrobacterium tumefaciens*. *Journal of Bacteriology* 169, 313–23.

Melchers, L. S. & Hooykaas, P. J. J. (1987) Virulence of *Agrobacterium*. *Oxford Surveys of Plant Molecular and Cell Biology* 4, 167–220.

Michelmore, R., Marsh, E., Seeley, S. & Landry, B. (1987) Transformation of lettuce (*Lactuca sativa*) mediated by *Agrobacterium tumefaciens*. *Plant Cell Reports* 6, 439–42.

Moloney, M. M., Walker, J. M. & Sharma, K. K. (1989) High efficiency transformation of *Brassica napus* using *Agrobacterium* vectors. *Plant Cell Reports* 8, 238–42.

Morris, J. W., Castle, L. A. & Morris, R. O. (1989) Efficacy of different *Agrobacterium tumefaciens* strains in transformation of pinaceous gymnosperms. *Physiological and Molecular Plant Pathology* 34, 451–61.

Muller, A. J., Mendel, R. R., Schiemann, J., Simoens, C. & Inze, D. (1987) High meiotic stability of a foreign gene introduced into tobacco by *Agrobacterium*-mediated transformation. *Molecular and General Genetics* 207, 171–5.

Naina, N. S., Gupta, P. K. & Mascarenhas, A. F. (1989) Genetic transformation and regeneration of transgenic *Azadirachta indica* plants using *Agrobacterium tumefaciens*. *Current Science* 58(4), 184–7.

Nehra, N. S., Chibbar, R. N., Kartha, K. K., Datla, R. S., Crosby, W. L. & Stushnoff, C. (1990) *Agrobacterium* mediated transformation of strawberry calli and recovery of transgenic plants. *Plant Cell Reports* 9, 10–13.

OECD (1990) *Database File: Field Releases of Genetically Modified Organisms*. Organization for Economic Co-operation and Development, Environmental Directorate, Paris.

Owens, L. D. (1981) Characterization of kanamycin-resistant cell lines of *Nicotiana tabacum*. *Plant Physiology* 67, 1166–8.

Peralta, E. G., Hellmiss, R. & Ream, W. (1986) *Overdrive* a T-DNA transmission enhancer on the *A. tumefaciens* tumour inducing plasmid. *EMBO Journal* 5, 1137–42.

Petit, A. & Tempé, J. (1985) The function of T-DNA in nature. In: Van Vloten-Doting, L., Groot, G. & Hall, T. (eds), *Molecular Form and Function of the Plant Genome*. Plenum Press, New York, pp. 625–36.

Potrykus, I. (1990) Gene transfer to cereals: an assessment. *Bio/Technology* 8, 535–41.

Puonti-Kaerlas, J., Eriksson, T. & Engstrom, P. (1990) Production of transgenic pea (*Pisum sativum*) plants by *Agrobacterium tumefaciens*-mediated gene transfer. *Theoretical and Applied Genetics* 80, 246–52.

Rao, S., Lippincott, B..& Lippincott, J. (1982) *Agrobacterium* adherence involves the pectic portion of the host cell wall and is sensitive to the degree of pectin methylation. *Plant Physiology* 56, 574–80.

Sahi, S., Chilton, M.-D. & Chilton, W. (1990) Corn metabolites affect growth and virulence of *Agrobacterium tumefaciens*. *Proceedings of the National Academy of Sciences USA* 87, 3879–83.

Schafer, W., Gorz, A. & Kahl, G. (1987) T-DNA integration and expression in a monocot crop plant after induction of *Agrobacterium*. *Nature* 327, 529–31.

Schrammeijer, B., Sijmons, P. C., van den Elzen, P. J. M. & Hoekema, A. (1990) Meristem transformation of sunflower via *Agrobacterium*. *Plant Cell Reports* 9, 55–60.

Scott, R. J. & Draper, J. (1987) Transformation of carrot tissues derived from proembryogenic suspension cells: a useful model system for gene expression studies in plants. *Plant Molecular Biology* 8, 265–74.

Shahin, E. A. & Yashar, M. (1986) Transformation of cabbage by a binary vector in *Agrobacterium tumefaciens*. In: *Crucifer Genetics Workshop III Proceedings*. University of Guelph, Guelph Canada, pp. 41.

Shahin, E. A., Spielmann, A., Sukhapinda, K., Simpson, R. B. & Yasher, M. (1986) Transformation of cultivated alfalfa using disarmed *Agrobacterium tumefaciens*. *Crop Science* 26, 1235–9.

Sheikholeslam, S. N. & Weeks, D. P. (1987) Acetosyringone promotes high efficiency transformation of *Arabidopsis thaliana* explants by *Agrobacterium tumefaciens*. *Plant Molecular Biology* 8, 291–8.

Shen, W. H., Petit, A., Guern, J. & Tempe, J. (1988) Hairy roots are more sensitive to auxin than normal roots. *Proceedings of the National Academy of Sciences USA* 85, 3417–21.

Spielmann, A. & Simpson, R. B. (1986) T-DNA structure in transgenic tobacco plants with multiple independent integration sites. *Molecular and General Genetics* 205, 34–41.

Stachel, S. E., Messens, E., van Montagu, M. & Zambryski, P. (1985) Identification of the signal molecules produced by wounded plant cells that activate T-DNA transfer in *Agrobacterium tumefaciens*. *Nature* 318, 624–9.

Stalker, D. M., McBride, K. E. & Malyj, L. D. (1988) Herbicide resistance in transgenic plants expressing a bacterial detoxification gene. *Science* 242, 419–23.

Suseelan, K. N., Bhagwat, A., Mathews, H. & Bhatia, C. R. (1987) *Agrobacterium tumefaciens*-induced tumour formation on some tropical dicot and monocot plants. *Current Science* 56, 888–9.

Tepfer, D. (1989) Ri T-DNA from *Agrobacterium rhizogenes:* a source of genes having applications in rhizosphere biology and plant development, ecology, and evolution. In: Kosuge, T. & Nester, E. W. (eds), *Plant–Microbe Interactions: Molecular and Genetic Perspectives,* vol. 3. McGraw-Hill, New York, pp. 294–342.

Thomashow, M. F., Karlinsey, J. E., Marks, J. R. & Hulbert, R. E. (1987) Identification of a new virulence locus in *Agrobacterium tumefaciens* that affects polysaccharide composition and plant cell attachment. *Journal of Bacteriology* 169, 3209–16.

Tinland, B., Rohfritsch, O., Michler, P. & Otten, L. (1990) *Agrobacterium tumefaciens* T-DNA gene *6b* stimulates rol-induced root formation, permits growth at high auxin concentrations and increases root size. *Molecular and General Genetics* 223, 1–10.

Trulson, A., Simpson, R. & Shahin, E. (1986) Transformation of cucumber (*Cucumis sativus* L.) plants with *Agrobacterium rhizogenes. Theoretical and Applied Genetics* 73, 11–15.

Van de Elzen, P. J. M., Townsend, J., Lee, K. Y. & Bedbrook, J. R. (1985) A chimeric hygromycin resistance gene as a selectable marker in plant cells. *Plant Molecular Biology* 5, 299–302.

Veluthambi, K., Jayaswal, R. K. & Gelvin, S. B. (1987) Virulence genes A, G, and D mediate the double-stranded border cleavage of T-DNA from the *Agrobacterium* Ti plasmid. *Proceedings of the National Academy of Sciences USA* 84, 1881–5.

Veluthambi, K., Krishnan, M., Gould, J. H., Smith, R. H. & Gelvin, S. B. (1989) Opines stimulate induction of the *vir* genes of the *Agrobacterium tumefaciens* Ti plasmid. *Journal of Bacteriology* 171, 3696–703.

Wallroth, M., Gerats, A. G. M., Rogers, S. G., Fraley, R. T. & Horsch, R. B. (1986) Chromosomal localization of foreign genes in *Petunia hybrida. Molecular and General Genetics* 202, 6–15.

Wang, K., Genetello, C., van Montagu, M. & Zambryski, P. C. (1987) Sequence context of the T-DNA border repeat element determines its relative activity during T-DNA transfer to plant cells. *Molecular and General Genetics* 210, 338–46.

White, D. W. R. & Greenwood, D. (1987) Transformation of the forage legume *Trifolium repens* L. using binary *Agrobacterium* vectors. *Plant Molecular Biology* 8, 461–9.

Zambryski, P. (1988) Basic processes underlying *Agrobacterium*-mediated DNA transfer to plant cells. *Annual Review of Genetics* 22, 1–30.

Zambryski, P., Joos, H., Genetello, C., Leemans, J., van Montagu, M. & Schell, J. (1983) Ti plasmid vector for the introduction of DNA into plant cells without alteration of their normal regeneration capacity. *EMBO Journal* 2(12), 2143–50.

Chapter 4
Electroporation for Direct Gene Transfer into Plant Protoplasts

Carl Rathus[1] & Robert G. Birch[2]

[1]*Bureau of Sugar Experiment Stations, PO Box 86, Indooroopilly, Queensland 4068, Australia.*
Present Address: CSIRO Division of Tropical Crops and Pastures, 306 Carmody Rd, St. Lucia, Queensland 4067, Australia [2] *Department of Botany, The University of Queensland, Queensland 4072, Australia*

Introduction

A major problem for the introduction of foreign genes into plant cells is the presence of a cell wall, which is an effective barrier to uptake of DNA from the surrounding solution. *Agrobacterium* species and some plant viruses can transfer DNA across the cell wall and then into the plant genome, but only for *Agrobacterium* has it been possible to separate genetic transformation capability from pathogenicity. Unfortunately, *Agrobacterium* has a limited natural host range primarily restricted to dicots, and as discussed in Chapter 3, major economic plant groups including the cereals, grasses and many legumes remain resistant to *Agrobacterium*-mediated gene transfer.

Efforts have therefore been made to develop direct gene transfer (DGT) techniques to introduce DNA into plant cells, without the need of an intermediate biological host. One DGT method (discussed in Chapter 5) involves firing small particles coated with DNA through the cell wall and the plasma membrane. Most other DGT methods require the removal of the plant cell wall to form protoplasts. Unfortunately, when the objective is the recovery of transgenic plants, removal of the wall to facilitate DNA entry creates the additional problem of regenerating intact cells and then plants from single protoplasts. This remains an important hurdle in many plant species, especially in the Gramineae. However, direct gene transfer has also proved to be a powerful tool for tailoring foreign genes for expression in plants, using a method that allows evaluation within days without the need to use cells capable of further divisions.

Techniques for direct gene transfer into plant protoplasts

Many techniques developed for gene transfer into animal cells have been tested with some success on plant protoplasts. Liposomes, spheroplasts, poly-L-ornithine or poly-L-lysine complexed genes, calcium phosphate co-precipitation and microinjection (Steinbiss & Broughton, 1983) have, however, been largely overtaken by the use of polyethylene glycol (PEG) and electroporation. Originally used in protoplast fusion studies, PEG has also been developed into an extremely simple transformation technique for some plant species (Krüger-Lebus & Potrykus, 1987). In combination with divalent cations, PEG 6000 at a concentration of 15–25% (w/v) minimizes charge repulsion between DNA and membrane, protects DNA from nuclease activity and stimulates endocytosis. PEG treatment has facilitated production of transgenic tobacco (Negrutiu *et al.*, 1987), moth bean (Köhler *et al.*, 1987) and rice (Zhang & Wu, 1988) plants; stable transformation of sugarcane (Chen *et al.*, 1987) and various model species (e.g. Schocher *et al.*, 1986), and transient expression of foreign genes in several grasses and cereals including rice, maize, barley, rye and Guinea grass (Junker *et al.*, 1987; Vasil *et al.*, 1988). Transformation frequencies approaching 1% of surviving cells have been achieved for a few species, but they are generally much lower (Table 4.1). For some species, protoplast survival following PEG treatment can be unacceptably low (Kao & Saleem, 1986).

In many laboratories, electroporation is the method of choice when high transformation frequencies are required. Apparatus to deliver appropriate electric pulses safely need not be expensive, and the electroporation treatment is both reproducible and very simple to apply. This chapter outlines the theoretical basis for electroporation and reviews its use as a method for obtaining both transient and stable transformation in plants.

Electroporation

Terminology and equipment

Electroporation is the application of high-voltage electric pulses to cells to induce transient membrane pores, allowing entry of macromolecules including DNA. Synonyms include electroinjection, electroinfection and electro-transfection. Electroporation has proved to be a broadly applicable gene transfer technique, facilitating foreign gene expression in intact bacterial (Dower *et al.*, 1988), yeast (Hashimoto *et al.*, 1985) and mammalian cells (Neumann *et al.*, 1982), and in protoplasts of fungi (Goldman *et al.*, 1990) and plants (Fromm *et al.*, 1985). However, different electrical conditions are required for efficient permeabilization of different cell types and sizes. As

Table 4.1. Examples of stable transformation in plants by direct gene transfer (DGT) into protoplasts.

DGT technique	Protoplasts required	Plant species	Transgenic plant recovered	RTF[a]	ATF[b]	References
Electroporation	+	Maize	−	$\sim 10^{-2}$	$\sim 10^{-4}$	Fromm et al., 1986
Electroporation	+	Maize	+	$1\text{-}5\times10^{-2}$	—	Rhodes et al., 1988
Electroporation	+	Maize	−	8×10^{-4}	2.4×10^{-5}	Huang & Dennis, 1989
Electroporation	+	Rapeseed	+	—	5×10^{-5}	Guerche et al., 1987
Electroporation	+	Soybean	−	—	—	Christou et al., 1987
Electroporation	+	Lettuce	+	—	10^{-3}	Chupeau et al., 1989
Electroporation	+	Rice	+	10^{-3}	4×10^{-6}	Toriyama et al., 1988
Electroporation	+	Rice	−	2.6×10^{-1}	2×10^{-4}	Yang et al., 1988
Electroporation	+	Rice	+	2.5×10^{-1}	3×10^{-5}	Zhang et al., 1988
Electroporation	+	Rice	+	$1\text{-}6\times10^{-3}$	$6\text{-}40\times10^{-6}$	Shimamoto et al., 1989
Electroporation/PEG	+	Tobacco	−	2×10^{-2}	2×10^{-3}	Shillito et al., 1985
Electroporation/PEG	+	Orchard grass	+	—	—	Horn et al., 1988
PEG	+	Moth bean	+	5.8×10^{-3}	7×10^{-6}	Köhler et al., 1987
PEG	+	T. monococcum	−	—	10^{-6}	Lörz et al., 1985
PEG	+	Arabidopsis thaliana	+	5×10^{-2}	10^{-4}	Damm et al., 1989
PEG	+	Rice	−	1.5×10^{-2}	3×10^{-5}	Uchimiya et al., 1986
PEG	+	Rice	+	1.5×10^{-1}	—	Zhang & Wu, 1988
PEG	+	Sugarcane	−	1.6×10^{-4}	8×10^{-7}	Chen et al., 1987
Liposomes	+	Tobacco	+	5×10^{-5}	1.5×10^{-5}	Bellini et al., 1989

[a]Relative transformation frequency (survival basis).
[b]Absolute transformation frequency (original basis).

the plant cell wall provides an effective barrier to macromolecules (Jones *et al.*, 1987), a protoplast isolation and culture system is required (Fig. 4.1).

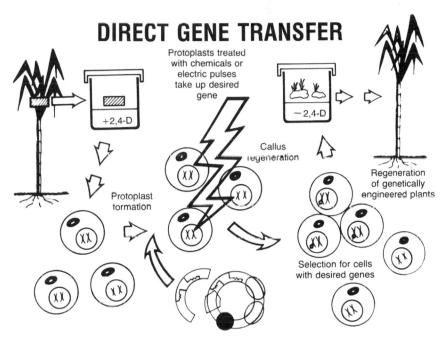

DIRECT GENE TRANSFER

Protoplasts treated with chemicals or electric pulses take up desired gene

+2,4-D

−2,4-D

Callus regeneration

Protoplast formation

Regeneration of genetically engineered plants

Selection for cells with desired genes

Fig. 4.1. Schematic representation of the steps in direct gene transfer.

The range of commercial and custom-built machines on the market and in the literature makes it difficult to generalize about the operation of an 'electroporator'. Plant protoplasts have been successfully electroporated using conditions ranging from a single, square wave, high-voltage pulse of short duration, to a single, exponential-decay pulse of lower voltage and longer duration, to customized pulse shapes or multiple pulses (Draper *et al.*, 1988; Larkin *et al.*, 1990; Table 4.2).

Capacitor discharge devices capable of supplying a reasonably wide range of voltages and pulse lengths safely and reproducibly can be constructed in an electronics workshop from parts costing less than A$2000. A simple circuit, without the peripheral circuits controlling pulse number and length, is shown in Fig. 4.2. High-capacitance discharge coupled with a timer circuit results in an approximately square wave, whereas a low-capacitance discharge gives an exponential decay curve.

Table 4.2. Examples of transient expression of genes introduced into plant protoplasts by electroporation.

Source of protoplasts	Electric field strength (V/cm)	Pulse shape	Pulse length[a]	Pulse number	Buffer conductivity	Reference
N. tabacum leaf mesophyll	300	Exponential	120 ms	20	Low	Guerche et al., 1987
Tobacco leaf mesophyll[c]	1000–1250	Exponential	10 µs	1	Medium	Shillito et al., 1985
Tobacco suspension culture	875	Exponential	54 ms	1	High	Fromm et al., 1985
Tobacco suspension culture	1500	Exponential	5 ms	6	High	Taylor & Larkin, 1988
Carrot suspension culture	550	Exponential	~ 50 ms	1	High	Boston et al., 1987
Carrot suspension culture	1300–3800	Square	6 µs	6	High	Langridge et al., 1985
Carrot suspension culture	750	Exponential	8 ms	1	High	Bates et al., 1988
Sugarbeet suspension culture	250–750	Square	100 µs	3	Low	Lindsey & Jones, 1987
Betulaceae suspension culture	500	Exponential	71 ms	1	Medium	Séguin & Lalonde, 1988
Maize suspension culture	875	Exponential	54 ms	1	High	Fromm et al., 1985

[a] For exponential decay pulses, pulse length is usually defined as the time taken for the EFS to drop to approximately one-third of its initial value.

[b] High – salt concentration > 50 mM.
Medium – salt concentration between 10 and 50 mM.
Low – salt concentration < 10 mM.

[c] Stable transformation.

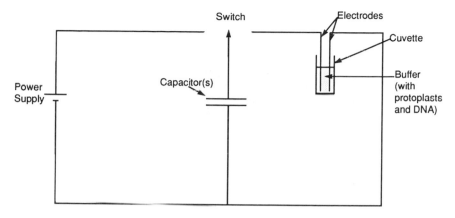

Fig. 4.2. Simple circuit used for electroporation. The capacitor is charged using a power supply, and then switched to discharge through the electroporation cuvette.

Model for electropermeabilization of cell membranes

Cell membranes are composed of a lipid bilayer interspersed with proteins. It has been suggested that, in electroporation, pore formation is a result of a 90° block rotation of two nearest-neighbour lipids in the bilayer (Sugar & Neumann, 1984) (Fig. 4.3). For a sphere, the transmembrane potential, ΔV, induced by an externally applied electric field, E_0, is given by:

$$\Delta V = 1.5\, RE_0 \cos \theta$$

where R is the radius of the sphere, E_0 is the initial electric field strength, and θ is the angle between a given membrane site and the field direction (Zimmermann, 1982). Therefore as protoplast radius increases a proportional decrease in E_0 is necessary to achieve the same transmembrane potential.

Pore formation has been shown to result at a transmembrane potential, ΔV_c, in the order of 1 V in animal cells and plant protoplasts (Kinoshita & Tsong, 1977; Steinbiss, 1978; Zhelev *et al.*, 1988). For a given protoplast type, ΔV_c decreases with increasing pulse duration, reaching a constant for pulse durations greater than approximately 1 ms. It has also been shown that there is an inverse relationship between electroporation temperature and both the EFS and the pulse duration required for pore formation (Zimmermann, 1986; Zhelev *et al.*, 1988).

The times taken for membrane breakdown and resealing are important for both DNA transfer and cell survival following electroporation. Pulsed-laser fluorescence microscopy of sea urchin eggs revealed that pore formation takes less than 1 μs in a high-salt medium (Kinoshita *et al.*, 1988). Pores

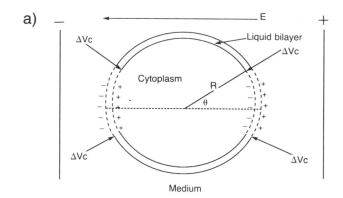

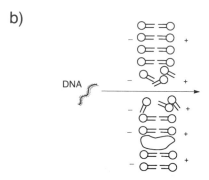

Fig. 4.3. Pore formation in lipid bilayer. (a) Pore formation in region of plasma membrane subjected to transmembrane potential greater than or equal to the transmembrane potential ΔV_c. (b) Close-up of disrupted lipid bilayer under the same conditions. See text for definition of symbols used. Adapted from Jones *et al.* (1987).

also recover rapidly (less than 2 s). However, with intense and long pulses some leakage continues, resulting in bursting, presumably from osmotic imbalance. Membrane disruption at lipid–protein junctions or in proteins recovers more slowly (Zimmermann & Vienken, 1982), taking from a few minutes at 25°C to 30 min at 5°C (Steinbiss, 1978). There is some evidence that PEG treatment induces more random pore sizes, resulting in slower and less efficient resealing (Hahn-Hägerdal *et al.*, 1986).

With pulse length in the ms range, EFS values of 250–1500 V/cm have typically been used to obtain transient expression. This is equivalent to transmembrane potentials of 0.5–2.8 V, assuming an average protoplast diameter of 25 μm. For shorter pulse lengths, EFS values equivalent to transmembrane potentials of up to 7 V have been used. (See Table 4.2 for some examples of conditions.) These values agree reasonably with the predictions from the model above. Mammalian cells, which are of similar size, have typically been transformed using EFSs of 500–1000 V/cm (Shigekawa & Dower, 1989). Bacteria, which are approximately an order of magnitude smaller than protoplasts, require an applied EFS of the order of 5000–10,000 V/cm (Shigekawa & Dower, 1989). Thus the model may be useful in selecting parameter ranges for various cell types, highlighting the importance of EFS, pulse length and temperature for pore formation and resealing.

However, differences in membrane composition affect the response of various cell types, and other factors, such as osmolarity and conductivity of the buffer, are also important. Thus, conditions must be experimentally optimized for each cell type, and, bearing the principles in mind, this can be accomplished efficiently using transient expression methodology.

Transient expression of introduced genes

Expression of introduced genes may be detected in protoplasts 1 to 3 days following electroporation, even for plant species where protoplasts fail to divide. Transient expression, defined as the expression of introduced genes that have not integrated into the host genome, provides a valuable experimental system for rapid preliminary studies of foreign genes in plant cells.

In order to obtain transient expression of a foreign gene in a cell, it is necessary for the DNA to move across the plasma membrane into the cytoplasm and then across the nuclear membrane into the nucleus, where transcription takes place. The resulting mRNA must move as usual to the cytoplasm to be translated into the protein gene product. Expression of the introduced gene is generally measured indirectly as activity of the gene product. This measure of 'expression' is affected by the stability of both the mRNA and the protein produced.

Reporter genes

Genes whose products are conveniently assayed are termed reporter genes as they are used to demonstrate expression of introduced genes. Some important properties of reporter genes are:

1. stability of the expressed protein both *in vivo* and *in vitro* under a range of conditions;
2. low background from endogenous enzyme activity;
3. a sensitive, quick, simple, clean and quantifiable assay.

Table 4.3 describes a range of reporter genes that have been used in transient expression studies. Perhaps the most versatile of the commonly used

Table 4.3. Reporter genes in plant transformation.

Gene	Origin	Assay type	Comments	References
Nopaline synthase (NOS)	T-DNA	Paper chromatography	Unquantifiable	e.g. Zambryski et al., 1983; Otten & Schilperoort, 1978
Octopine synthase (OCS)	T-DNA	Paper chromatography	Unquantifiable	e.g. De Greve et al., 1982
Neomycin phosphotransferase II (NPTII)	Tn5	Phosphorylation (^{32}P) autoradiography ELISA	Can also be used as selectable marker Difficult to quantify Quantifiable	e.g. Töpfer et al., 1988; Reiss et al., 1984
Chloramphenicol acetyltransferase (CAT)	Tn9	Acetylation (^{14}C) (autoradiography scintillation counting)	Quantifiable Endogenous CAT activity in some species	e.g. Taylor & Larkin, 1988
β-glucuronidase (GUS)	E. coli	Fluorometric Spectrophotometric Histochemical	Easily quantifiable Can be localized in situ	e.g. Jefferson et al., 1987
Firefly luciferase (LUC)	Photinus pyralis	Light emission Luminometer	Easily quantifiable Can be localized in situ	Ow et al., 1986; Gallie et al., 1989
Bacterial luciferase (LUX)	Vibrio harveyi	Light emission Luminometer	Easily quantifiable	Koncz et al., 1987
LacZ	E. coli	β-galactosidase activity Fluorometric Histochemical	Unquantifiable in several species (due to high background of endogenous β-galactosidase)	Teeri et al., 1989
Anthocyanin biosynthesis (Lc)	Maize	Cell pigmentation (visual)	Unquantifiable (number of transformed cells can be quantified)	Ludwig et al., 1990
Plant viral genomes	Tobacco mosaic virus	Inclusion bodies Coat protein	Pathogenicity limits application to demonstrating transformation	Nishiguchi et al., 1986

reporter genes at present is GUS. GUS activity can be quantified using either fluorometric or spectrophotometric assays, both of which are reasonably cheap and simple. Using a histochemical assay, gene expression can be visualized *in situ*. However, this assay is destructive, as the product is toxic to plant cells (Jefferson *et al.*, 1987). Luciferase reporter genes permit sensitive and non-toxic *in situ* assays, but at present this requires very expensive low-light camera equipment (Olsson *et al.*, 1990).

Transient expression assays using reporter genes can be powerful tools to develop and optimize gene transfer methods, to determine rapidly the relative strength of gene expression controlled by various promoter sequences, and to study constitutive and in some systems tissue-specific and environmentally inducible gene expression. Reporter genes can also be used to study gene regulation in transgenic plants.

Gene constructs for electroporation

An advantage of DGT techniques is that genes for transfer need not be engineered between *Agrobacterium* T-DNA border sequences, or into a Ti or broad-host-range plasmid. Almost any small, high-copy-number *E. coli* cloning vector containing an *E. coli* origin of replication and a bacterial selectable marker gene can be used for delivery (Draper *et al.*, 1988). However, expression of any introduced gene requires appropriate control sequences upstream (5') and downstream (3') of the coding region (Fig. 4.4).

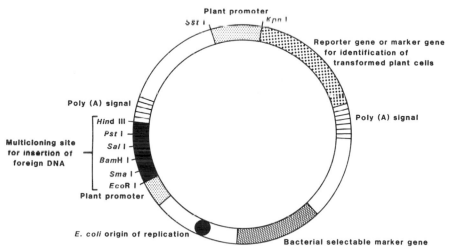

Fig. 4.4. Generalized vector for electroporation. The reporter or marker gene allows detection or selection of transformed cells. The multicloning site allows insertion of a foreign gene for expression in plant cells. The top plant promoter can be replaced with other sequences to be tested as promoters, based on reporter gene expression (adapted from Draper *et al.*, 1988).

The proximal 5' sequences correspond to the promoter region, an oriented DNA sequence that binds RNA polymerase immediately adjacent to the start site for RNA synthesis. A signal for polyadenylation of the mRNA is essential downstream of the coding region.

The NOS and CaMV 35S promoters are commonly used to drive constitutive expression in plant cells. Both are effective in dicots for transient and stable expression (Fromm *et al.*, 1985; Ecker & Davis, 1986), whereas the CaMV 35S promoter gives higher transient expression in monocots (Fromm *et al.*, 1985; Howard *et al.*, 1987; Junker *et al.*, 1987). Reports of transient expression in Gramineae show that the CaMV 35S promoter also gives greater gene expression than the mannopine synthase, 2' and 1' promoters from T-DNA, and the maize Adh1 promoter (Howard *et al.*, 1987; Junker *et al.*, 1987; Harpster *et al.*, 1988; Planckaert & Walbot, 1989). Enhancers (Ellis *et al.*, 1987) and introns (Callis *et al.*, 1987) can be used to increase gene expression, and an increasing range of tissue-specific or inducible promoters is becoming available.

Optimization of gene transfer using transient expression assays

Protoplast quality

Factors influencing the level of transient expression of genes introduced into cells by electroporation include the source of the protoplasts, the electrical conditions and the buffer. As other parameters are optimized, protoplast survival generally becomes limiting. 'Optimum' conditions are a compromise between increasing the transfer of DNA across the plasma membrane and decreasing protoplast viability.

Protoplasts are preferably isolated from suspension cultures. Mesophyll protoplasts appear to be more sensitive than those derived from suspension cells, so that applied voltage is more critical for gene transfer (Lindsey & Jones, 1987; Larkin *et al.*, 1990). Because of the relationship between protoplast diameter and critical transmembrane potential, the greater uniformity of suspension cells also results in more consistent expression (Jones *et al.*, 1989).

Although there is evidence that the physiological state of protoplasts is important, ability to define or reproduce the ideal state has proved elusive. Strict adherence to a standardized procedure, and use of fresh media stocks, especially vitamins, are important in reducing variability, whilst choice of cell wall degrading enzymes can affect plasma membrane properties (Nea & Bates, 1987). Caution is required when comparing the degree of expression in protoplasts isolated at different times, because of the large day-to-day variation (Knutson & Yee, 1987). If such comparisons are required, it is important to include one treatment (internal control) in all experiments, and relate expression to that treatment.

Cell cycle phase at electroporation affects transient expression in synchronized cultures. M-phase cells give three- to fourfold greater expression of the CAT gene than cells in other phases, possibly because delivery of the transferred DNA to the nucleus is facilitated by the absence of a nuclear membrane in M phase (Okada *et al.*, 1986). Meyer *et al.* (1985) observed apparently increased stable transformation frequencies for both S- and M-phase protoplasts. Cell cycle phase may account for the existence of a competent subpopulation in unsynchronized protoplasts (Bower & Birch, 1990). However, it is difficult to maintain good protoplast yield and quality during synchronization, so that optimization of cell cycle is impractical.

Electroporation buffer

Two divalent cations, Ca^{2+} and Mg^{2+}, have been examined for their role both as membrane stabilizers and in increasing gene transfer efficiency. Membrane stability is increased by 10 mM Ca^{2+}, but not by Mg^{2+} (Boss & Mott, 1980). Fromm *et al.* (1985) and Taylor & Larkin (1988) found sharp optima for Ca^{2+} (4 and 6 mM respectively), but a neutral to detrimental effect with Mg^{2+}. Bates *et al.* (1988) found that, if Ca^{2+} was not included, complete protoplast disruption followed electroporation. In all three cases high-salt buffers were used. In contrast, Mg^{2+} proved superior to Ca^{2+} in the enhancement of PEG-mediated gene transfer (Negrutiu *et al.*, 1987).

The conductivity of an electroporation medium is determined by the salt concentration. NaCl and KCl are often used to increase conductivity, although the high $CaCl_2$ and NaCl medium of Krens *et al.* (1982) has also been used (e.g. Langridge *et al.*, 1985). As buffer conductivity is increased, a smaller proportion of the electric charge will pass through the protoplasts.

A low-salt buffer and short, high-voltage pulses allowed efficient infection of tobacco protoplasts with viral RNA (Nishiguchi *et al.*, 1986). However, changes in protoplast density, nucleic acid concentration and divalent cation concentration all have significant effects on the conductivity of low-salt buffers, making optimization difficult (Hibi *et al.*, 1986).

High-salt buffers have been satisfactory in most systems. Survival is improved by higher-conductivity buffers for maize, *N. plumbaginifolia* and lucerne (Fromm *et al.*, 1985; Larkin *et al.*, 1990). Although NaCl was superior to KCl, their optima (150 and 100 mM respectively) correspond to approximately the same buffer conductivity (Taylor & Larkin, 1988).

pH has an effect on gene transfer, possibly due to membrane effects (Taylor & Larkin, 1988). Although a high pH (9.0) increases expression in *N. plumbaginifolia*, it appears to have an adverse effect on survival in more sensitive species such as lucerne (Larkin *et al.*, 1990).

PEG has been used in electroporation buffers to enhance transient

expression (Boston *et al.*, 1987; Séguin & Lalonde, 1988) and stable transformation frequencies (Shillito *et al.*, 1985) in tobacco. However, Guerche
et al. (1987) found no effect at either low (10%) or medium (20%) PEG
concentrations.

Electric pulse parameters

Four electrical conditions – pulse shape, EFS, pulse duration and number of
pulses – are central to electroporation. Square-wave pulses and exponential-
decay pulses are commonly used. As pulse shape is determined by the
apparatus used, direct comparisons have seldom been made. Exponential-
decay pulses are generated by capacitor discharge. Pulse duration is a function of the capacitance, the buffer conductivity and the applied voltage.
Pulse duration of square-wave pulses is set by a timer and is therefore
independent of other variables. Saunders *et al.* (1989) compared the two
pulse shapes and found that, although a higher percentage of transformants
could be obtained using exponential decay pulses, square pulses were effective over a broader range of pulse duration and EFS. The major contributing
factor was higher viability. It is possible that the choice of a low-conductivity
buffer favoured the shorter pulse lengths available in the square-wave
system.

The available electroporation apparatus determines the range of electrical parameters available. Some systematic experimentation will yield an
optimum for a given machine and plant species or genotype. However, the
results are rarely useful for other electroporator designs, and only a few
examples will be provided here. Using a capacitor discharge unit that
delivered exponential-decay pulses from a 24 μF capacitor, Taylor & Larkin
(1988) found increasing activity with increasing EFS, pulse length and pulse
number, until decreasing viability became limiting. With 4 pulses, activity
increased rapidly to a pulse length of 5 ms and then plateaued, whereas EFS
gave a sharp optimum at 1500–1750 V/cm.

Some attempts have been made to investigate interactions between the
various electrical parameters. Using a single exponential-decay pulse, Bates
et al. (1988) found that at 250 V/cm expression increased linearly with pulse
lengths up to 50 ms, whereas at 750 V/cm a clear optimum of 8 ms was obtained. Lindsey & Jones (1987) found that increasing the duration of a
square pulse from 50 to 99.9 μs while reducing the EFS from 1500 to 750 V/
cm gave increased expression. By increasing the delay between pulses from
0.5 to 9.9 s, the effect on viability of increasing both pulse duration and
number of pulses was reduced, resulting in higher expression. For electrical
parameters in particular, the compromise between gene transfer and protoplast viability must be borne in mind. High EFS, long pulses, multiple
pulses and low-conductivity buffers all contribute to increased membrane
permeabilization and therefore gene transfer, but also to reduced protoplast

viability, which will ultimately limit transient expression and stable transformation frequencies. Figure 4.5 illustrates this point.

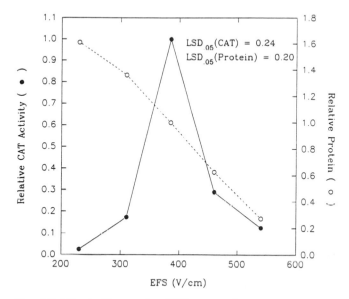

Fig. 4.5. Effect of increasing EFS on transient gene expression (relative CAT activity) and viability (relative protein in intact protoplasts), 18 hours after electroporation of sugarcane protoplasts. (Rathus & Birch, 1991).

Temperature and heat-shock treatments

Although temperature has an effect on pore formation and resealing, few transient expression studies have examined this variable. Electroporation is generally carried out on ice. Bates *et al.* (1988) found that electroporation at room temperature gave a threefold higher level of CAT expression than did electroporation in ice-cold media. As the protocol had been optimized at room temperature, this is not surprising. Where there are strong interactions between parameters, the order in which the parameters are optimized can have a large effect on the final optimum conditions arrived at.

Several groups have tested the effect of a heat shock (5 min at 45°C) prior to electroporation. Shillito *et al.* (1985) found a 5- to 20-fold increase in (stable) transformation efficiency, but others have obtained no benefit (Boston *et al.*, 1987; Guerche *et al.*, 1987; Bates *et al.*, 1988), and Larkin *et al.* (1990) found that the extent of expression decreased as the length of the heat shock (45°C) was increased.

Form of introduced DNA

Plasmids can be used in three forms: supercoiled, linear or relaxed. Shillito *et al.* (1985) found that the linear form gave a tenfold greater efficiency in stable transformation experiments. This may be a result of more efficient integration into the plant genome. In transient expression experiments the results have been variable. Okada *et al.* (1986) and Séguin & Lalonde (1988) found no difference between supercoiled and linear plasmid, while Bates *et al.* (1988) found supercoiled to be better. The most thorough study showed five- to tenfold higher CAT and LUC transient expression with linear, compared with supercoiled, plasmid (Ballas *et al.,* 1987). This was due to more efficient transcription of the linear form, rather than to a differential efficiency in transfer of the plasmids into the nucleus.

The use of carrier DNA such as salmon sperm or calf thymus DNA has increased expression in some cases (Guerche *et al.,* 1987; Bates *et al.,* 1988). The carrier DNA may protect the plasmid from nucleases to some extent. Increasing plasmid concentration also increases expression although a plateau will be reached (Lindsey & Jones, 1987; Bates *et al.,* 1988).

Optimization strategies

Provided the electroporation apparatus used is flexible, a reasonable compromise between expression and viability can be achieved for most plant systems. Because of the number of interacting parameters influencing gene transfer and protoplast survival, a sequential analysis of selected variables (pulse characteristics, buffer pH and composition, and DNA form and concentration) is recommended in order to achieve a high degree of transient expression. Subsequent analysis of interactions among electrical parameters (EFS, pulse length and pulse number) may lead to higher levels of expression. Table 4.2 illustrates the diversity of electrical conditions determined for several plant species after some optimization. No clear-cut pattern emerges. The 'optimum' conditions determined for any new system will depend on plant genotype, the source of protoplasts, the electroporation apparatus, and the order in which the parameters are optimized. The importance placed on protoplast survival, for subsequent development of a stable transformation system, will also have an effect.

Electroporation into intact plant cells

In 1986, Morikawa *et al.* demonstrated introduction of viral RNA into tobacco mesophyll cells, isolated with macerase, after electroporation at low voltage (650 V/cm) and very low capacitance (2 nF). Because of the ability of viruses to replicate within the host cell, detection of viral coat protein is an extremely sensitive assay for the introduction of viral RNA. These conditions are unlikely to yield detectable expression of introduced non-infectious DNA. Low transient expression of a CAT gene introduced into sugarbeet suspension cells was later demonstrated by Lindsey & Jones (1987). Pretreatment of the cells with pectinase led to a sixfold increase in expression to nearly one-half the amount obtained using suspension culture-derived protoplasts. These early reports showed that the pectinase-treated cell wall is not an insurmountable barrier to nucleic acids.

However, no improvements to these methods were reported until 1990, when Dekeyser *et al.* demonstrated transient expression of introduced genes in electroporated rice, maize, barley and wheat leaf bases. The protocol involved elimination of explant-released nucleases from the electroporation buffer prior to addition of the DNA, increasing the DNA/explant incubation time and extending the pulse length. Histochemical GUS assays revealed gene transfer into cells throughout the tissue. Tissue-specific GUS expression was also driven by an appropriate promoter, indicating the potential of the method for rapid evaluation of tissue-specific promoters and enhancers. Very importantly, gene transfer into regenerable tissue was demonstrated. Although the results to date are confined to transient expression, early confirmation of stable transformation can be expected. Thus the technique may facilitate production of transgenic plants in species for which protoplast culture has proved difficult.

Stable incorporation and expression of introduced genes

To obtain stable transformation, a functional copy of the introduced gene must be incorporated into the plant genome so that it is replicated and transmitted in mitotic divisions. The gene must also be expressed in daughter cells. A high stable transformation frequency following electroporation paves the way for recovery of transgenic plants, provided there is no block to regeneration of plants from protoplast-derived callus.

Transformation frequency

Transformation frequency can be expressed as the number of transformed calli per unit surviving protoplast (relative transformation frequency) or per unit initial protoplast (absolute transformation frequency) (Paszkowski *et al.*, 1988). High absolute transformation frequency (ATF) requires gene transfer into a high proportion of protoplasts, and a high proportion of protoplasts dividing and forming callus. For most applications, absolute transformation frequency is more important. Using electroporation, ATFs of between 10^{-3} and 10^{-6} have been obtained (see Table 4.1). This is similar to the ATFs obtained in *E. coli* using standard $CaCl_2$ co-precipitation methods, the difference being the number of cells that can be treated and the efficiency per μg of DNA used.

When early selection cannot be applied for a desired trait, a high relative transformation frequency (RTF) may be more important. RTFs have generally been below 1%, but values of greater than 10% have been obtained using both electroporation (Yang *et al.*, 1988; Zhang *et al.*, 1988) and PEG (Zhang & Wu, 1988) (see Table 4.1). In fact, Zhang & Wu (1988) transformed rice protoplasts with a plasmid containing only the GUS gene, and detected transformed plants by using the GUS histochemical assay on callus at the base of regenerated plants. Sixty-one out of 378 calli tested were positive, as were the plants derived from these calli.

Although some electroporation treatments can enhance DNA synthesis, frequency of divisions and plant regeneration from protoplasts (Ochatt *et al.*, 1988; Rech *et al.*, 1988), conditions used to produce high transient expression usually reduce protoplast survival. In order to adapt electroporation conditions optimized for transient expression to stable transformation it may be necessary to reduce the severity of the electroporation conditions (Fromm *et al.*, 1985), or improve the protoplast culture method (Huang & Dennis, 1989), in order to obtain divisions and subsequent callus regeneration.

Selection of transformed protoplasts

Because of the generally low transformation frequencies and the inability to screen rapidly for many desirable traits, a general strategy is to link the gene for the desired trait to a selectable marker gene. For example, the NPTII gene allows transformants to be selected during regeneration on the basis of resistance to kanamycin. Incorporation and expression of the linked gene can be confirmed by DNA or RNA hybridization, and sometimes by assaying for activity of the gene product, to guide selection of regenerants for slower agronomic testing.

Two problems can arise during the primary selection for antibiotic resistance. If selection is not stringent enough, non-transformed cells will fail to survive. Reconstruction experiments, where protoplasts from already transformed cells are seeded into non-transformed protoplasts, can be used to improve the plating protocol, to allow the recovery of a low density of transformed cells (Larkin *et al.*, 1990). Because of the wide variation in tolerance to different antibiotics found in different plant species, it is important when attempting to develop a new transformation system to establish a kill curve for non-transformed tissue and to choose an antibiotic and concentration that ensures tight selection, but is not directly or indirectly toxic to transformed tissue.

Some selectable marker genes used in plant transformation are listed in Table 4.4. Selection for NPTII expression on media containing kanamycin has been successful for a range of species including maize (e.g. Rhodes *et al.*,

Table 4.4. Selectable marker genes in plant transformation.

Gene	Origin	Resistance	Comments	References
Neomycin phosphotransferase II (NPT II)	*Tn5*	Kanamycin Neomycin G418 Paromomycin	Kanamycin tolerance in some (monocot) species. Can also be used as a reporter gene	e.g. Bevan *et al.*, 1983 Herrera-Estrella *et al.*, 1983 Guerche *et al.*, 1987 Okada *et al.*, 1986
Hygromycin phosphotransferase (Hpt)	*E. coli*	Hygromycin B	No enzyme assay available	Van den Elzen *et al.*, 1985 Waldron *et al.*, 1985
Mouse dihydrofolate reductase (DHFR)	Mouse	Methotrexate	No enzyme assay available	Herrera-Estrella *et al.*, 1983; Eichholtz *et al.*, 1987
Bleomycin resistance	*Tn5*	Bleomycin	Enzyme activity unknown	Hille *et al.*, 1986
Phosphinotricin acetyltransferase (PAT)	*S. hygroscopicus bar* gene	Bialaphos	Assay available	Spencer *et al.*, 1990
Acetolactate synthase (ALS)	Mutant *Arabidopsis* ALS gene	Sulphonylurea herbicides	Assay available	Haughn *et al.*, 1988

1988), rice (Yang *et al.*, 1988) and sugarcane (Chen *et al.*, 1987), although the level and timing of selection can be critical in avoiding escapes (Hauptmann *et al.*, 1987; Yang *et al.*, 1988). G418 and paromomycin have both been used as alternatives to kanamycin to improve the tightness of selection in dicots (Okada *et al.*, 1986; Guerche *et al.*, 1987) and rice (Dekeyser *et al.*, 1989).

Tight selection can be attained using the Hpt gene for hygromycin resistance (van den Elzen *et al.*, 1985; Waldron *et al.*, 1985), but the lack of an effective enzyme assay (Waldron *et al.*, 1985; Huang & Dennis, 1989) and its greater phytotoxicity (Huang & Dennis, 1989) are drawbacks.

Integration pattern of introduced DNA

In order for a foreign gene to be expressed efficiently, it is necessary for at least one copy to integrate intact in a potentially transcribed region of the genome. In contrast to *Agrobacterium*-mediated gene transfer, DGT methods tend to result in complex integration patterns including concatemers and rearrangements of the integrated DNA, and multiple insertion sites (Czernilofsky *et al.*, 1986). This occurs with both linear (e.g. Horn *et al.*, 1988) and supercoiled plasmid (e.g. Huang & Dennis, 1989), although simple patterns have been observed in both animal (Boggs *et al.*, 1986; Gusew *et al.*, 1987) and plant systems. Meyer *et al.* (1988) found that by using a supercoiled plasmid that included a specific 2 kb genomic fragment from *Petunia hybrida*, a 20-fold increase in transformation efficiency was obtained while retaining a simple integration pattern. Although Southern hybridization banding patterns may reveal complex integration patterns of an introduced gene, simple inheritance of the trait results if only one copy of the gene is functional (Riggs & Bates, 1986; Uchimiya *et al.*, 1986; Hara *et al.*, 1988).

Directed integration of foreign DNA into a predicted location in the plant genome would be useful as a means both of controlling gene expression and of avoiding the inactivation of endogenous genes. For yeast, including sequences homologous to parts of the host genome on the plasmid can determine the site of integration (see Sturley & Young, 1986, for review). However, this is not the case in plants (Lurquin & Paszty, 1988). It has been shown that integration via illegitimate recombination is approximately 100-fold more frequent than homologous recombination in both animal (Lin *et al.*, 1985; Jasin & Berg, 1988) and plant cells (Paszkowski *et al.*, 1988). Homologous recombination between co-transformed plasmids also occurs at a low frequency relative to other rearrangements (Wirtz *et al.*, 1987).

Integration and expression of non-selected DNA

The production of agronomically useful transgenic plants will in most cases require the integration and expression of non-selectable genes. Co-transformation of a non-selected gene with a selected gene has been demonstrated many times with both PEG-mediated (e.g. Damm *et al.*, 1989) and electroporation-mediated transfer (e.g. Shimamoto *et al.*, 1989). In most cases the non-selectable gene was carried on a separate plasmid, but Hara *et al.* (1988) and Uchimiya *et al.* (1986) obtained co-expression frequencies of 4/16 and 3/17 using single constructs containing both genes. Using genes introduced on separate plasmids, co-expression has ranged from low (Toriyama *et al.*, 1988: 0/5; Matsuki *et al.*, 1989: 2/26) to high (Shimamoto *et al.*, 1989: 41/57). Adjusting the ratios of supplied plasmids to obtain a high co-transformation ratio for genes on separate plasmids may save considerable recombinant DNA work in inserting non-selectable genes in plasmids suitable for transfer. Schocher *et al.* (1986) electroporated into tobacco protoplasts either a single plasmid containing the NPTII gene and the NOS gene, or separate plasmids each containing one of the genes. Using a tenfold excess of the NOS gene on a linearized plasmid resulted in approximately 50% of Km^R-resistant clones expressing NOS, equal to the frequency of co-expression of the genes supplied on a single plasmid.

Southern hybridization has revealed integration at a common site of genes supplied on separate plasmids (Tagu *et al.*, 1988; Lyznik *et al.*, 1989), or introduced genes and carrier DNA (Peerbolte *et al.*, 1985; Jongsma *et al.*, 1987). These results suggest that introduced DNA tends to undergo recombination before integration. There may be cases where such linkage of an introduced agronomic gene to a selectable marker gene could greatly facilitate subsequent selection for the agronomic gene in conventional breeding.

However, the reverse pattern also occurs. Based on segregation analysis, Peerbolte *et al.* (1985) found that one selected and two non-selected genes, supplied on a single plasmid, were integrated on to separate chromosomes in at least one transformed plant. This implies considerable fragmentation prior to integration. A potential benefit of unlinked genes for crop improvement is the ease with which the selectable gene can be eliminated in the F_1 generation, while retaining the non-selected gene. Elimination of the selectable marker will:

1. eliminate the metabolic cost of a gene expressing a protein of no use to the plant;
2. allow the same selectable marker to be used in a subsequent transformation of the same plant;
3. eliminate any pleiotropic effects;
4. avoid unforeseen dangers of a ubiquitous marker gene, analogous to the link between maize *tms* and susceptibility to *Helminthosporium maydis* (Strobel, 1982).

These considerations may ultimately favour DGT over *Agrobacterium* for practical gene manipulation of plants.

Can DGT proponents learn from *Agrobacterium*?

Evidence from work with mammalian, fungal and plant cells indicates that, during direct gene transfer, plasmid DNA is first degraded or fragmented, followed by ligation to a more stable intermediate and then integrated into the host genome (Perucho *et al.*, 1980; Wernars *et al.*, 1985; Jongsma *et al.*, 1987). It appears that integration is commonly into only one or a few sites, but it is unknown whether these sites are random.

Neither DGT nor *Agrobacterium*-mediated gene transfer allows the DNA to be targeted into a specific genomic site at high frequency. For this reason, correlation between copy number and degree of expression is often poor (Sanders *et al.*, 1987; Shirsat *et al.*, 1989), as the expression of genes varies in different parts of the genome (position effect). However, there is now clear evidence that *Agrobacterium*-mediated gene transfer targets DNA for integration into potentially transcribed regions of the plant genome (Koncz *et al.*, 1989), possibly mediated by proteins bound to single-stranded T-DNA (Howard & Citovsky, 1990). Inspired by the possibility of single-stranded (ss) DNA transfer by *Agrobacterium*, Rodenburg *et al.* (1989) found that electroporation of ssDNA into plant protoplasts led to a three- to tenfold higher frequency of stable transformation than did double-stranded DNA. When the mechanism employed by *Agrobacterium* is clarified, it will be worth while to experiment with artificial T-DNA constructs linked to essential proteins to increase the efficiency of stable genetic transformation by direct gene transfer.

Future prospects for electroporation

Electroporation is already a powerful tool for the study of gene expression and regulation, and the production of transgenic plants containing novel agronomically useful genes. Some recent advances in related areas have enhanced its potential. The development of sensitive non-toxic *in situ* assays (Olsson *et al.*, 1990) will enable gene regulation to be studied throughout the life of a plant, and allow the early selection and subculture of transformed cells. It may also be possible to adapt the mechanisms evolved by *Agrobacterium* (Koncz *et al.*, 1989; Howard & Citovsky, 1990) to increase integration and expression of genes transferred using DGT methods.

Electroporation has several advantages over other gene transfer methods. Provided good-quality protoplasts can be produced, electroporation is a simple and reproducible procedure, ideally suited to rapid quantitative studies of gene expression using transient assays. Unlike *Agrobacterium*-mediated gene transfer, there are no host range limitations, and no need to use high concentrations of antibiotics to eliminate bacteria from the plant cell culture. Simple high-copy-number plasmids can be used, and selectable and non-selectable genes can be incorporated on separate chromosomes, facilitating the elimination of unwanted selectable genes in the F_1 generation.

The major limitation to the general applicability of electroporation is the difficulty of regenerating plants from protoplasts in many plant species. Development of techniques for establishing and maintaining embryogenic suspension cultures and their use as a source of protoplasts has led to regeneration in some of the important cereals and grasses, although the protocols are often genotype-specific and difficult to reproduce. The possibility of increased undesired somaclonal variation due to prolonged tissue culture required for protoplast work may also prove problematic for direct cultivar improvement.

Electroporation into organized plant tissue (Dekeyser *et al.*, 1990) opens up new and exciting possibilities on two fronts. Firstly, tissue-specific gene expression can be studied using transient expression systems, dramatically reducing the time from gene transfer to result, because plant regeneration is not required. The power of this technique is enhanced because the plasmid DNA can penetrate into the tissue through several layers of cells. Secondly, and most importantly for immediate cultivar improvement work, regenerable tissue can be used. It should therefore be possible to produce transgenic plants by electroporation, without the need for protoplast culture. Although transgenic plants regenerated from intact tissues are likely to be chimeric, uniformly transformed plants should be attainable by segregation through tissue culture or conventional breeding.

References

Ballas, N., Zakai, N. & Loyter, A. (1987) Transient expression of the plasmid pCaMVCAT in plant protoplasts following transformation with polyethylene glycol. *Experimental Cell Research* 170, 228–34.

Bates, G. W., Piastuch, W., Riggs, C. D. & Rabussay, D. (1988) Electroporation of DNA delivery to plant protoplasts. *Plant Cell, Tissue and Organ Culture* 12, 213–18.

Bellini, C., Guerche, P., Spielmann, A., Goujaud, J., Lesaint, C. & Cabouche, M. (1989) Genetic analysis of transgenic tobacco plants obtained by liposome-mediated transformation: absence of evidence for the mutagenic effect of inserted sequences in sixty characterized transformants. *Journal of Heredity* 80, 361–7.

Bevan, M., Flavell, R. B. & Chilton, M.-D. (1983) A chimeric antibiotic resistance gene as a selectable marker for plant cell transformation. *Nature* 304, 184–7.

Boggs, S. S., Gregg, R. G., Borenstein, N. & Smithies, O. (1986) Efficient transformation and frequent single-site, single-copy insertion of DNA can be obtained in mouse erythroleucemia cells transformed by electroporation. *Experimental Hematology* 4, 988–94.

Boss, W. F. & Mott, R. L. (1980) Effects of divalent cations and polyethylene glycol on the membrane fluidity of protoplasts. *Plant Physiology* 66, 835–7.

Boston, R. S., Becwar, M. R., Ryan, R. D., Goldsborough, P. B., Larkins, B. A. & Hodges, T. K. (1987) Expression from heterologous promoters in electroporated carrot protoplasts. *Plant Physiology* 83, 742–6.

Bower, R. & Birch, R. G. (1990) Competence for gene transfer by electroporation in a subpopulation of protoplasts from uniform carrot cell suspension cultures. *Plant Cell Reports* 9, 368–89.

Callis, J., Fromm, M. & Walbot, V. (1987) Introns increase gene expression in cultured maize cells. *Genes and Development* 1, 1183–200.

Chen, W. H., Gartland, K. M. A., Davey, M. R., Sotak, R., Gartland, J. S., Mulligan, B. J., Power, J. B. & Cocking, E. C. (1987) Transformation of sugarcane protoplasts by direct uptake of a selectable chimaeric gene. *Plant Cell Reports* 6, 297–301.

Christou, P., Murphy, J. E. & Swain, W. F. (1987) Stable transformation of soybean by electroporation and root formation from transformed callus. *Proceedings of the National Academy of Sciences USA* 84, 3962–6.

Chupeau, M.-C., Bellini, C., Guerche, P., Maisonneuve, B., Vastra, G. & Chupeau, Y. (1989) Transgenic plants of lettuce (*Lactuca sativa*) obtained through electroporation of protoplasts. *Bio/Technology* 7, 503–7.

Czernilofsky, A. P., Hain, R., Herrera-Estrella, L., Goyvaerts, E., Baker, B. J. & Schell, J. (1986) Fate of selectable marker DNA integrated into the genome of *Nicotiana tabacum. DNA* 5, 101–13.

Damm, B., Schmidt, R. & Willmitzer, L. (1989) Efficient transformation of *Arabidopsis thaliana* using direct gene transfer to protoplasts. *Molecular and General Genetics* 217, 6–12.

De Greve, H., Leemans, J., Hernalsteens, J.-P., Thia-Toongh, L., DeBeuckeleer, M., Willmitzer, L., Otten, L., van Montagu, M. & Schell, J. (1982) Regeneration of normal and fertile plants that express octopine synthase, from tobacco crown galls after deletion of tumour controlling functions. *Nature* 300, 752–5.

Dekeyser, R., Claes, B., Marichal, M., van Montagu, M. & Caplan, A. (1989) Evaluation of selectable markers for rice transformation. *Plant Physiology* 90, 217–23.

Dekeyser, R. A., Claes, B., de Recke, R. M. U., Habits, M. E., van Montagu, M. C. & Caplan, A. B. (1990) Transient gene expression in intact and organized rice tissues. *Plant Cell* 2, 591–602.

Dower, W. J., Miller, J. F. & Ragsdale, C. W. (1988) High efficiency transformation of *E. coli* by high voltage electroporation. *Nucleic Acids Research* 16, 6127–45.

Draper, J., Scott, R., Kumar, A. & Dury, G. (1988) Transformation of plant cells by DNA-mediated gene transfer. In: Draper, J., Scott, R., Armitage, P. & Walden, R. (eds), *Plant Genetic Transformation and Gene Expression: a Laboratory Manual*. Blackwell Scientific Publications, Oxford, pp. 161–98.

Ecker, J. R. & Davis, R. W. (1986) Inhibition of gene expression in plant cells by expression of antisense RNA. *Proceedings of the National Academy of Sciences USA* 83, 5372–6.

Eichholtz, D. A., Rogers, S. G., Horsch, R. B., Klee, H. J., Hayford, M., Hoffmann, N. L., Bradford, S. B., Fink, C., Flick, J., O'Connell, K. M. & Fraley, R. T. (1987) Expression of mouse dihydrofolate reductase gene confers methotrexate resistance in transgenic petunia plants. *Somatic Cell and Molecular Genetics* 13, 67–76.

Ellis, J. G., Llewellyn, D. J., Dennis, E. S.& Peacock, W. J. (1987) Maize *Adh-1* promoter sequences control anaerobic regulation: addition of upstream promoter elements from constitutive genes is necessary for expression in tobacco. *EMBO Journal* 6, 11–16.

Fromm, M., Taylor, L. P. & Walbot, V. (1985) Expression of genes transferred into monocot and dicot plant cells by electroporation. *Proceedings of the National Academy of Sciences USA* 82, 5824–8.

Fromm, M. E., Taylor, L. P. & Walbot, V. (1986) Stable transformation of maize after gene transfer by electroporation. *Nature* 319, 791–3.

Gallie, D. R., Lucas, W. J. & Walbot, V. (1989) Visualizing mRNA expression in plant protoplasts: Factors influencing efficient mRNA uptake and translation. *Plant Cell* 1, 301–11.

Goldman, G. H., van Montagu, M. & Herrera-Estrella, A. (1990) Transformation of *Trichoderma harzianum* by high-voltage electric pulse. *Current Genetics* 7, 1169–74.

Guerche, P., Bellini, C., Le Moullec J.-M. & Caboche, M. (1987) Use of a transient expression assay for the optimization of direct gene transfer into tobacco mesophyll protoplasts by electroporation. *Biochimie* 69, 621–8.

Gusew, N., Nepven, A. & Chartrand, P. (1987) Linear DNA must have free ends to transform rat cells efficiently. *Molecular and General Genetics* 206, 121–5.

Hahn-Hägerdal, B., Hosono, K., Zachrisson, A. & Bornman, C. H. (1986) Polyethylene glycol and electric field treatment of plant protoplasts: characterization of some membrane properties. *Physiologia Plantarum* 67, 359–64.

Hara, A., Hashimoto, H., Morota, H., Harada, H. & Uchimiya, H. (1988) Evidence for the location of foreign genes in three different chromosomes in a plant obtained from direct DNA transfer. *Botanical Magazine Tokyo* 101, 131–40.

Harpster, M. H., Townsend, J. A., Jones, J. D. G., Bedbrook, J. & Dunsmuir, P. (1988) Relative strengths of the 35S cauliflower mosaic virus, 1', 2', and nopaline synthase promoters in transformed tobacco, sugarbeet, and oilseed rape callus tissue. *Molecular and General Genetics* 212, 182–90.

Hashimoto, H., Morikawa, H., Yamada, Y. & Kimura, A. (1985) A novel method for transformation of intact yeast cells by electroinjection of plasmid DNA. *Applied Microbiology and Biotechnology* 21, 336–9.

Haughn, G. W., Smith, J., Mazur, B. & Somerville, C. (1988) Transformation with a mutant *Arabidopsis* acetolactate synthase gene renders tobacco resistant to sulfonylurea herbicides. *Molecular and General Genetics* 211, 266–71.

Hauptmann, R. M., Vasil, V., Ozias-Akins, P., Tabaeizadeh, Z., Rogers, S. G., Horsch, R. B., Vasil, I. K. & Fraley, R. T. (1987) Transient expression of electroporated DNA in monocotyledonous and dicotyledonous species. *Plant Cell Reports* 6, 265–70.

Herrera-Estrella, L., Depicker, A., van Montagu, M. & Schell, J. (1983) Expression of chimaeric genes transferred into cells using Ti-plasmid-derived vector. *Nature* 303, 209–13.

Hibi, T., Kano, H., Sugiura, M., Kazami, T. & Kimura, S. (1986) High efficiency electro-transfection of tobacco mesophyll protoplasts with tobacco mosaic virus DNA. *Journal of General Virology* 67, 2037–42.

Hille, J., Verheggen, F., Roelvink, P., Franssen, H., van Kammen, A. & Zabel, P. (1986) Bleomycin resistance: a new dominant selectable marker for plant cell transformation. *Plant Molecular Biology* 7, 171–6.

Horn, M. E., Shillito, R. D., Conger, B. V. & Harms, C. T. (1988) Transgenic plants of Orchard grass (*Dactylis glomerata* L.) from protoplasts. *Plant Cell Reports* 7, 469–72.

Howard, E. & Citovsky, V. (1990) The emerging structure of the *Agrobacterium* T-DNA transfer complex. *BioEssays* 12, 103–8.

Howard, E. S., Walker, J. C., Dennis, E. S. & Peacock, W. J. (1987) Regulated expression of an alcohol dehydrogenase 1 chimeric gene introduced into maize protoplasts. *Planta* 170, 535–40.

Huang, Y.-W. & Dennis, E. S. (1989) Factors influencing stable transformation of maize protoplasts by electroporation. *Plant Cell, Tissue and Organ Culture* 18, 281–96.

Jasin, M. & Berg, P. (1988) Homologous integration in mammalian cells without target gene selection. *Genes and Development* 2, 1353–63.

Jefferson, R. A., Kavanagh, T. A. & Bevan, M. W. (1987) GUS fusions: β-glucuronidase as a sensitive and versatile gene fusion marker in higher plants. *EMBO Journal* 6, 3901–7.

Jones, H., Templelaar, M. J. & Jones, M. G. K. (1987) Recent advances in plant electroporation. *Oxford Surveys of Plant Molecular and Cell Biology* 4, 347–59.

Jones, H., Ooms, G. & Jones, M. G. K. (1989) Transient gene expression in electroporated *Solanum* protoplasts. *Plant Molecular Biology* 13, 503–11.

Jongsma, M., Koornneef, M., Zabel, P. & Hille, J. (1987) Tomato protoplast DNA transformation: physical linkage and recombination of exogenous DNA sequences. *Plant Molecular Biology* 8, 383–94.

Junker, B., Zimny, J., Lührs, R. & Lörz, H. (1987) Transient expression of chimaeric genes in dividing and non-dividing cereal protoplasts after PEG-induced DNA uptake. *Plant Cell Reports* 6, 329–32.

Kao, K. N. & Saleem, M. (1986) Improved fusion of mesophyll and cotyledon protoplasts with PEG and high pH-Ca^{2+} solutions. *Plant Physiology* 122, 217–25.

Kinoshita, K. & Tsong, T. Y. (1977) Voltage-induced pore formation and hemolysis of human erythrocytes. *Biochimica et Biophysica Acta* 471, 227–42.

Kinoshita, K., Ashikawa, I., Saita, N., Yoshimura, H., Itoh, H., Nagayama, K. & Ikegami, A. (1988) Electroporation of cell membrane visualized under a pulsed-laser fluorescence microscope. *Biophysics Journal* 53, 1015–19.

Knutson, J. C. & Yee, D. (1987) Electroporation: parameters affecting transfer of DNA into mammalian cells. *Analytical Biochemistry* 164, 44–52.

Köhler, F., Golz, C., Eapen, S., Kohn, H. & Schieder, O. (1987) Stable transformation of moth bean *Vigna aconitifolia* via direct gene transfer. *Plant Cell Reports* 6, 313–17.

Koncz, C., Olsson, O., Langridge, W. H. R., Schell, J. & Szalay, A. A. (1987) Expression and assembly of functional bacterial luciferase in plants. *Proceedings of the National Academy of Sciences USA* 84, 131–5.

Koncz, C., Martini, N., Mayerhofer, R., Koncz-Kalman, Z., Korber, H., Redei, G. P. & Schell, J. (1989) High-frequency T-DNA-mediated gene tagging in plants. *Proceedings of the National Academy of Sciences USA* 86, 8467–71.

Krens, F. A., Molendijk, L., Wullems, G. J. & Schilperoort, R. A. (1982) *In vitro* transformation of plant protoplasts with Ti-plasmid DNA. *Nature* 296, 72–4.

Krüger-Lebus, S. & Potrykus, I. (1987) A simple and efficient method for direct gene transfer to *Petunia hybrida* without electroporation. *Plant Molecular Biology Reports* 5, 289–94.

Langridge, W. H. R., Bao, J. L. & Szalay, A. A. (1985) Electric field-mediated stable transformation of carrot protoplasts with naked DNA. *Plant Cell Reports* 4, 355–9.

Larkin, P. J., Taylor, B. H., Gersmann, M. & Brettel, R. I. S. (1990) Direct gene transfer to protoplasts. *Australian Journal of Plant Physiology* 17, 291–302.

Lin, F.-L., Sperle, K. & Sternberg, N. (1985) Recombination in mouse L cells between DNA introduced into cells and homologous chromosomal sequences. *Proceedings of the National Academy of Sciences USA* 82, 1391–5.

Lindsey, K. & Jones, M. G. K. (1987) Transient gene expression in electroporated protoplasts and intact cells of sugar beet. *Plant Molecular Biology* 10, 43–52.

Lörz, H., Baker, B. & Schell, J. (1985) Gene transfer to cereal cells mediated by protoplast transformation. *Molecular and General Genetics* 199, 178–82.

Ludwig, S. R., Bowen, B., Beach, L. & Wessler, S. R. (1990) A regulatory gene as a novel visible marker for maize transformation. *Science* 247, 449 50.

Lurquin, P. F. & Paszty, C. (1988) Electroporation of tobacco protoplasts with homologous and nonhomologous transformation vectors. *Journal of Plant Physiology* 133, 332–5.

Lyznik, L. A., Ryan, R. D., Ritchie, S. W. & Hodges, T. K. (1989) Stable co-transformation of maize protoplasts with *gusA* and *neo* genes. *Plant Molecular Biology* 13, 151–61.

Matsuki, R., Onodera, H., Yamaguchi, T. & Uchimiya, H. (1989) Tissue-specific expression of the *rolC* promoter of the Ri plasmid in transgenic rice plants. *Molecular and General Genetics* 220, 12–16.

Meyer, P., Walgenbach, E., Bussmann, K., Hombrecher, G. & Saedler, H. (1985) Synchronized tobacco protoplasts are efficiently transformed by DNA. *Molecular and General Genetics* 201, 513–18.

Meyer, P., Kartzke, S., Niedenkof, I., Heidmann, I., Bussmann, K. & Saedler, H. (1988) A genomic DNA segment from *Petunia hybrida* leads to increased transformation frequencies and simple integration patterns. *Proceedings of the National Academy of Sciences USA* 85, 8568–72.

Morikawa, H., Iida, A., Matsui, C., Ikegami, M. & Yamada, Y. (1986) Gene transfer into intact plant cells by electroinjection through cell walls and membranes. *Gene* 41, 121–4.

Nea, L. J. & Bates, G. W. (1987) Factors affecting protoplast electrofusion efficiency. *Plant Cell Reports* 6, 337–40.

Negrutiu, I., Shillito, R., Potrykus, I., Biansini, G. & Sala, F. (1987) Hybrid genes in the analysis of transformation conditions. I. Setting up a simple method for direct gene transfer in plant protoplasts. *Plant Molecular Biology* 8, 363–73.

Neumann, E., Schaefer-Ridder, M., Wang, Y. & Hofschneider, P. H. (1982) Gene transfer into mouse lyoma cells by electroporation in high electric fields. *EMBO Journal* 1, 841–5.

Nishiguchi, M., Langridge, W. H. R., Szalay, A. A. & Zaitlin, M. (1986) Electroporation-mediated infection of tobacco leaf protoplasts with tobacco mosaic virus RNA and cucumber mosaic virus RNA. *Plant Cell Reports* 5, 57–60.

Ochatt, S. J., Chand, P. K., Rech, E. L., Davey, M. R. & Power, J. B. (1988) Electroporation-mediated improvement of plant regeneration from colt cherry (*Prunus avium*×*pseudocerasus*) protoplasts. *Plant Science* 54, 165–9.

Okada, K., Takebe, I. & Nagata, T. (1986) Expression and integration of genes introduced into highly synchronised plant protoplasts. *Molecular and General Genetics* 205, 398–403.

Olsson, O., Nilsson, O. & Koncz, C. (1990) Engineered luciferase proteins as tools to study plant gene regulation *in vivo*. *Journal of Bioluminescence and Chemiluminescence* 5, 79–87.

Otten, L. A. B. M. & Schilperoort, R. A. (1978) A rapid micro scale method for the detection of lysopine and nopaline dehydrogenase activities. *Biochimica et Biophysica Acta* 527, 497–500.

Ow, D. W., Wood, K. V., DeLuca, M., De Wet, J. R., Helinski, D. R. & Howell, S. H. (1986) Transient and stable expression of the firefly luciferase gene in plant cells and transgenic plants. *Science* 234, 856–9.

Paszkowski, J., Bauer, M., Bogucki, A. & Potrykus, I. (1988) Gene targeting in plants. *EMBO Journal* 57, 4021–6.

Peerbolte, R., Kresas, F. A., Mans, R. M. W., Floor, M., Hoge, J. H. C., Wullems, G. J. & Schilperoort, R. A. (1985) Transformation of plant protoplasts with DNA: cotransformation of non-selected calf thymus carrier DNA and meiotic segregation of transforming DNA sequences. *Plant Molecular Biology* 5, 235–46.

Perucho, M., Hanahan, D. & Wigler, M. (1980) Genetic and physical linkage of exogenous sequences in transformed cells. *Cell* 22, 309–17.

Planckaert, F. & Walbot, V. (1989) Transient gene expression after electroporation of protoplasts derived from embryogenic maize callus. *Plant Cell Reports* 8, 144–7.

Rathus, C. & Birch, R. G. (1991) Optimization of conditions for electroporation and transient expression of foreign genes in sugarcane protoplasts. *Plant Science,* in press.

Rech, E. L., Ochatt, S. J., Chand, P. K., Davey, M. R., Mulligan, B. J. & Power, J. B. (1988) Electroporation increases DNA synthesis in cultured plant protoplasts. *Bio/Technology* 6, 1091–3.

Reiss, B., Springel, R., Will, H. & Schaller, H. (1984) A new sensitive method for qualitative and quantitative assay of neomycin phosphotransferase in crude cell extracts. *Gene* 30, 211–18.

Rhodes, C. A., Pierce, D. A., Mettler, I. J., Mascarenhas, D. & Detmer, J. J. (1988) Genetically transformed maize plants from protoplasts. *Science* 240, 204–7.

Riggs, C. D. & Bates, G. W. (1986) Stable transformation of tobacco by electroporation: evidence for plasmid concatenation. *Proceedings of the National Academy of Sciences USA* 83, 5602–6.

Rodenburg, K. W., de Groot, M. J. A., Schilperoort, R. A. & Hooykaas, P. J. J. (1989) Single-stranded DNA used as an efficient new vehicle for transformation of plant protoplats. *Plant Molecular Biology* 13, 711–19.

Sanders, P. R., Winter, J. A., Barnason, A. R., Rogers, S. G. & Fraley, R. T. (1987) Comparison of cauliflower mosaic virus 35S and nopaline synthase promoters in transgenic plants. *Nucleic Acids Research* 15, 1543–58.

Saunders, J. A., Smith, C. R. & Kaper, J. M. (1989) Effects of electroporation pulse wave on the incorporation of viral RNA into tobacco protoplasts. *Bio Techniques* 7, 1124–31.

Schocher, R. J., Shillito, R. D., Saul, M. W., Paszkowski, J. & Potrykus, I. (1986) Co-transformation of unlinked foreign genes into plants by direct gene transfer. *Bio/Technology* 4, 1093–6.

Séguin, A. & Lalonde, M. (1988) Gene transfer by electroporation in Betulaceae protoplasts: *Alnus incana*. *Plant Cell Reports* 7, 367–70.

Shigekawa, K. & Dower, W. J. (1989) Electroporation: a general approach to the introduction of macromolecules into prokaryotic and eukaryotic cells. *Australian Journal of Biotechnology* 3, 56–64.

Shillito, R. D., Saul, M. W., Paszkowski, J., Miller, M. & Potrykus, I. (1985) High efficiency direct gene transfer to plants. *Bio/Technology* 3, 1099–103.

Shimamoto, K., Terada, R., Izawa, T. & Fujimoto, H. (1989) Fertile transgenic rice plants regenerated from transformed protoplasts. *Nature* 338, 274–6.

Shirsat, A. H., Wilford, N. & Croy, R. R. D. (1989) Gene copy number and levels of expression in transgenic plants of a seed specific gene. *Plant Science* 61, 75–80.

Spencer, T. M., Gordon-Kamm, W. J., Daines, R. J., Start, W. G. & Lemaux, P. G. (1990) Bialaphos selection of stable transformants from maize cell culture. *Theoretical and Applied Genetics* 79, 625–31.

Steinbiss, H.-H. (1978) Dielektrischer Durchbruch des Plasmalemmas von *Valerianella locusta*–Protoplasten. *Zeitschrift für Pflanzenphysiologie* 88, 95–102.

Steinbiss, H.-H. & Broughton, W. J. (1983) Methods and mechanisms of gene uptake in protoplasts. *International Review of Cytology* Supplement 16, 191–208.

Strobel, G. A. (1982) Phytotoxins. *Annual Review of Biochemistry* 51, 309–33.

Sturley, S. L. & Young, T. W. (1986) Genetic manipulation of commercial yeast strains. *Biotechnology and Genetic Engineering Reviews* 4, 1–38.

Sugar, I. P. & Neumann, E. (1984) Stochastic model for electric field-induced membrane pores: electroporation. *Biophysical Chemistry* 19, 211–25.

Tagu, D., Bergounioux, C., Cretin, C., Perennes, C. & Gadal, P. (1988) Direct gene transfer in *Petunia hybrida* electroporated protoplasts: evidence for co-transformation with a phosphoenolpyruvate carboxylase cDNA from sorghum leaf. *Protoplasma* 146, 101–5.

Taylor, B. H. & Larkin, P. J. (1988) Analysis of electroporation efficiency in plant protoplasts. *Australian Journal of Biotechnology* 1, 52–7.

Teeri, T. H., Lehväslaiho, H., Franck, M., Uotila, J., Heino, P., Palva, E. T., Van Montagu, M. & Herrera-Estrella, L. (1989) Gene fusions to *lacZ* reveal new expression patterns of chimeric genes in transgenic plants. *EMBO Journal* 8, 343–50.

Töpfer, R., Pröls, M., Schell, J. & Steinbiss, H.-H. (1988) Transient gene expression in tobacco protoplasts. II. Comparison of the reporter gene systems for CAT, NPTII and GUS. *Plant Cell Reports* 7, 225–8.

Toriyama, K., Arimoto, Y., Uchimiya, H. & Hinata, I. (1988) Transgenic rice plants after direct gene transfer into protoplasts. *Bio/Technology* 6, 1072–4.

Uchimiya, H., Hirochika, H., Hashimoto, H., Hara, A., Masuda, T., Kasumimoto, T., Harada, H., Ikeda, J.-E. & Yoshioka, M. (1986) Co-expression and inheritance of foreign genes in transformants obtained by direct DNA transformation of tobacco protoplasts. *Molecular and General Genetics* 205, 1–8.

van den Elzen, P. J. M., Townsend, J., Lee, K. Y. & Bedbrook, J. R. (1985) A chimaeric hygromycin resistance gene as a selectable marker in plant cells. *Plant Molecular Biology* 5, 299–302.

Vasil, V., Hauptmann, R. M., Morrish, F. M. & Vasil, I. K. (1988) Comparative analysis of free DNA delivery and expression into protoplasts of *Panicum maximum* Jacq. (Guinea grass) by electroporation and polyethylene glycol. *Plant Cell Reports* 7, 499–503.

Waldron, C., Murphy, E. B., Roberts, J. L., Gustafson, G. D., Armour, S. L. & Malcolm, S. K. (1985) Resistance to hygromycin B: a new marker for plant transformation studies. *Plant Molecular Biology* 5, 103–8.

Wernars, K., Goosen, T., Wennekes, L. M. J., Visser, J., Bos, C. J., Van den Boek, H. W. J., Van Gorcum, R. F. M., Van den Hondel, C. A. M. J. J. & Pouwels, P. H. (1985) Gene amplification in *Aspergillus nidulans* by transformation with vectors containing the *amd*S gene. *Current Genetics* 9, 361–8.

Wirtz, U., Schell, J. & Czernilofsky, A. P. (1987) Recombination of selectable marker DNA in *Nicotiana tabacum*. *DNA* 6, 245–53.

Yang, H., Zhang, H. M., Davey, M. R., Mulligan, B. J. & Cocking, E. C. (1988) Production of kanamycin resistant rice tissues following DNA uptake into protoplasts. *Plant Cell Reports* 7, 421–5.

Zambryski, P., Joos, H., Genetello, C., Leemans, J., Van Montagu, M. & Schell, J. (1983) Ti plasmid vector for the introduction of DNA into plant cells without alteration of their normal regeneration capacity. *EMBO Journal* 2, 2143–50.

Zhang, H. M., Yang, H., Rech, E. L., Golds, T. J., Davis, A. S., Mulligan, B. J., Cocking, E. C. & Davey, M. R. (1988) Transgenic rice plants produced by electroporation-mediated plasmid uptake into protoplasts. *Plant Cell Reports* 7, 379–84.

Zhang, W. & Wu, R. (1988) Efficient regeneration of transgenic plants from rice protoplasts and correctly regulated expression of the foreign gene in the plants. *Theoretical and Applied Genetics* 76, 835–40.

Zhelev, D. V., Dimitrov, D. S. & Doinov, P. (1988) Correlation between physical parameters in electrofusion and electroporation of protoplasts. *Bioelectrochemistry and Bioenergetics* 20, 155–67.

Zimmermann, V. (1982) Electric field-mediated fusion and related electrical phenomena. *Biochimica et Biophysica Acta* 694, 227–77.

Zimmermann, V. (1986) Electric breakdown, electropermeabilization and electrofusion. *Reviews of Physiology Biochemistry and Pharmacology* 10, 175–256.

Zimmermann, V. & Vienken, J. (1982) Electric field-induced cell-to-cell fusion. *Journal of Membrane Biology* 67, 165–82.

Chapter 5
Microprojectile Techniques for Direct Gene Transfer into Intact Plant Cells

Tricia Franks & Robert G. Birch
Department of Botany, The University of Queensland, Queensland 4072, Australia

Introduction

Development of *Agrobacterium* as a vehicle for routine genetic transformation of model plant species (Chapter 3) and the ability to rapidly assay expression of genes transferred into protoplasts of many plant species (Chapter 4) have underpinned the recent, explosive growth of plant molecular genetics. However, because of host-range limitations of *Agrobacterium* and difficulties with regeneration from protoplasts, reliable production of transgenic plants has been achieved for relatively few economic plant species. In this context, bombardment of intact plant cells and tissues with high-velocity, DNA-coated microprojectiles can be considered crude, but effective. Results to date suggest that the technique can be developed to facilitate production of transgenic plants of virtually any species.

Plant virologists have for decades employed high-velocity micropro-jectiles to wound plant cells and facilitate entry of viral particles or nucleic acids (MacKenzie *et al.*, 1966). However, routine virological techniques (50 mg/ml of 600 mesh carborundum in inoculum sprayed from an air-brush at 5 kg/cm^2) are not effective for transfer of non-infectious nucleic acids. The key steps to develop a gene transfer technique were taken by researchers at Cornell University who developed a range of devices to accelerate tungsten microprojectiles (1–4 μm in diameter) to velocities (approx. 250 m/s) sufficient to penetrate plant cell walls and membranes. Onion epidermal cells penetrated by a small number of projectiles remained viable, and up to 40% of cells penetrated by tungsten microprojectiles coated with TMV RNA developed distinctive crystalline inclusions, indicating expression of the viral nucleic acid. The Cornell researchers recognized that particle

Table 5.1. Transient and stable expression of microprojectile-delivered DNA in various species and tissues.

Species	Transient expression in tissue types	Stable expression[a]	R$_1$ progeny	References
Higher plants				
Arabidopsis	Seedlings			Bruce *et al.*, 1989
Barley	Anther culture			Creissen *et al.*, 1990
	Callus, immature embryos			Kartha *et al.*, 1989
	Seedlings			Bruce *et al.*, 1989
	Suspension cell culture			Kartha *et al.*, 1989; Mendel *et al.*, 1989
Cotton		Plants (from embryogenic suspension culture)	−	Finer & McMullen, 1990
Cucumber	Seedlings			Bruce *et al.*, 1989
Eggplant	Adventitious shoots of hypocotyls			Morikawa *et al.*, 1989
Maize	Aleurone, embryos			Goff *et al.*, 1990; Klein *et al.*, 1989b
	Callus			Goff *et al.*, 1990
	Seedlings			Ludwig *et al.*, 1990
	Suspension cell culture			Klein *et al.*, 1988a,b; Oard *et al.*, 1990
		Callus (from suspension culture)		Klein *et al.*, 1989a
		Plants (from suspension culture)	+	Gordon-Kamm *et al.*, 1990
Oat	Seedlings			Bruce *et al.*, 1989
Onion	Epidermal cells (viral RNA)			Sanford *et al.*, 1987
Rice	Seedlings			Bruce *et al.*, 1989
	Suspension cell culture			Wang *et al.*, 1989

Organism	Material	Transformation[a]	References
Soybean	Suspension cell culture		Wang et al., 1989
	Callus (from callus)		Christou et al., 1988
	Plants (from immature shoot meristems)	+	Christou et al., 1989; McCabe et al., 1988b
Tobacco	Plastics of suspension cell culture		Daniell et al., 1990
	Pollen		Twell et al., 1989
	Seedlings		Bruce et al., 1989
	Suspension cell culture		Morikawa et al., 1989
	Callus and plants from intact leaves and suspension cell culture	+	Klein et al., 1988c
Wheat	Suspension cell culture		Wang et al., 1989
Algae			
Chlamydomonas reinhardtii	Chloroplast		Blowers et al., 1989; Boynton et al., 1988
	Nucleus		Day et al., 1990; Debuchy et al., 1989; Kindle et al., 1989
Fungi			
Neurospora crassa	Nucleus		Armaleo et al., 1990
Saccharomyces pombe	Nucleus		Armaleo et al., 1990
Saccharomyces cerevisiae	Nucleus		Armaleo et al., 1990; Fox et al., 1988; Johnston et al., 1988
	Mitochondria		

[a]Unless otherwise stated, stable transformation refers to integration into the nuclear genome.

bombardment could be a nearly universal mechanism for transporting substances such as biological stains, proteins (antibodies or enzymes), synthetic macromolecules and genetic material into any living cell, with particular significance for genetic transformation (Sanford *et al.*, 1987).

Following the demonstration of strong transient reporter gene expression in onion epidermal tissue bombarded with DNA-coated microprojectiles (Klein *et al.*, 1987) the technique was adopted and modified successfully by numerous research groups world-wide. To date, the targets of bombarding DNA have been primarily plant, algal and fungal cells. Nuclear and organelle genomes of plants, algae and fungi have been transformed by the technique. Many plant species have been shown to support transient expression of bombarding foreign DNA in various tissues. Transgenic plants generated by microprojectile techniques have already been reported for crop species as diverse as cotton, soybean and maize (see Table 5.1).

Dubbed the 'biolistic' process by its inventors (Sanford, 1988), the use of high-velocity, DNA-coated microprojectiles for gene transfer has also been referred to as particle bombardment, microprojectile bombardment, particle acceleration, the gene gun method or the particle gun method.

Here we present a summary of the microprojectile techniques currently used to deliver nucleic acids into plant cells. We wish to emphasize the need to tailor any microprojectile bombardment system to suit a particular species and tissue target. It is important firstly to achieve maximal DNA delivery rates to cells, and secondly to consider the best possible strategy for generating stably transformed plants. In addition to direct use of microprojectile-mediated gene transfer for plant improvement, we discuss the contribution of this technology to studies of gene regulation, gene function and basic plant biology.

Apparatus for microprojectile acceleration

Sanford *et al.* (1987) found that devices using gas discharge, transferred mechanical impulse, a macroprojectile and stopping plate, or a centripetal acceleration system all accelerated microprojectiles to velocities sufficient to penetrate plant cells. The most effective device used a gunpowder charge to accelerate a nylon macroprojectile down a barrel towards a stop plate. DNA-coated microprojectiles on the front of the macroprojectile continue at high velocity through an aperture in the stop plate, and travel to the target tissue in a chamber under partial vacuum (Klein *et al.*, 1987; Sanford *et al.*, 1987). Full details of a commercially available apparatus operating on this principle were subsequently provided in a patent application (Sanford *et al.*, 1989).

An alternative apparatus, also described in a patent application, used a high-voltage electric discharge through a water droplet to generate a shock wave which accelerates a carrier sheet into a retaining screen, allowing DNA-coated microprojectiles to continue at high velocity into the target tissue (McCabe *et al.*, 1988a). Other simple particle gun designs using either gunpowder charges, solenoid-controlled compressed gas pulses or a modified air rifle have been published (Morikawa *et al.*, 1989; Oard *et al.*, 1990; Zumbruun *et al.*, 1989), and numerous modifications are in use in laboratories around the world. Figure 5.1 shows a typical design from our laboratory, using either gunpowder charges or gas pulses for acceleration. Most devices in current use employ a macroprojectile and stop plate rather than a carrier sheet and screen. Electric discharge has not been widely used because of the expense of the large capacitors required.

Although results with microprojectile bombardment are notoriously inconsistent, similar transformation frequencies have been achieved with a

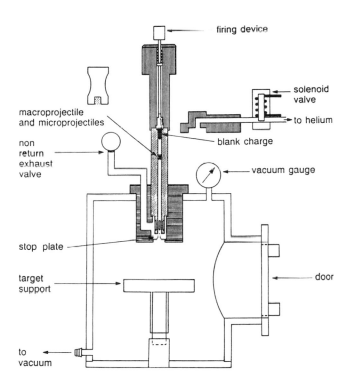

Fig. 5.1. Schematic diagram of a particle bombardment device. DNA-coated microprojectiles are placed on the surface of a plastic macroprojectile, which is propelled by either a gunpowder charge or a gas pulse towards a stop plate. The accelerated microprojectiles continue through a vacuum chamber into the target tissue.

variety of machines. In selecting or designing a microprojectile accelerator for gene transfer work, we suggest safety, transformation efficiency, reliability and convenience of use as major criteria. It seems likely that the next generation of devices may be capable of accurately targeting cells, in addition to achieving greater consistency between shots.

Optimizing DNA delivery to intact cells

Early reports of microprojectile-mediated delivery of DNA to intact cells were concerned with maximizing the number of cells which transiently express the incoming DNA for a given apparatus and cell type.

The GUS (β-glucuronidase) reporter system of Jefferson *et al.* (1987) has been most useful for determining the number of expressing cells per bombardment. One to two days after bombardment with the reporter gene, target cells from suspension culture, callus culture or intact tissue are incubated with the GUS histochemical substrate (5-bromo-4-chloro-3-indoyl-β-D-glucuronic acid). Those cells which express *gus* stain blue, and the number and distribution of the cells which have received and expressed the bombarding DNA are easily visualized.

At least five key factors interact to affect the frequency of transiently expressing cells in the bombarded tissue.

1. DNA attachment to the microprojectiles prior to bombardment.
2. Size of the microprojectiles.
3. Impact velocity of the microprojectile/DNA complex.
4. Degree of tissue damage suffered on bombardment.
5. Potential for expression of the bombarding DNA in the recipient tissue.

In a model study for optimizing microprojectile delivery protocols, Klein *et al.* (1988b) investigated factors influencing *gus* gene delivery into maize suspension culture cells. The parameters they tested included the size of the microprojectiles, the quantity of DNA loaded on to microprojectiles at co-precipitation, the concentration of co-precipitation factors ($CaCl_2$ and spermidine), the quantity of tungsten particles per bombardment, the distance at which the target tissue was placed from the macroprojectile stop plate, the means of anchoring the target cells and the vacuum in the chamber through which the microprojectiles were propelled towards the target tissue. Transient expression frequencies increased from 10 to 500 cells per bombardment as these parameters were optimized.

DNA attachment to microprojectiles

The features required in materials to be used for microprojectiles in gene transfer work include the following.

1. Availability in a range of defined particle sizes around 1 μm diameter.
2. High density to achieve the momentum required for cell wall penetration.
3. Sufficiently inert to reduce the likelihood of explosive oxidation of fine microprojectile powders which would be both dangerous and inconvenient for storage.
4. Non-reactivity with DNA and other components of the precipitating mixes.
5. Low toxicity to plant cells.

Tungsten and gold microprojectiles meet these requirements and have been widely used in gene transfer studies.

Any effective procedure for attaching DNA to microprojectiles is probably universally applicable to all cell types regardless of the other optimized bombardment conditions. DNA is commonly bound to tungsten particles by variations of the $CaCl_2$/spermidine co-precipitation method of Klein *et al.* (1988b). We use stock suspensions of 1–2 μm mean diameter tungsten particles stored frozen at 50 mg/ml in 50% glycerol and take care to separate aggregated particles by sonicating the suspension just prior to preparing a precipitation mix. The mix is prepared in an eppendorf tube, and contains 25 μl tungsten suspension, 2 μl DNA (1 mg/ml), 25 μl $CaCl_2$ (2.5 M) and 10 μl spermidine (0.1 M, free base form), added and mixed sequentially in that order. The mix is kept on ice for at least 15 min before bombardment. Just prior to bombardment, the eppendorf containing the mix is centrifuged momentarily, 25 μl of the supernatant is removed and discarded and the tube base is touched briefly to the tip of a sonicator horn tip to disrupt agglutinated particles. We typically load 6 μl of this 'wet' suspension of DNA-coated particles on to the macroprojectile surface. DNA-coated particles may dry as they are propelled through the vacuum chamber towards the target.

In contrast, DNA has generally been precipitated on to gold particles using ethanol, and then dried on to the carrier surface before acceleration (McCabe *et al.*, 1988a). Both techniques are convenient and effective, although most laboratories develop a preference for one or the other.

Microprojectile size

The optimal size and impact velocity of the microprojectiles depend on properties of the target cells such as size, penetrability and resilience. Those working with large plant cells generally use microprojectiles 0.8–1.2 μm in diameter. For example, Klein *et al.* (1988b) found highest transient expression when maize suspension culture cells were bombarded with DNA-coated tungsten particles with an average diameter of 1.2 μm compared with 0.6 μm and 2.4 μm particles. Armaleo *et al.* (1990) transformed smaller fungal cells with microprojectiles in the size range 0.5–0.65 μm after finding that projectiles larger than one-tenth of the yeast cell diameter tended to destroy the cell structure. Paradoxically, yeast mitochondria have been transformed using 1 μm microprojectiles, a diameter comparable to that of the organelles themselves (Johnston *et al.*, 1988).

Microprojectile velocity and tissue damage

Even for a single plant species, it may be necessary to alter microprojectile velocity for optimal transformation rates with different tissue types, depending on cell wall thickness and the need to penetrate through several cell layers. Although there may be substantial variation between consecutive shots, velocity can generally be controlled to some degree by altering the accelerating force (gunpowder charge size, gas pulse pressure, electrical discharge energy, etc.), the vacuum in the target chamber or the distance travelled by the microprojectiles. If necessary, macroprojectile velocity can be measured with a chronograph device, used to measure firearm muzzle velocities. Klein *et al.* (1988b) found the highest vacuum they tested (28 in of Hg) was optimal and that transient expression rates were greatest when the tissue was placed at the shortest tested distance (6 cm) from the stop plate.

Using macroprojectile and stop-plate devices, velocities sufficient for gene transfer frequently cause target tissue damage from macroprojectile fragmentation, air blasts and perhaps acoustic shock. This results in a central area of tissue damage, sometimes referred to as the 'zone of death', fringed by an area in which there are high frequencies of transiently expressing cells (Klein *et al.*, 1988c). We have found that nylon 66 and UHMW polyethylene macroprojectiles cause fewer fragmentation problems than many other plastics including PVC, teflon, polycarbonate and polyurethanes. It is also important for the well in the macroprojectile to be slightly wider than the stop-plate aperture, and the optimal diameter for this aperture may decrease as macroprojectile impact velocity is increased. Damage can also be decreased by interposing a mesh screen to shelter the target tissue from macroprojectile fragments. Although blocking some microprojectiles, such a screen may also increase transient expression frequencies by deflecting particles to penetrate cells over a wider area of the target tissue.

Armaleo *et al.* (1990) observed that yeast cells which received bombarding particles commonly suffered cytoplasmic extrusions. They achieved highest yeast transformation rates using plating media of high osmolarity, possibly due to protection against osmotic disruption following cell wall damage. Experiments are warranted to determine whether media of high osmolarity can similarly protect plant tissues from cell death upon bombardment.

Expression of the bombarding DNA in the recipient tissue

When optimizing a microprojectile bombardment system, it is important to choose a genetic construct that will be expressed at reasonably high rates in the bombarded tissue. The GUS histochemical assay is relatively insensitive, and cells expressing the *gus* gene to a relatively low degree may not be revealed. In our work with sugarcane suspension cultures, a change from CaMV 35S to a stronger monocot promoter resulted in a 100-fold increase in the apparent frequency of transiently expressing cells based on GUS histochemical staining (Franks & Birch, 1991). Genotype, metabolic capacity and developmental state of the target tissue will also influence the expression rate of the incoming DNA. Bruce *et al.* (1989) have clearly demonstrated the effects of the gene regulatory sequences, the genetic background of the target tissue and environmental stimuli on the extent of transient expression in bombarded tissues.

Microprojectile techniques for plant transformation

It is very important to achieve a high frequency of transiently expressing cells following microprojectile bombardment because, at best, only a few percent of transiently expressing cells will integrate and stably express the introduced gene. Estimates for rates of conversion from transient expression to stable incorporation of microprojectile-delivered DNA in plant cells range from less than 1% (Finer & McMullen, 1990) up to almost 5% (Klein *et al.*, 1988c). Ideally, we aim to determine conditions yielding at least 10^3 transiently expressing cells per bombardment in regenerable tissues, for a high probability of regeneration of transgenic plants. Although data on this aspect are scant in the literature, Klein *et al.* (1988c) regenerated transgenic tobacco plants from leaves showing *c.* 80 transiently expressing cells per bombardment whereas Finer & McMullen (1990) regenerated transgenic cotton plants from embryogenic suspension cultures showing over 4000 transiently expressing cells per bombardment.

Strategies for generating stable transformants

To date, regeneration of transgenic plants following microprojectile bombardment has been reported for four plant species. McCabe *et al.* (1988b) and Christou *et al.* (1989) bombarded meristems from immature seeds of soybean with GUS and NPTII gene constructs. Plants were regenerated from bombarded embryos without selection, and leaf segments assayed for GUS and NPTII activity revealed chimeric and uniformly transformed plants. Three of the transformants were tested further. One chimeric plant and two uniformly transformed plants were fertile and transmitted the marker gene to the R_1 generation (Christou *et al.*, 1989).

Using a different strategy, Klein *et al.* (1988c), Finer & McMullen (1990) and Gordon-Kamm *et al.* (1990) have transformed plants of tobacco, cotton and maize respectively. In each case regenerable tissue was bombarded with a selectable gene. The bombarded cells were proliferated as non-differentiated tissue without selection for a period of 7–14 days, and then under selection for the appropriate resistance. Plants were subsequently regenerated from selected callus. Southern analyses confirmed that foreign DNA had integrated into the nuclear genomes of the cotton and maize plants. The progeny from self-fertilization of a transformed tobacco plant inherited NPTII resistance in a 15:1 ratio, suggesting that the plant was heterozygous and carried the transforming DNA at two independent loci (Klein *et al.*, 1988c). Some of the maize plants were fertile and transmitted the transforming DNA to progeny. Preliminary data from backcrossing one transgenic maize plant are consistent with a 1:1 segregation ratio of a single dominant gene. However, morphological variation was observed, possibly derived from somatic rearrangements during tissue culture (Gordon-Kamm *et al.*, 1990).

Choice of target tissue

When the objective of microprojectile bombardment is recovery of transgenic plants, the choice of target tissues will be limited to those which can be regenerated to plants, and which show satisfactory rates of penetration by microprojectiles as evidenced by transient expression studies. For some species, these criteria may be met by both organized meristematic tissues (such as embryos from seed or proembryoids or shoots produced through tissue culture) and dedifferentiated totipotent cells (such as cell suspensions, callus or tissue explants which regenerate via a callus stage).

An advantage of meristematic target tissue is that extended tissue culture, with attendant costs and risk of somaclonal variation, may be avoided. The disadvantage is that a high proportion of transformed regenerants is likely to be chimeric, as observed with soybean (Christou *et al.*, 1989). Direct antibiotic selection would be lethal to many chimeric plantlets, so numerous regenerating plants must be screened, using an assay for reporter gene expression. It may be possible to streamline this process using rapid, non-toxic luciferase assays to scan large numbers of regenerating plants (Olsson *et al.*, 1990).

By contrast, regeneration from bombarded dedifferentiated cells permits early selection for antibiotic resistance and increases the likelihood of uniformly transformed plants as observed with cotton and maize (Finer & McMullen, 1990; Gordon-Kamm *et al.*, 1990). However, depending upon the plant species and selectable marker gene used, cross-protection between cells may still result in chimeric primary regenerants.

When chimeric plants are obtained, it may be possible to derive uniformly transformed progeny from lateral buds, through proliferation in tissue culture or by conventional breeding, depending upon the location of transformed sectors and the cell lineage patterns of the species concerned (see McDaniel & Poethig, 1988).

Microprojectile-delivered DNA has been expressed transiently in barley anther culture cells (Creissen *et al.*, 1990). Such haploid tissues are potentially useful targets for introducing foreign genes into a species gene pool via microprojectiles. Genes could be directly introduced to established cultivars of homozygous crops, by regenerating diploid plants from stably transformed haploid tissue. Heterozygous crops would require additonal classical plant breeding measures.

Pollen of tobacco has also been shown to support the transient expression of bombarding foreign genes (Twell *et al.*, 1989). A patent application has been filed for a method to produce transgenic seed through fertilization with pollen which has been bombarded with DNA-coated microprojectiles (McCabe *et al.*, 1988a). This approach has the potential to quickly generate large numbers of transformants without the need for tissue culture manipulations. However, confirmation of the production of transgenic plants using transformed pollen has not appeared in the scientific literature.

Choice of a selectable marker

Tailoring a system for effective transgenic plant production also requires a careful choice of an appropriate selectable marker. Hauptmann *et al.* (1988) addressed this factor with particular reference to grass species, which tend to have high natural resistance to aminoglycoside antibiotics, such as kanamycin, commonly used for selecting transformed plants. Gordon-Kamm *et al.* (1990) found bialaphos herbicide to be an effective selective agent for embryonic maize cultures. They used the *bar* gene encoding PAT-mediated bialaphos resistance in microprojectile-mediated transformation to isolate the first reported fertile transgenic maize plants. No chimeric callus or plants were detected in numerous independent lines regenerated under bialaphos selection. The importance of the choice of a selectable marker is also demonstrated by the continued difficulty of developing transformation systems for eukaryotic algae. For the algae in particular, there has been little success with heterologous selectable marker genes, possibly due to problems with promoter specificity or strongly biased codon usage. However, a shift to selection systems involving complementation of mutations in algal genes has been successful (Day *et al.*, 1990). This strategy has also been successful for demonstrating fungal transformation using microprojectiles (Armaleo *et al.*, 1990), but the necessity to work with recipient cells carrying a defined mutation makes the strategy unattractive for plant cultivar improvement.

Optimizing a stable transformation system

In summary, when planning a strategy for transgenic plant production using microprojectile techniques, consider the following points:

1. identify penetrable target tissues capable of plant regeneration;
2. maximize the success of initial DNA delivery based on transient expression frequency;
3. minimize tissue culture manipulations to reduce undesired somaclonal variation;
4. be aware of the need for appropriate gene regulatory sequences;
5. (a) carefully choose a selectable marker and fine tune the matching selection conditions, or

   (b) adopt a non-toxic, *in situ* reporter gene assay to reveal chimeric regenerants.

Stable transformation by microprojectile techniques may benefit from future experimentation with different structural forms of bombarding DNA and mechanisms for stimulating DNA repair/recombination systems, as for other direct gene transfer techniques (see Chapter 4).

Analysis of stable nuclear transformants

Patterns of integration

Southern analyses of DNA extracted from tissues stably transformed by microprojectile techniques reveal trends in integration patterns analogous to those resulting from other direct gene transfer techniques (see Chapter 4). For transformants generated by microprojectiles, integration patterns vary markedly from one individual to the next. Most transformed plant tissues have one or more (up to 20) intact copies of the marker gene inserted at a single site or multiple sites in the genome. In many cases there are also multiple rearranged forms of the introduced DNA at various sites in the genome (Christou *et al.*, 1988; Klein *et al.*, 1988c, 1989a; Finer & McMullen, 1990; Gordon-Kamm *et al.*, 1990). In algal and fungal nuclear transformants selected by complementation of mutant strains with wild-type genes, the introduced DNA was inserted at sites either linked or unlinked to the native locus (Debuchy *et al.*, 1989; Kindle *et al.*, 1989; Armaleo *et al.*, 1990; Day *et al.*, 1990).

   When individual transformants are reanalysed after repeated subcultures, the integration patterns are unchanged, indicating that no genomic rearrangements have occurred subsequent to transformation (Kindle *et al.*, 1989; Day *et al.*, 1990; Gordon-Kamm *et al.*, 1990). This implies that the concatenation and rearrangement of foreign DNA occur prior to the integration event. Evidence in favour of pre-integration rearrangement comes with the observation that different genes on different plasmids which are co-precipitated and co-transformed are often linked in the genome (see for example: Day *et al.*, 1990; Gordon-Kamm *et al.*, 1990; Spencer *et al.*, 1990).

This has implications for plant improvement when the trait to be introduced is not selectable in dedifferentiated tissue culture. For non-selectable traits, one can identify transformants based on expression of a co-introduced marker gene, with an increased probability of co-transformation with the desired non-selectable gene.

Co-transformation and co-expression

In microprojectile-mediated gene transfer experiments, stable transformants have been generated from tissues bombarded with two marker genes on the same or different constructs (see Table 5.2). For the single report of transformation with genes on the same construct, co-transformation frequencies were 100% and the degree of expression of the non-selected gene was variable but always detectable. However, co-expression did not necessarily follow co-transformation of genes on different plasmids (see Table 5.2). From the limited available data, it seems probable that the patterns of co-transformation and co-expression for microprojectile-mediated gene transfer will parallel those of other direct gene transfer techniques (see Chapter 4).

Failure to detect co-expression of apparently intact genes following direct gene transfer may result from subtle gene rearrangements, or from effects on expression of the position of integration of the transforming DNA in the genome. There may also be a mechanism for deactivating gene expression from transferred genes other than the one originally selected. The extent of expression of a selected gene does not necessarily correlate with the gene copy number in the genome (e.g. Gordon-Kamm *et al.*, 1990) and methylation-dependent gene deactivation has been observed in *Agrobacterium*-mediated transformants (Matzke *et al.*, 1989). Attempts to over-express plant genes by introducing additional copies have, in several cases, resulted in a dramatic co-ordinate suppression of the transgene and its endogenous homologue (Jorgensen, 1990; Napoli *et al.*, 1990). The mechanism of this unexpected *trans* interaction is unknown, but the phenomenon could clearly be important in plant genetic manipulation work using direct gene transfer techniques which tend to introduce multiple gene copies.

Special applications of microprojectile techniques

Organelle transformation

Most chloroplast and mitochondrial proteins are encoded in the nuclear genome whereas others are encoded in the small multicopy genomes of the

Table 5.2. Co-transformation and co-expression of non-selected markers following microprojectile bombardment.

Species	Transformed tissue	First screenable/ selectable marker	Second marker	Marker genes on same construct	Co-transformation frequency	Co-expression frequency	Number of transformants analysed	References
Maize	plant	PAT	GUS	–	77%	18%	39	Gordon-Kamm et al., 1990
Maize	callus	NPTII	GUS	+	100%	100%[a]	11	Klein et al., 1989a
Maize	callus	PAT	GUS	–	69%	50%	16	Spencer et al., 1990
Soybean	plant	GUS	NPTII	–	50%	50%	2	Christou et al., 1989

[a]Note that expression rates were not uniform between individuals and within individual cell populations.

organelles themselves (Chapters 8, 9). For mitochondria in particular, evolutionary transfer of genetic information from the organelle to the nucleus has been extreme. Genetic manipulation to modify some organelle traits can therefore be accomplished most directly by changes to the nuclear genome. In some cases, transfer of novel gene products into chloroplasts or mitochondria has been achieved by fusing a sequence encoding an organelle transit-peptide to the gene of interest prior to nuclear transformation (see Cheung *et al.*, 1988; Howe, 1988). However, some traits of substantial agronomic importance (e.g. cytoplasmic male sterility and some forms of herbicide resistance) are organelle-encoded, and the energy-generating semi-autonomous organelles have long been considered potential sites for production of novel and valuable products in genetically engineered plant cells. Thus organelle transformation was attempted prior to the development of microprojectile techniques, but without verified success.

Failure to detect transformed chloroplasts and mitochondria following transfer of foreign DNA into the cytosol may result from impermeability of the organelle double membranes to DNA, or from inability to detect expression of a single introduced gene copy in a cell containing numerous chloroplasts or mitochondria, each containing multiple copies of the corresponding organelle genome (Howe, 1988). Microprojectile-mediated gene transfer has now been used successfully to directly transform the chloroplast genome of *Chlamydomonas* (Boynton *et al.*, 1988; Blowers *et al.*, 1989), the mitochondrial genome of yeast (Fox *et al.*, 1988; Johnston *et al.*, 1988) and the plastids of tobacco (Svab *et al.*, 1990). This is presumably because the bombarding tungsten particles are able to penetrate the organelle membranes and deliver multiple copies of the transforming DNA to the multiploid genome.

The first example of *Chlamydomonas* chloroplast transformation using microprojectiles was accomplished by complementing carbon-requiring mutant strains to photosynthetic competence with microprojectile delivered wild-type DNA. Most often transformation was by gene replacement and in 25% of the transformants, the mechanism for mutant complementation was homologous recombination, which did not involve gene replacement. Unintegrated donor plasmid persisted in 25% of transformants due to the operation of a plasmid replication origin in the chloroplast (Boynton *et al.*, 1988). Subsequently, Blowers *et al.* (1989) transformed *Chlamydomonas* chloroplast genomes with a chimeric foreign gene by flanking the foreign DNA with wild-type chloroplast DNA homologous to the recipient mutant genome. The foreign DNA was integrated by homologous recombination and stably maintained in the chloroplast chromosome in the absence of selection.

Chloroplast transformation of *Chlamydomonas* may be particularly straightforward because of the single large cup-shaped chloroplast which lies adjacent to the plasma membrane along most of the cell boundary and surrounds the nucleus (Boynton *et al.*, 1988). The ability to transform the chloroplast of this relatively simple organism immediately opens the way for plastid gene regulation studies because numerous mutants and cloned chloroplast genes of *Chlamydomonas* are available. The system is a good *in vivo* model for higher plants because many chloroplast gene sequences are known to be highly conserved across the plant and algal kingdoms (see Howe, 1988).

Biolistic gene transfer can potentially facilitate direct engineering of chloroplast genomes of cultivated plants, to convey herbicide resistance for example. Where a novel protein is required in the plastid, chloroplast transformation should be more energy-efficient and more universally applicable than adding a transit peptide to a gene expressed from the nuclear genome. Even with a transit peptide, not all proteins will be correctly transported across the organelle double membrane (Lubben *et al.*, 1988).

Daniell *et al.* (1990) first demonstrated transient expression of microprojectile-delivered chloroplast expression vectors in the plastids of cultured tobacco cells. Stable transformation of higher plant chloroplasts is complicated because each cell contains multiple plastids with several hundred plastid genomes (Chapter 9). Svab *et al.* (1990) used a nonlethal spectinomycin resistance selection system to identify microprojectile generated 'transplastomic' tobacco callus lines which were subsequently regenerated to plants with transformed plastids. Recovery of organelle transformants was possible because the selection system allowed the nondestructive scoring of green callus as resistant and white callus as susceptible. Hence there was sufficient time for the resistant plastid genome to increase copy number to an extent necessary for effective phenotypic expression. However, the transformation frequency of 1 per 50 bombardments was low compared to frequencies of 2 to 6 nuclear transformants per bombardment. These authors proposed that successful plastid transformation occurs by a three step process of DNA recombination, copy correction and 'sorting out' of the ptDNA copies towards a state of homoplasmy (Svab *et al.*, 1990).

For mitochondria, evidence also suggests that provided a proportion of organelles are transformed with a selectable marker, the resulting phenotype can be detected due to an enrichment of transformed organelles during selection. Johnston *et al.* (1988) worked with a yeast strain carrying a nuclear gene mutation resulting in auxotrophy, and a mitochondrial gene mutation resulting in deficient respiration and hence a requirement for a fermentable carbon source for growth. As efforts to select directly for organelle transformants were unsuccessful, they co-bombarded the mutant strain with DNAs capable of complementing both the nuclear and mitochondrial mutations and selected first for nuclear transformants, subsequently screening for mitochondrial transformation. Using this strategy they produced one mitochondrial transformant per 1000–2000 nuclear transformants (Johnston *et al.*, 1988).

Using a similar strategy, Fox *et al.*, (1988) have successfully transferred plasmid DNA carrying a wild-type mitochondrial respiratory gene (*oxi*1) to the mitochondria of a respiratory-deficient yeast strain completely lacking mitochondrial DNA (*rho*^{0}). After first selecting for complementation of a nuclear auxotrophic mutant by a co-bombarding nuclear gene, they identified new strains in which mitochondria had taken up the plasmid containing *oxi*1. This was done by performing matings of various nuclear transformants with a second respiratory-deficient strain (*oxi*1$^-$). Those nuclear transformants able to complement the *oxi*1$^-$ to respiratory competence by mitochondrial fusions following mating were designated mitochondrial transformants. The mitochondrial transformants contained plasmid DNA in reiterated concatamers equivalent in length to native mtDNA, indicating that the replicon of the transforming plasmid was functional in the yeast mitochondria. This ability to transfer plasmid DNA into *rho*^{0} mitochondria by microprojectile techniques, and then to test these introduced sequences in various strains following mating, should be very useful to study mitochondrial gene regulation by *trans* analysis of deliberately modified genes (see Fox *et al.*, 1988).

It is interesting to note that for all reported cases of integrative organelle transformation, DNA integrated into the genome by homologous recombination (Boynton *et al.*, 1988; Johnston *et al.*, 1988; Blowers *et al.*, 1989; Svab *et al.*, 1990). As discussed above, this is less commonly the case for nuclear transformation.

Blowers *et al.* (1989) experimented with the forms of bombarding DNA in an effort to improve *Chlamydomonas* chloroplast transformation efficiencies. Linear DNA molecules resulted in higher transformation rates. Chloroplast sequences at the free ends of the input DNA stimulated homologous recombination, and transformation frequencies diminished when single-stranded transforming DNA was used (Blowers *et al.*, 1989). With an improved understanding of recombination mechanisms in the future, it may prove possible to target transforming genes to precise genomic locations.

Analysis of plant gene regulatory elements

Given the right 'biolistic' parameters and an appropriate genetic construct for expression, almost any species and cell type is amenable to gene transfer by microprojectile bombardment. Table 5.1 summarizes species and tissues which have already been shown to accommodate transient expression of microprojectile-delivered genes. Thus microprojectile-mediated gene transfer provides an excellent system for assaying patterns of gene expression as influenced by the environment, tissue type, developmental stage and genotypic background.

Although transgenic plants provide a bench-mark for studies of plant gene regulation (Benfey & Chua, 1989), transient expression systems are a valuable adjunct because of their speed and relative ease of use (see Chapter 4). Both *Agrobacterium* and protoplast transformation systems have been successfully applied to the analysis of regulatory sequences derived from the transformed plant species as well as heterologous gene sequences (Willmitzer, 1988; Chapter 4). However, not all cell types are readily transformed at high frequency by *Agrobacterium*, and protoplast expression systems are not always representative of the tissue from which they were derived (for example, Vernet *et al.*, 1982).

Microprojectile-mediated gene transfer can facilitate analysis of DNA regulatory sequences in intact tissues of any species, with the speed and ease of transient assay systems. For example, Bruce *et al.* (1989) exploited microprojectile-mediated gene transfer to study DNA sequences involved in the regulation of oat phytochrome genes (*phy*), using a rapid assay for light induction of transient expression of a *phy*–CAT (chloramphenicol acetyltransferase) reporter construct in bombarded intact rice seedlings. They first bombarded seedlings with increasing amounts of the oat *phy*–CAT construct to check that the rate of expression was only influenced by the action of cellular factors and not by the amount of bombarding DNA. Given the variable nature of microprojectile techniques, they monitored the success of each bombardment and standardized the expression of the CAT construct for each treatment, by co-bombarding with a second reporter gene (LUC, firefly luciferase) under the control of a constitutive promoter. Introduced *phy*–CAT gene expression in response to light induction paralleled the light induction pattern of endogenous rice phytochrome mRNA concentrations. This confirmed that regulatory properties of the *phy* sequences fused to the CAT reporter were retained in the bombarded cells. Although these oat *phy* sequences were functional in other monocots, bombarded dicot species did not exhibit CAT expression. Hence regulation of these genes by the two plant groups must be considerably divergent (Bruce *et al.*, 1989).

This example of the use of the CAT reporter system to assess gene expression *in vivo* required homogenization of the bombarded tissue prior to assaying for CAT. Visually detectable regulated gene expression allows an assessment of regulation as it occurs *in situ* and this has been used to advantage in the investigations of genes which are involved in regulation of the biosynthesis of anthocyanin pigment in maize (Klein *et al.*, 1989b; Goff *et al.*, 1990; Ludwig *et al.*, 1990). A gene (*Lc*) from a large gene family (*R*) which is known to be involved in tissue specificity of pigmentation was placed under the control of a constitutive promoter and introduced to cells of various maize tissues by particle bombardment. The constitutively expressed *Lc* gene induced transient anthocyanin biosynthesis in tissues that are not normally pigmented by that gene. Thus *Lc* and the other genes of the *R* family cause diverse tissue-specific patterns of pigmentation through differential gene expression which is controlled by the various promoters, rather than differences in the gene products themselves (Ludwig *et al.*, 1990).

Klein *et al.* (1989b) used rapid complementation tests to further analyse the genetic and tissue-specific regulation of anthocyanin-related gene expression. They showed that anthocyanin production in cells of mutant genotypes was restored following bombardment of the tissue with complementing wild-type structural genes. They then fused 5' and 3' regulatory signals from the anthocyanin biosynthesis genes to a *luc* reporter gene and bombarded the construct into tissues carrying either dominant or recessive alleles of genes known to regulate anthocyanin production. Luciferase assays indicated that transient expression of the introduced gene fusion was controlled as expected by the endogenous regulatory genes.

Goff *et al.* (1990) used microprojectile bombardment for further analysis of *trans*-activation of maize anthocyanin structural genes by regulatory genes. They bombarded intact maize tissues sequentially with *luc* under the control of the 5' regulatory sequences of an anthocyanin biosynthetic gene, and with various constructs of constitutively expressed regulatory gene sequences. Stimulated *luc* expression indicated *trans*-activation of bio synthetic gene 5' regulatory sequences by the product of the regulatory gene.

Together, these studies represent a marvellous exploitation of micro projectile-mediated gene transfer to examine gene expression in response to environment, developmental state, genotypic background and *trans*-activation by the products of co-transformed genes. The simplicity of this approach to assaying regulatory properties of DNA sequences suggests that it will be extensively applied in the future. The system is immediately applicable to even those species which are not yet transformable. Towards the ultimate production of transgenic plants, it offers a quick intermediate means for checking that a transforming construct will be expressed in the desired fashion.

Genetic mosaics for cell lineage analysis

Although microprojectile techniques have not been applied to cell lineage analysis in plants, the properties of this DNA delivery system seem ideally suited to such investigations. Genetic mosaics have been studied since the turn of the century, yielding considerable information about cell lineage patterns and the mechanisms for morphogenesis in plants (for review, see Poethig, 1989). Some of the genetic mosaics used in such studies have arisen spontaneously whereas others have been generated artificially. Somatic segregation of mutant and normal chloroplasts, interspecific mosaics and changes in the nuclear genome of somatic cells have provided traceable cell lineages.

For example, some cell lineage studies have in the past followed chemically induced ploidy differences using tedious cytological analyses. In other cases, irradiating stocks heterozygous for one or more cell marker mutations has deactivated an allele, revealing the marker and hence initiating a traceable lineage in the developing plant.

Traits visible in shoots, such as chlorophyll or anthocyanin pigmentation, are especially easy to discern and have been used to advantage in cell line studies. However, root cell lineages are difficult to examine using genetic mosaics because of a shortage of traits that can serve as cell markers. The only reported analysis of the cell lineage of the root was based on largely impractical x-ray-induced chromosomal rearrangements (see Poethig, 1989).

Microprojectile bombardment can easily deliver dominant visual marker genes to various intact tissues of developing plants. A proportion of these genes will become stably incorporated in the genome of cells and give rise to cell lineages which can be followed. Although biolistic techniques have not yet been deliberately applied for cell lineage analysis, the recovery of uniform transgenic soybean plants from bombarded shoot meristems is relevant. This result indicates that whole plants were ultimately derived from single cells by a *de novo* organogenic pathway, which contradicts the widely held view that organogenesis follows a multicellular regeneration pathway (Christou *et al.,* 1989). More work is warranted in this area, not only because the experimental results appear to conflict with current theory, but also because of the practical importance of designing appropriate selection and regeneration strategies to obtain uniform transgenic plants following microprojectile bombardment.

Limitations of the technique

The current major technical limitation to microprojectile-mediated gene transfer is the inconsistency of results between successive, replicated treatments. Most groups using the technique find that results are repeatable, but not consistent. For the production of transgenic plants, this need not be a serious problem provided that high transformation rates can be achieved, and supplies of regenerable target tissue or DNA for transfer are not severely limited. However, for quantitative studies of gene regulation it is an inconvenience necessitating increased replicates and increased assays for internal controls. Even experiments to 'optimize' parameters for maximal transient expression frequencies in a new target tissue type can become extremely frustrating because of the large number of replicates required to detect real improvements against a background of random variation.

A major factor contributing to inconsistency may be an inability to precisely and reproducibly control the velocity of the bombarding microprojectiles. This problem may be reduced through ongoing research into factors such as irregularities in microprojectile shape and size; DNA coating methodology to generate uniformly covered particles which do not agglutinate rapidly; and engineering of devices to provide more uniform acceleration and distribution of the bombarding particles (see Herman, 1990; Sanford, 1990).

Conclusions

Perhaps the ultimate aim of any plant gene transfer technique is to improve the agronomic performance of a species by extending its gene pool or by precisely correcting genetic deficiencies of cultivars. In preliminary work towards the same goal, gene transfer techniques are powerful tools for tagging and characterizing potentially useful plant genes and gene regulatory sequences required for tailored expression of introduced genes in transgenic plants. Indeed, the availability of transient expression systems following gene transfer is rapidly increasing our understanding of plant gene regulation, the knowledge base on which much of future plant improvement will be founded.

Microprojectile-mediated gene transfer has been used to develop transient and stable gene expression systems for a broad spectrum of plant species. The technique has circumvented the host range constraints of *Agrobacterium* and the tissue culture limitations of protoplast transformation, and, at present, it uniquely facilitates organelle transformation. In addition to applications in plant improvement, microprojectile-mediated gene transfer into plant, algal and fungal cells should be valuable in other biotechnologies, such as production of pharmaceuticals and biochemicals from improved cell lines in fermentation systems. The unique features of the technique will certainly be exploited in ingenious approaches by other biological researchers. For example, bombardment of infected plant tissues may provide an opportunity for genetic manipulation of biotrophic fungal phytopathogens *in situ*, in research to elucidate the mechanisms of pathogenicity and plant disease resistance.

It is ironic that the future application of microprojectile-mediated gene transfer will be limited, not by technology, but by those who hold the relevant patents. Although it represents a major breakthrough toward practical genetic manipulation of important crop groups such as the cereals, microprojectile-mediated gene transfer is certainly not the only possible approach to this goal. Enforcement of excessive demands by patent holders, such as substantial royalties on any plants transformed using the technique, will have the effect of turning research and development away from microprojectiles and into alternatives such as electroporation of intact plant tissues or extending the *Agrobacterium* host range. We suggest that devices for microprojectile bombardment deserve the benefits of patent protection to reward and encourage the research and engineering work necessary to improve the consistency of bombardments. Only if microprojectile bombardment is then applicable in the research community without additional commercial constraints will its potential as a nearly universal gene transfer technique in fact be realized.

Acknowledgements

We thank Ms Lyn Jessup for drawing Fig. 5.1 and Dr John Manners for helpful discussions. We are grateful to the Sugar Research Council (Australia) for support of our research into microprojectile-mediated transformation of sugarcane.

References

Armaleo, D., Ye, G.-N., Klein, T. M., Shah, K. B., Sanford, J. C. & Johnston, S. A. (1990) Biolistic nuclear transformation of *Saccharomyces cerevisiae* and other fungi. *Current Genetics* 17, 98–103.

Benfey, P. N. & Chua, N.-H. (1989) Regulated genes in transgenic plants. *Science* 244, 174–81.

Blowers, A. D., Bogorad, L., Shark, K. B. & Sanford, J. C. (1989) Studies on *Chlamydomonas* chloroplast transformation: foreign DNA can be stably maintained in the chromosome. *Plant Cell* 1, 123–32.

Boynton, J. E., Gilham, N. W., Harris, E. H., Hosler, J. P., Johnson, A. M., Jones, A. R., Randolph-Anderson, B. L., Robertson, P., Klein, T. M., Shark, K. B. & Sanford, J. C. (1988) Chloroplast transformation of *Chlamydomonas* with high velocity microprojectiles. *Science* 240, 1534–7.

Bruce, W. B., Christensen, A. H., Klein, T., Fromm, M. & Quail, P. H. (1989) Photoregulation of a phytochrome gene promoter from oat transferred into rice by particle bombardment. *Proceedings of the National Academy of Sciences USA* 86, 9692–6.

Cheung, A. Y., Bogorad, L., van Montagu, M. & Schell, J. (1988) Relocating a gene for herbicide tolerance: a chloroplast gene is converted into a nuclear gene. *Proceedings of the National Academy of Sciences USA* 85, 391–5.

Christou, P., McCabe, D. E. & Swain, W. F. (1988) Stable transformation of soybean callus by DNA-coated gold particles. *Plant Physiology* 87, 671–4.

Christou, P., Swain, W. F., Yang, N.-S. & McCabe, D. E. (1989) Inheritance and expression of foreign genes in transgenic soybean plants. *Proceedings of the National Academy of Sciences USA* 86, 7500–4.

Creissen, G., Smith, C., Francis, R., Reynolds, H. & Mullineaux, P. (1990) *Agrobacterium*- and microprojectile-mediated viral DNA delivery into barley microspore-derived cultures. *Plant Cell Reports* 8, 680–3.

Daniell, H., Vivekananda, J., Nielsen, B. L., Ye, G. N., Tewari, K. K. & Sanford, J. C. (1990) Transient foreign gene expression in chloroplasts of cultured tobacco cells after biolistic delivery of chloroplast vectors. *Proceedings of the National Academy of Sciences USA* 87, 88–92.

Day, A., Debuchy, R., van Dillewijn, J., Purton, S. & Rochaix, J. D. (1990) Studies on the maintenance and expression of cloned DNA fragments in the nuclear genome of the green alga *Chlamydomonas reinhardtii*. *Physiologia Plantarum* 78, 254–60.

Debuchy, R., Purton, S. & Rochaix, J. D. (1989) The arginosuccinate lyase gene of *Chlamydomonas reinhardtii*: an important tool for nuclear transformation and for correlating the genetic and molecular maps of the *ARG7* locus. *EMBO Journal* 8(10), 2803–9.

Finer, J. J. & McMullen, M. D. (1990) Transformation of cotton (*Gossypium hirsutum* L.) via particle bombardment. *Plant Cell Reports* 8, 586–9.

Fox, T. D., Sanford, J. C. & McMullin, T. W. (1988) Plasmids can stably transform yeast mitochondria lacking endogenous mtDNA. *Proceedings of the National Academy of Sciences USA* 85, 7288–92.

Franks, T. & Birch R. G. (1991) Gene transfer into intact sugarcane cells using microprojectile bombardment. *Australian Journal of Plant Physiology* 18 (in press).

Goff, S. A., Klein, T. A., Roth, B. A., Fromm, M. E., Cone, K. C., Radicella, J. P. & Chandler, V. L. (1990) Transactivation of anthocyanin biosynthetic genes following transfer of B regulatory genes into maize tissues. *EMBO Journal* 9(8), 2517–22.

Gordon-Kamm, W. J., Spencer, T. M., Mangano, M. L., Adams, T. R., Daines, R. J., Start, W. G., O'Brien, J. V., Chambers, S. A., Adams Jr, W. R., Willetts, N. G., Rice, T. B., Mackey, C. J., Krueger, R. W., Kausch, A. P. & Lemaux, P. G. (1990) Transformation of maize cells and regeneration of fertile transgenic plants. *Plant Cell* 2, 603–18.

Hauptmann, R. M., Vasil, V., Ozias-Akins, P., Tabaeizadeh, Z., Rogers, S. G., Fraley, R. T., Horsch, R. B. & Vasil, I. K. (1988) Evaluation of selectable markers for obtaining stable transformants in the gramineae. *Plant Physiology* 86, 602–6.

Herman, E. B. (1990) Microprojectile bombardment and the 'zone of death'. *Agricell Report* 15(1), 1–4.

Howe, C. J. (1988) Organelle transformation. *Trends in Genetics* 4(6), 150–3.

Jefferson, R. A., Kavanagh, T. A. & Bevan, M. W. (1987) GUS fusions: β-glucuronidase as a sensitive and versatile gene fusion marker in higher plants. *EMBO Journal* 6(13), 3901–7.

Johnston, S. A., Anzianno, P. Q., Shark, K., Sanford, J. C. & Butow, R. A. (1988) Mitochondrial transformation in yeast by bombardment with microprojectiles. *Science* 240, 1534–41.

Jorgensen, R. (1990) Altered gene expression in plants due to *trans* interactions between homologous genes. *Trends in Biotechnology* 8, 340–4.

Kartha, K. K., Chibbar, R. N., Georges, F., Leung, N., Caswell, K., Kendall, E. & Zureshi, J. (1989) Transient expression of chloramphenicol acetyltransferase (CAT) gene in barley cell cultures and immature embryos through microprojectile bombardment. *Plant Cell Reports* 8, 429–32.

Kindle, K. L., Schnell, R. A., Fernandez, E. & Lefebvre, P. A. (1989) Stable nuclear transformation of *Chlamydomonas* using the *Chlamydomonas* gene for nitrate reductase. *Journal of Cell Biology* 109, 2589–601.

Klein, T. M., Wolf, E. D., Wu, R. & Sanford, J. C. (1987) High velocity microprojectiles for delivering nucleic acids into living cells. *Nature* 327, 70–3.

Klein, T. M., Fromm, M., Weissinger, A., Tomes, D., Schaff, S., Sletten, M. & Sanford, J. C. (1988a) Transfer of foreign genes into intact maize cells with high velocity microprojectiles. *Proceedings of the National Academy of Sciences USA* 85, 4305–9.

Klein, T. M., Gradziel, T., Fromm, M. E. & Sanford, J. C. (1988b) Factors influencing gene delivery into *Zea mays* cells by high velocity microprojectiles. *Bio/Technology* 6, 559–63.

Klein, T. M., Harper, E. C., Svab, Z., Sanford, J. C., Fromm, M. E. & Maliga, P. (1988c) Stable genetic transformation of intact *Nicotiana* cells by the particle bombardment process. *Proceedings of the National Academy of Sciences USA* 85, 8502–5.

Klein, T. M., Kornstein, L., Sanford, J. C. & Fromm, M. E. (1989a) Genetic transformation of maize cells by particle bombardment. *Plant Physiology* 91, 440–4.

Klein, T. M., Roth, B. A. & Fromm, M. E. (1989b) Regulation of anthocyanin biosynthetic genes introduced into intact maize tissues by microprojectiles. *Proceedings of the National Academy of Sciences USA* 86, 6681–5.

Lubben, T. H., Theg, S. M. & Keegstra, K. (1988) Transport of proteins into chloroplasts. In: Govindjee, Bohnert, H. J., Bottomley, W., Bryant, D. A., Mullet, J. E., Ogren, W. L., Pakrasi, H. & Somerville, C. R. (eds), *Molecular Biology of Photosynthesis*. Kluwer Academic Publishers, Dordrecht, Boston, London, pp. 713–34.

Ludwig, S. R., Bower, B., Beach, L. & Wessler, S. R. (1990) A regulatory gene as a novel visible marker for maize transformation. *Science* 247, 449–50.

McCabe, D. E., Swain, W. F. & Martinell, B. J. (Agracetus) (1988a) Pollen-mediated plant transformation. *European Patent Application* no. 87310612.4. Published 8 June 1988. Bulletin 88/23.

McCabe, D. E., Swain, W. F., Martinell, B. J. & Christou, P. (1988b) Stable transformation of soybean (*Glycine max*) by particle acceleration. *Bio/Technology* 6, 923–6.

McDaniel, C. N. & Poethig, R. S. (1988) Cell lineage patterns in the shoot apical meristem of the germinating maize embryo. *Planta* 175, 13–22.

MacKenzie, D. R., Anderson, P. M. & Wernham, C. C. (1966) A mobile air blast inoculator for plot experiments with maize dwarf mosaic virus. *Plant Disease Reporter* 50, 363–7.

Matzke, M. A., Primig, M., Trnovsky, J. & Matzke, A. J. M. (1989) Reversible methylation and inactivation of marker genes in sequentially transformed tobacco plants. *EMBO Journal* 8(3), 643–9.

Mendel, R. R., Muller, B., Schulze, J., Kolesnikov, V. & Zelenin, A. (1989) Delivery of foreign genes to intact barley cells by high-velocity microprojectiles. *Theoretical and Applied Genetics* 78, 31–4.

Morikawa, H., Iida, A. & Yamada, Y. (1989) Transient expression of foreign genes in plant cells and tissues obtained by a simple biolistic device (particle-gun). *Applied Microbiological Biotechnology* 31, 320–2.

Napoli, C., Lemieux, C. & Jorgensen, R. (1990) Introduction of a chimeric chalcone synthase gene into petunia results in reversible co-suppression of homologous genes *in trans*. *Plant Cell* 2, 279–89.

Oard, J. H., Paige, D. F., Simmonds, J. A. & Gradziel, T. M. (1990) Transient gene expression in maize rice and wheat cells using an airgun apparatus. *Plant Physiology* 92, 334–9.

Olsson, O., Nilsson, O. & Koncz, C. (1990) Engineered luciferase proteins as tools to study plant gene regulation *in vivo*. *Journal of Bioluminescence and Chemiluminescence* 5, 79–87.

Poethig, S. (1989) Genetic mosaics and cell lineage analysis in plants. *Trends in Genetics* 5(8), 273–7.

Sanford, J. C. (1988) The biolistic process. *Trends in Biotechnology* 6, 299–302.

Sanford J. C. (1990) Biolistic plant transformation. *Physiologia Plantarum* 79, 206–9.

Sanford, J. C., Klein, T. M., Wold, E. D. & Allen, N. (1987) Delivery of substances into cells and tissues using a particle bombardment process. *Particulate Science Technology* 5, 27–37.

Sanford, J. C., Wolf, E. D. & Allen, N. K. (Biolistics Inc.) (1989) Biolistic apparatus for delivering substances into cells and tissues in a non-lethal manner. *US Patent Application* no. 23743/88. Published 31 August 1989. Document no. 161807.

Spencer, T. M., Gordon-Kamm, W. J., Daines, R. J., Start, W. G. & Lemaux, P. G. (1990) Bialaphos selection of stable transformants from maize cell culture. *Theoretical and Applied Genetics* 79, 625–31.

Svab, Z., Hajdukiewicz, P. & Maliga, P. (1990) Stable transformation of plastids in higher plants. *Proceedings of the National Academy of Sciences USA* 87, 8526–30.

Twell, D., Klein, T. M., Fromm, M. E. & McCormick, S. (1989) Transient expression of chimaeric genes delivered into pollen by microprojectile bombardment. *Plant Physiology* 91, 1270–4.

Vernet, T., Fleck, J., Durr, A., Fritsch, C., Pinck, M. & Hirth, L. (1982) Expression of the gene coding for the small subunit of ribulosebisphosphate carboxylase during differentiation of tobacco plant protoplasts. *European Journal of Biochemistry* 126, 489–94.

Wang, Y.-C., Klein, T. M., Fromm, M., Cao, J., Sanford, J. C. & Wu, R. (1989) Transient expression of foreign genes in rice, wheat and soybean cells following particle bombardment. *Plant Molecular Biology* 11, 433–9.

Willmitzer, L. (1988) The use of transgenic plants to study plant gene expression. *Trends in Genetics* 4(1), 13–18.

Zumbruun, G., Schneider, M. & Rochaix, J.-D. (1989) A simple particle gun for DNA mediated cell transformation. *Technique 1*, 204–16.

Chapter 6
Localization of Transferred Genes in Genetically Modified Plants

A. Mouras[1], S. Hinnisdaels[2], C. Taylor[3] & K. C. Armstrong[4]

[1]*Université de Bordeaux II, Laboratoire Biologie Cellulaire, Avenue des Facultés, 33405 Talence Cedex, France*
[2]*Université de Bruxelles, Laboratoire Génétique des Plantes-Institut Biologie Moleculaire, Paardenstraat 65-B-1640, Rode-St Genèse Belgium*
[3]*Centre for Cereal Biotechnology, Waite Agricultural Research Institute, University of Adelaide, Glen Osmond, South Australia 5064*
[4]*Plant Research Centre, Central Experimental Farm, Ottawa, Ontario, Canada, K1AOC6*

Introduction

Advances in plant cytogenetics have occurred as new and efficient methods for the analysis of chromosome morphology have been devised. The application of these techniques requires high-quality metaphase spreads so that reliable and accurate chromosome identification and classification can be performed. This system is time-consuming, however, and can be accomplished only by experienced researchers.

The development of new biotechnological methods, particularly the introduction of foreign genes into plants (transgenic plants) and gene tagging using transposable elements, requires the development of a new and efficient gene mapping system. Molecular biology and molecular cytology methods have been devised, such as restriction fragment patterns of nuclear and organelle DNA (RFLP = restriction fragment length polymorphism) (Tanksley *et al.*, 1989) and *in situ* hybridization (Hutchinson *et al.*, 1981; Mouras *et al.*, 1987b). In the latter, the use of radioactively or non-radioactively labelled nucleic acids provides a very sensitive means for the detection of complementary nucleic acid sequences, on both interphase nuclei and metaphase chromosomes.

The localization of transferred genes, irrespective of the transformation system used, is of great interest. It is now well established that the copy number and the insertion site(s) are directly dependent on the transformation method used: in agrotransformed plants, molecular analyses have generally shown one and occasionally more than one insertion site (Schell, 1987), resulting in one copy per cell; in contrast, plants transformed by direct gene transfer (DGT) may exhibit an integration of alien DNA in several loci and inserts may exist in multiple copies (Paszkowski *et al.*, 1984). Therefore, with the use of gene transfer methods in plants, gene localization by *in situ* hybridization combined with classical genetic analyses becomes an essential technique for gene mapping.

In genetic modification experiments via somatic hybridization (symmetric or asymmetric) the hybrid may contain the total or only part of the chromosome set of the donor parent; in some cases reported (Bates *et al.*, 1987; Gleba *et al.*, 1988), the genetic information from the donor may be reduced to as little as a chromosome fragment. This situation has also been found in some sexual hybrids. Le *et al.* (1989), for example, have recently demonstrated the translocation of a rye chromosome arm on to a wheat chromosome in a wheat–rye hybrid by using total genomic rye DNA as a probe. Thus cytogenetic and molecular cytological analyses are essential in order to define precisely the new set of chromosomes in hybrid plants. Moreover, it is important to note that up until now, recombination at the nuclear level in somatic hybrids has been suspected (Dudits *et al.*, 1980; Bates *et al.*, 1987) but not demonstrated, as cytogenetic analyses have either not been carried out or were not conclusive.

In this chapter we describe the cytogenetic techniques currently used to:

1. characterize the chromosome set from species to species;
2. characterize chromosomes within a particular species;
3. localize gene(s) or alien chromatin by *in situ* hybridization using as probe unique gene sequences, repetitive DNA or total genomic DNA.

Plant chromosome analyses

During mitosis and meiosis, chromosomes can be visualized, after staining, in the light microscope. Almost all cytogenetic work has been carried out with condensed chromosomes, i.e. metaphase chromosomes obtained from dividing cells (somatic cells in mitosis, gametic cells in meiosis). The number, size and shape of metaphase chromosomes constitute the so called karyotype, the set of chromosomal features being distinctive for each organism but theoretically identical between the cells within the same organism.

In order to study chromosome structure, it is necessary to arrest the dividing cells in a suitable stage of the cell cycle. This is usually metaphase, although there may be some advantage in using chromosomes from another stage of the cell cycle. In metaphase, the chromosomes are well contracted and are very amenable to cytological manipulations. It should be noted that the manner in which the cells are arrested should cause minimal distortion of the structure of the chromosomes.

Prerequisites for cytogenetic studies

Prefixation

Prefixation is performed to stop the cell cycle in mitosis, particularly in metaphase. Antimitotic agents are most commonly used for this purpose. Generally these chemicals prevent either the polymerization of the spindle fibres or the attachment of the kinetochore to the spindle, thereby preventing the chromosomes from moving towards the poles of the dividing cell. The most commonly used chemical is colchicine; however, this agent does not provide reproducible results with plant chromosomes. In our experiments the chemicals α-chloronaphthalene or α-bromonaphthalene have proved to be very suitable for plant chromosome analyses.

Fixation

Fixation may be defined as the process by which the cellular components are fixed selectively to a desired extent. Farmer's or Carnoy's solution (Sharma & Sharma, 1965), containing both ethanol (95 or 100%) and acetic acid in the ratio of 3:1, represents the most suitable fixative solution for plant chromosomes. It should be freshly prepared and chilled prior to use.

Cytological slide preparation

The preparation of cytological slides (chromosome preparation) represents the most critical point in the cytogenetic method; upon this the quality of chromosome spreading and subsequent chromosome morphology depends. These methods will be developed in a later section.

Staining and slide mounting

The choice of staining compound is dependent to a large extent on the type of analyses to be carried out. For instance, fuchsin (Feulgen reaction) and carmine are most commonly used for plant chromosome observation and counting. As with mammalian cytogenetics, however, Giemsa staining is also very suitable and commonly used. The staining of chromosomes is an important step in chromosomal analysis and the relevant literature should be consulted prior to undertaking extensive projects.

Once slides have been prepared and stained, it is usual to cover the cellular material with a coverslip prior to observation. Depending on the material, method of staining and type of analysis to be carried out, however, the application of a coverslip may not be essential. For example, metaphase spreads prepared from protoplasting root tips when stained with Giemsa require no coverslip. The immersion oil may be placed directly on to the preparation without any damage to the chromosomes and nuclei. Moreover, this preparation may be destained and the chromosomes banded (by incubation with barium hydroxide) and restained without any deleterious effects.

Plant material

Cytogenetic studies, particularly chromosome feature analyses, require many metaphase plates to be examined. Thus plant meristems, i.e. root and shoot tips, are the most suitable material, as they contain many dividing cells. Alternatively, cell suspension cultures can be used.

Meristematic cells

In plants, root tips as opposed to shoot tips are the most suitable material for chromosome analyses: (i) the mitotic index is in the order of 2–4%, which is generally twofold (or more) higher than in shoot tips; (ii) plants have many root tips while in general only one shoot tip; (iii) the taking of roots rarely causes injury to the plant; and (iv) cuttings can generally be induced to root under appropriate conditions.

Agrobacterium rhizogenes – *transformed root cultures*

When a quantity of root meristems is required, especially for *in situ* hybridization, root cultures present an alternative to cuttings or germination of seeds. Under normal conditions, roots are unable to grow in axenic cultures. However, it is possible, upon transformation with *A. rhizogenes*, to induce the growth of so-called hairy roots, which are capable of profuse growth in axenic culture. The process of root induction from leaves of a selected plant takes approximately 3 weeks with a further 3–4 weeks to obtain a large quantity of meristems. Thus within 2 months the material is ready for use. This compares favourably with the 3–4 months required to obtain roots in sufficient quantity from cuttings.

Cell suspension cultures

Liquid cell culture represents another method for obtaining dividing cells in large quantities. When subcultured frequently (once or twice a week according to the plant species), the mitotic index of cell suspension in log phase is around 0.05% and up to 1% in some cases. Several authors have reported synchronization of cell division using plant growth regulators (Jouanneau, 1971), or chemicals, such as aphidicolin and hydroxyurea, which are known to inhibit DNA polymerase (Eriksson, 1966; Nagata *et al.*, 1982). Mitotic indices of cell cultures ranging from 15 to 30% (Erickson, 1966; Huang *et al.*, 1988) and 60% (Nagata *et al.*, 1982) have been reported. The synchronization of cell cultures is, however, not always reproducible, and moreover, cell culture *per se*, associated with certain chemical treatments, is known to be a source of chromosomal aberrations (Mouras, 1981; Karp, 1989).

Metaphase plate preparation

When appropriate material is available, the cells are arrested in metaphase with an antimitotic agent. At this stage the shape of chromosomes is characteristic for each plant species. The cells or tissues may then be fixed directly or transformed into protoplasts prior to fixation. To obtain metaphase plates with chromosomes which are well spread and flat, several methods are available.

Squash method

Probably the first procedure devised for the analyses of chromosomes, this method is very efficient and gives excellent results. However, for *in situ* hybridization, the presence of the cell wall represents a barrier for the accessibility of the probe to the target DNA. This method, the most widely used up to date, is probably responsible for the difficulty experienced in the detection of single-copy genes in plants (Mouras *et al.*, 1987b; Huang *et al.*, 1988; Simpson *et al.*, 1988; Clark & Karp, 1989; Gustafson *et al.*, 1990).

Protoplast preparation from fixed cells

Two principal methods, albeit similar, have been developed. Both methods result in the transformation of fixed root meristem cells into protoplasts via enzymatic digestion of cell walls with cellulase and pectinase.

In the method devised by Pijnacker & Ferwerda (1984), enzymatic digestion is stopped just before the cells are spontaneously released from the meristematic region; the disaggregation of root tips is carried out on the surface of the glass slide in a drop of 60% acetic acid. The chromosome spreading is induced by the addition of a drop of fresh cold fixative (Farmer's solution). This method yields excellent results, particularly with potato plants, and has been applied successfully for *in situ* hybridization (Pijnacker & Ferwerda, 1984).

In the method used by Ambros *et al.* (1986), the enzymatic digestion is performed until the protoplasts from root tips are spontaneously and completely released into solution. After washing, the protoplast suspension is placed in fresh fixative solution and dropped on to clean, dry slides and air-dried under a permanent light air flow. This method has been used extensively, mainly with monocots (Simpson *et al.*, 1988); however, the quality of metaphase plates varies considerably within and between plant species.

Protoplast preparation from unfixed cells

This method produces protoplasts from meristematic cells (root meristems, cell culture) which have been arrested in metaphase stage pirior to the protoplasting step (Mouras *et al.*, 1978, 1987b). Once protoplasts have been formed (via enzymatic digestion), the suspension is treated in the same manner as for mammalian cells, first being subjected to a hypotonic shock followed by fixation. During the protoplast swelling (without bursting), the chromosomes are able to separate from each other and to disperse within the protoplast. This allows for an even and reproducible distribution of chromosomes (in metaphase plates) upon spreading. This method has proved to be reliable and has been applied successfully to root meristems of several species, including some monocots, and to cell suspensions (Mouras, 1981, 1984; Mouras *et al.*, 1978, 1986, 1987b, 1989). It has also been extensively used for karyotyping plant species (Mouras, 1984; Mouras *et al.*, 1986; Chevre *et al.*, 1985) and for *in situ* hybridization experiments (Mouras *et al.*, 1987b, 1989; Mouras & Negrutiu, 1989).

The major disadvantage with this method is that it requires large amounts of material (roots) in order to produce workable quantitites of protoplasts. Because of this the use of *A. rhizogenes*-transformed root cultures or cuttings, when available, is recommended.

Plant karyotype

Principles

The precise localization of transferred genes in transgenic plants or the identification of translocation and chromosomal rearrangements within or between plant genotypes requires a precise knowledge of the karyotype of the plant species studied. The number, size and shape of chromosomes constitute the karyotype. Each chromosome can be individually characterized by both the relative length and the centromeric index (Essad *et al.*, 1966). Additional characteristics such as satellites, secondary constrictions and banding pattern allow a more precise chromosome identification.

Following the commonly used nomenclature of Essad *et al.* (1966) for determining plant karyotype, individual chromosomes within metaphase plates are measured and the following data obtained:

1. absolute length (T);

2. relative length (T/θ): the ratio between the absolute length (T) of one chromosome and the chromosome mean length (θ) of the chromosome set for each metaphase plate;
3. centromeric index (C/L): the ratio between the short (C) and long (L) arm of each chromosome or alternatively the ratio between the short arm (C) and the absolute length (T) for each chromosome.

It follows that chromosomal identification based on absolute length, relative length and centromeric index requires that a large number of metaphase spreads be analysed (at least 25) in order to obtain statistically accurate data for each chromosome. Once chromosomes have been identified accurately, they may be arranged into groups based on the following classification:

1. metacentric chromosomes: $0.75 < C/L < 1.0$
2. submetacentric chromosomes: $0.5 < C/L < 0.75$
3. subtelocentric chromosomes: $0.25 < C/L < 0.5$
4. telocentric chromosomes: $C/L < 0.25$

For presentation, each chromosome pair within each group should be ordered according to decreasing relative length.

Chromosome banding

Identification of plant chromosomes has in the past been based on the analysis of chromosome size, arm length ratios, pairing and, more recently, chromosome banding. To date, however, banding on plant chromosomes has essentially been limited to C-banding and N-banding. Unlike the complex banding patterns observed with human and mammalian chromosomes, these methods have in general revealed only relatively simple banding patterns, apparently correlating to the presence of constitutive heterochromatin (Jones & Flavell, 1982a, b). Nevertheless, the application of these cytogenetic methods to plant species has provided a wealth of information on the characterization, evolution and organization of plant genomes. This is particularly true for cereals. It is now possible to identify all rye, wheat and barley chromosomes on the basis of C-banding (Darvey & Gustafson, 1975; Lelley *et al.*, 1978; Lapitan *et al.*, 1986; Lange, 1988). Moreover, this technique is sensitive enough to allow the identification of various alien chromosomes within hybrid cereal genomes. For instance, C-banding has been used extensively to analyse the chromosome complement of triticale; it is possible to identify all rye chromosomes within this wheat–rye hybrid. The sensitivity of this technique is such that translocation and other chromosomal rearrangements can be identified also (Lange, 1988; Jouve *et al.*, 1989).

C-banding techniques, however, suffer a number of severe limitations.

The most critical is a general absence of interstitial bands on chromosomes (Lange, 1988). This is particularly true for all seven rye chromosomes. A large degree of inter- and intravarietal polymorphism has also been observed for C-banded cereal chromosomes, making standardization of karyotypes difficult (Lelley *et al.*, 1978; Jouve *et al.*, 1989). A further drawback with this method lies in the fact that C-banding results in the staining of all constitutive heterochromatin irrespective of its nucleic acid composition (Rayburn & Gill, 1986). Hence, all constitutive heterochromatin appears the same. Clearly, then, the use of C-banding analysis is limited. This is particularly true when the objective of breeding programmes is stable introgression of small segments of alien chromatin. To this end, it is important that techniques for the reliable identification and characterization of alien DNA become available for breeders.

Gene mapping on metaphase chromosomes by *in situ* hybridization

In situ hybridization may be defined as the annealing of an RNA or DNA probe to nucleic acid within cytological preparations. However, in contrast to the Southern blot method, where DNA fixed on to nitrocellulose or nylon membranes is present as a large heterogeneous population of single-stranded molecules, equivalent to 10^5–10^6 genomes, in cytological preparations the DNA is present for each cell, in few loci and in the form of interphase nuclei or condensed chromosomes, both highly complex structures of varying degrees of DNA condensation. Thus it is obvious that:

1. gene localization on chromosomes requires a highly sensitive hybridization method;
2. the accessibility of gene or target DNA must be optimized. This is important when using plant material as plant cells present a number of additional problems for the accessibility of the probe to the target DNA.

The following steps have proved important in increasing the sensitivity of *in situ* hybridization to plant chromosomes:

1. removing the cell wall, i.e. protoplasting plant cells;
2. releasing proteins and RNA from cytological preparations by appropriate RNase and alternatively protease treatments;
3. efficient denaturation of both probe and chromosomal DNA.

Additional increases in the sensitivity of the method may be gained through the use of probes of high specific activity.

The general methodology for *in situ* hybridization may be summarized as follows:

1. cytological preparations and processing;

2. prehybridization;
3. hybridization
   (a) probe labelling
   (b) DNA denaturation (probe and chromosomes)
   (c) hybridization;
4. post-hybridization;
5. autoradiography.

Here we have made comments on only the critical points of the methodology.

Cytological preparation and processing

Cytological procedure

There is no single method for the processing and preparation of meristematic cells from different plant species. The cytogenetic technique used will vary according to the plant species and the aim of the experiment. It is advisable to select the least disruptive method for karyotyping and to note that the method used for the preparation of cytological material is also dependent on the nature of the probe being used. For example, for the localization of repetitive DNA sequences, the constraints are lower than those required for the localization of single- and low-copy genes. In both cases, it is necessary to remove the cell wall. Therefore, the cytological procedure is defined according to the plant species, the plant material (organ type), the quantity available and the nature of the probe.

Slide preparation

The use of subbed slides is generally recommended for the preparation of cytological material for *in situ* hybridization when combined with autoradiography. This process generally leads to a significant decrease in the background caused by non-specific binding of nucleic acid probes to the slides (Gerhard *et al.*, 1981).

Slide mounting

Protoplast suspensions represent the most suitable material for making cytological preparations as they provide large numbers of cells. Generally, the use of this method results in well-spread metaphase plates clear of debris. Spreading of cellular material is carried out in essentially the same manner as for mammalian cell suspensions. The cytological preparations can be immediately used or stored at −70°C for several months.

Probe labelling

Recombinant DNA probes

Recombinant DNA technology provides the opportunity to obtain DNA or RNA (riboprobes) probes from genomic or cDNA clones of virtually any desired sequence. Both DNA and RNA probes have been successfully used for the localization of high- or low-copy DNA sequences on metaphase chromosomes; however, DNA probes are probably most commonly used.

Choice of probe labelling

Both radioactive and non-radioactive methods are presently available for the labelling of DNA and RNA probes. Commonly, both systems make use of the modified nick-translation or random priming reaction.

^{3}H and ^{35}S are the most commonly used radio-isotopes for labelling probes for *in situ* hybridization. ^{3}H is particularly suitable, due to its relatively low energy. ^{3}H-labelled probes provide very good resolution and allow precise location of sequences on chromosomes. ^{35}S-labelled probes are rarely used, although of higher specific activity than ^{3}H-labelled probes, due to poor resolution caused by the higher-energy beta emissions.

In recent years, non-isotopic methods for labelling DNA probes have become available. These now include, among others, sulphonation and the use of biotin or digoxigenin.

Both isotopic and non-isotopic labelling methods have been used for the detection of high and middle repetitive sequences (Hutchinson *et al.*, 1981; Rayburn & Gill, 1985; Mouras *et al.*, 1987b, 1989). The detection of low-copy or single-copy genes has essentially been limited to ^{3}H-labelled probes (Mouras *et al.*, 1987b, 1989; Huang *et al.*, 1988, 1989). The improvement in sensitivity and efficiency of non-radioactive labelling methods should, however, result in a wide application of this technology to the localization of low-copy sequences by *in situ* hybridization. Indeed, several groups have recently reported the localization of low-copy sequences on plant metaphase chromosomes using non-isotopically labelled probes (Ambros *et al.*, 1986; Simpson *et al.*, 1988; Clark *et al.*, 1989; Gustafson *et al.*, 1990).

Characteristics of labelled probes

The modified nick-translation method (Rigby *et al.*, 1977) and the random priming labelling method (Feinberg & Vogelstein, 1983) are the most commonly used methods for probe labelling. Both methods result in the efficient incorporation of isotopic and non-isotopic nucleotide analogues. Using nick-translation methods, probes with medium to high specific activity (10^7–10^8 cpm/µg) are usually produced. The random priming method, however, regularly produces probes with a specific activity an order of magnitude higher.

In our experience, the increase in specific activity of a probe labelled by random priming is often accompanied by an increase in background in cytological preparations. This background is most likely to be a result of the small size (100–300 bp in average) of the probe fragments produced during the random priming reaction. They are more able to permeate cellular structures but also, due to their small size, to bind aspecifically (to DNA, proteins or other charged chemicals groups). Random primed DNA appears to be suitable to detect high or middle repetitive sequences; however, as previously mentioned, the increase in background often accompanying the use of these probes makes them unsuitable for the detection of low- or single-copy sequences.

Nick-translated probes have, in our experience, proved to be more useful for the detection of low-copy and single-copy sequences. This is probably due to the larger size of these probes (> 500 bp on average) and amplification of the hybridization signal through the increased ability of these probes to form networks. For this purpose, labelling of the entire plasmid molecule carrying the sequence of interest is usually carried out and dextran sulphate is included in the hybridization mixture to favour the formation of probe networks. These modifications of the *in situ* hybridization protocol have been estimated to increase the rate of hybridization by 20–30-fold (Gerhard *et al.*, 1981). Thus, network formation should compensate for the relatively low specific activity of nick-translated probes and allow the use of this type of probe for single-copy gene localization (Gerhard *et al.*, 1981; Mouras *et al.*, 1987b).

Hybridization

General assessment

The protocol used for the preparation and hybridization of different plant cells, in order to localize DNA sequences, particularly low-copy sequences on plant metaphase chromosomes, should be tested for each source of material. The preparation of plant cytological material is obviously extremely important. The production of well-spread, flat metaphase chromosome plates, clear of cell wall, cytoplasm and other cellular debris, aids chromosomal identification as well as hybridization. The enzymatic removal of RNA prior to hybridization has also been found to increase the efficiency and specificity of hybridization through the removal of competition between cellular RNA and genomic sequences.

The accessibility of the target sequence (DNA) to the probe is another critical step. This is facilitated by an efficient denaturation step.

The hybridization process (reannealing of probe with target DNA) must also be carried out in a manner such that network formation between probe molecules is facilitated on small (single-copy) DNA gene targets.

Denaturation of probe and chromosomal DNA

In most protocols probe and chromosomal DNA are denatured separately (Gerhard *et al.*, 1981). Chromosomal DNA denaturation is accomplished at 70°C for 2–3 min, in 70% formamide/2×SSC,[1] pH 7.0. Cytological preparations are then quickly immersed in 70% and then 95% ethanol at 0°C or −20°C, to prevent reannealing of chromosomal DNA. The probe, either in water or in the hybridization mixture (containing 50% formamide), is denatured by boiling for 5 min or heating at 70°C for 5–15 min respectively, followed by rapid cooling in ice-water. The probe mixture is then applied to chromosomal DNA and incubated at the desired temperature, generally for 15 h. Finally it is washed to remove excess and non-specifically bound DNA prior to autoradiography. With this protocol the hybridization efficiency is generally in the range of 10–20%. In our comparative experiments the efficiency varied between 4 and 9%. We have found, however, that simultaneous denaturation of probe and chromosomal DNA on to cytological preparations increases the efficiency of hybridization to 40%. Moreover, prior denaturation of probe DNA in probe mixture at 70°C for 15 min, followed by rapid cooling in ice-water, and subsequent re-denaturation concomitantly with chromosomal denaturation (80°C for 30 s), increases the hybridization efficiency to more than 80% in the case of single-copy genes and 95–99% for repetitive sequences of DNA such as rDNA (Mouras *et al.*, 1987b). The high hybridization efficiency observed in our experiments makes it unnecessary to examine more than 50 metaphase plates in order to localize the signal to any particular chromosome(s).

Thus the efficient denaturation of probe and target DNA combined with well-spread metaphase plates clear of debris and some expertise in plant karyotypes, makes gene mapping readily available and relatively easy.

Hybridization conditions

The presence of formamide in the probe mixture makes it possible to carry out the hybridization at temperatures ranging between 25 and 40°C. Some species, such as *Nicotiana tabacum*, appear not to be affected by the temperature of hybridization; the chromosomes of *Nicotiana plumbaginifolia*, however, appear to be very sensitive to the temperature of hybridization. At 40°C, although the hybridization efficiency remains very high, the chromo-

---

[1]SSC = saline sodium citrate: 0.15 M NaCl, 0.015 M trisodium citrate.

somes suffer great modifications, such as swelling, or more drastically, they sometimes appear as ghost chromosomes.

Therefore, it seems that from plant to plant, the temperature and in some cases the time of hybridization may need to be adjusted in order to find the optimum conditions.

To improve the accessibility of chromosomal DNA to the probe, cytological preparations have sometimes been treated prior to hybridization with a proteolytic enzyme such as proteinase K. This type of treatment is efficient with mammalian cells, but in our experiments it appeared ineffective and caused dramatic distortion of the chromosomes.

Post-hybridization

The post-hybridization process is necessary for the elimination of excess probe, poorly matched hybrids and aspecifically bound DNA. Efficient washing reduces the background, thereby increasing the signal/noise ratio. This process can be performed by extensive washes in a saline solution such as SSC with a concentration lower than that of the hybridization solution, at room temperature or at the same temperature as hybridization. In some cases, more stringent washes may be required; therefore additional steps such as 50% formamide/2×SSC at 42°C once or twice for 10 min (Huang *et al.*, 1988) or 1×SSC at 60°C for 15 min (Mouras *et al.*, 1987b) have often been included.

After washes and dehydration in an ethanol series, cytological slides are subjected to different treatments according to the type of probe used (radioactive or non-radioactive) in order to detect the hybridization signal (Rayburn & Gill, 1985; Ambros *et al.*, 1986; Mouras *et al.*, 1987b, 1989).

Chromosomal assignment of transferred genes

Transformation of plants using *Agrobacterium* vectors (Chapter 3) or direct gene transfer (Chapters 4, 5) has become a routine procedure for the study of gene function and expression as well as for modifying plant genotypes (Schell, 1987). In general, plant transformation vectors contain selectable markers such as aminoglycoside phosphotransferase or neomycin phosphotransferase, which can be used to select transformed plants (Fraley *et al.*, 1985). Once transformed plants have been selected, it is necessary to determine whether the gene under investigation has also been integrated, and in what form. With Southern analysis it is possible to characterize the presence and the integrity of introduced DNA in transgenic plants; accurate *in situ* hybridization, however, allows a precise localization of that sequence on metaphase chromosomes.

Localization of T-DNA in agrotransformed cells

Transformed tumour cells of tobacco

During cell transformation by *A. tumefaciens*, T-DNA from the Ti plasmid is introduced and integrated into the genome of host cells (Van Lijsebettens *et al.*, 1986). This stable transformation results in expression of a tumorous phenotype: unregulated synthesis of growth regulators (hormones), tumour formation after grafting on to healthy plants and loss of regeneration potential. These properties characterize the so-called crown-gall cells.

Cytogenetic analyses of the crown-gall line isolated by Morel (1948) and a series of its clones and subclones have been carried out (Mouras, 1981, 1984; Mouras & Lutz, 1983). We have shown that:

1. clones contain multiple translocated chromosomes (Fig. 6.1a);
2. within the set of rearranged chromosomes, some are found to be specific to tumour cells (Fig. 6.1b), while others appear to be common to normal (non-tumorous) and abnormal cells cultivated *in vitro*;
3. the presence and the number of specifically rearranged chromosomes (marker chromosomes) correlate very closely to the oncogenic potential of the crown-gall line (Fig. 6.1c);
4. when weak tumour lines that had been maintained in the absence of selective pressure were transferred to growth under selective pressure, they showed concomitantly and significantly increased chromosome marker number and tumour potential;
5. the loss of marker chromosomes is accompanied by a decrease in the oncogenic trait and sometimes in the ability to regenerate plants.

These results, unfortunately, have not been extended to different crown-gall lines and cannot be generalized.

For the localization of T-DNA on to chromosomes the following hypothesis was proposed: assuming that the mathematical correlation calculated is correct, then the oncogenic element (T-DNA) is expected to be present on the marker chromosomes. To demonstrate this correlation, these lines were analysed by Southern analyses and *in situ* hybridization.

Southern hybridizations were carried out using as probes the plasmid pGV0120 containing the 16 kb BamH1 fragment 2, corresponding to the right fragment (TR) of the T-DNA (De Vos *et al.*, 1981), and the plasmid pGV0153 containing the 6 kb BamH1 fragment 8 corresponding to the left fragment (TL) of the T-DNA (Akiyoshi *et al.*, 1982). The autoradiography showed a very faint or negative signal with the TR fragment of the T-DNA and a clear positive signal with the TL fragment (Fig. 6.1d). The intensity of the signal varied from clone to clone, in parallel with the copy number: two copies in weak tumour lines, five or six copies in strong tumour lines.

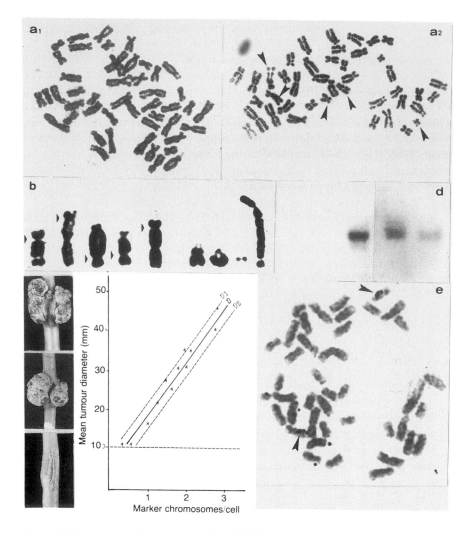

Fig. 6.1. Evidence for the presence of the T-DNA on marker chromosomes in clones derived from a crown-gall line of tobacco:

(a) Metaphase plates from untransformed (a1) and transformed (a2) cells of tobacco. Arrows indicate rearranged chromosomes.

(b) Rearranged chromosomes in tumour cells. Arrows indicate aberrant chromosomes characteristic of tumour cells (marker chromosomes).

(c) The relationship between tumour strength and the number of marker chromosomes per cell in clones derived from the crown-gall line of tobacco: least square regression line D lies between the 0.05% confidence limits, D1 and D2.

(d) Blot hybridization of the DNA from the tumour line of tobacco with the 6 kb Bam H1 fragment 8 of the TL region of the T-DNA of the plasmid pTiA6. From left to right: control, crown-gall line, and clone derived from the tumour line.

(e) Metaphase plate after *in situ* hybridization with the TL fragment of the T-DNA. The label is clearly shown on marker chromosomes (arrows).

In situ hybridization has been carried out using the TL fragment of T-DNA as probe. As shown in Fig. 6.1e, the hybridization signal has been effectively localized on marker chromosomes and/or on a small metacentric chromosome closely resembling that always present in marker chromosomes (Mouras, 1984; Mouras *et al.*, 1987a, 1989b). Thus, as proposed, the loci of T-DNA integration in this tobacco crown-gall line have been localized to rearranged marker chromosomes.

Transformed root cells of Crepis capillaris

Ambros and co-workers (1986) reported for the first time the localization of a unique sequence of transferred T-DNA in plants transformed with *A. rhizogenes*.

Southern blot analyses of DNA isolated from a transformed root line verified the presence of a single copy of T-DNA, as estimated by reconstruction experiments; the size of the transferred DNA represented a sequence of about 17 kb.

In situ hybridization experiments were performed with both ^{3}H and biotinylated probes. The probes (RNA and DNA) were hybridized on metaphase chromosomes according to a protocol where probe and chromosomal DNA were denatured together on cytological preparations. The hybridization efficiency ranged between 20 and 50% and was significantly higher than that generally reported for mammalian cells (Gerhard *et al.*, 1981). Hence the authors were able to show that with both ^{3}H and biotinylated probes, it was possible to localize a DNA fragment in the size range of around 10–20 kb.

Localization of a low-copy resistance gene in a transgenic plant

The plant material used in these experiments consisted of R_1 homozygous tobacco plant resistant to kanamycin. Plants were obtained by selfing R_0 kanamycin-resistant hemizygous primary transformants (Paszkowski *et al.*, 1984, Potrykus *et al.*, 1985). The transformation was carried out using direct gene transfer: protoplasts from the mesophyll of tobacco were incubated in the presence of the plasmid pABD I carrying the gene for kanamycin resistance. After division of protoplasts, resistant colonies were obtained on selective medium. Plants were regenerated on appropriate medium and subjected to genetic, cytogenetic and molecular analyses.

Genomic DNA of transformed plants probed with the pABD I plasmid or with the Eco R5 fragment of the kanamycin gene revealed a positive signal in R_0 plants as well as in progeny (R_1 plants). The resistant R_1 plants were found to contain, at one locus, the foreign gene within three to five rearranged copies of the pABD I plasmid (Paszkowski *et al.*, 1984), representing an insertion of approximately 15–25 kb of DNA.

The chromosomal localization of the transferred gene has been carried out using the *in situ* hybridization method devised for this purpose (Mouras *et al.*, 1987b):

- the cytogenetic method using protoplasts has been shown to be more reliable for the accessibility of target DNA to the probe;
- the hybridization efficiency has also been improved by using an efficient DNA denaturation protocol (see above);
- the sensitivity of the *in situ* hybridization method was tested in an initial experiment, using as probe a plasmid containing a repetitive sequence of DNA specific to tobacco.

We subsequently demonstrated (Mouras *et al.*, 1987b) that the localization of DNA sequences on chromosomes depends on:

- the size of the target;
- the copy number and the distribution of target DNA (either as dispersed sequences or as tandemly repeated sequences);
- the specific activity of the probe and the quantity of probe DNA placed on each cytological preparation.

Using a nick-translated probe with a specific activity of approximately 7×10^7 cpm/µg we observed after 10–12 weeks of autoradiography a clear and intense hybridization signal on interphase nuclei and on homologous chromosomes for the kanamycin gene (Fig. 6.2).

Although the method involves long time exposure (up to 12 weeks), the signal observed on both chromosomes and interphase nuclei was very clear and unequivocal, and therefore the method does not require the development of a histogram for the localization of an alien DNA sequence (Ambros *et al.*, 1986; Huang *et al.*, 1988, 1989).

The localization by *in situ* hybridization of transferred genes on chromosomes in plants is essentially limited by the sensitivity and the very long autoradiography exposure time of this technique. Nevertheless, progress has been made recently and it is now possible to detect a single-copy gene: the parsley 4-coumarate:CoA ligase gene (3.3 kb) (Huang *et al.*, 1989) and the *adh* gene (1.6 kb) in *Nicotiana plumbaginifolia* (Mouras *et al.*, 1989). More refinements will be required in order to detect target DNA of less than 1 kb, the size of most endogenous plant genes.

Chromosome-mediated gene transfer: identification of alien chromosomes

Plant breeding has been enriched with the new methods and tools of genetic engineering. However, many characters responsible for traits of importance, as well as not being characterized at molecular level, are often related

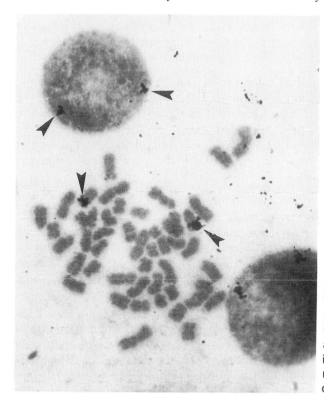

Fig. 6.2. Localization of a low-copy resistance gene in transgenic tobacco plants. The hybridization signals are clearly shown on interphase nuclei and metaphase chromosomes (arrows).

to several gene families. Consequently wide hybridization and/or somatic hybridization is of interest for genetic improvement of crop species (Chapters 7, 8). Extensive cytogenetic analyses are therefore required to characterize chromosome complements in hybrid lines where addition, substitution and translocation of chromosomes have occurred. Essentially, two strategies have been used:

- classical cytogenetic methods as the first approach for chromosome identification;
- *in situ* hybridization, when required, allowing a more precise characterization and location of alien genetic information.

Asymmetric somatic hybrids

One of the most attractive features of plant somatic cell hybridization is the production of highly asymmetric fusion products in which only a small part of the donor genome has been transferred. Although a number of interesting asymmetric clones have already been produced by fusing mitotically inactivated protoplasts with normal recipient protoplasts, highly asymmetric hybrids containing only one or two donor chromosomes (or fragments) have rarely been described (Dudits *et al.*, 1980; Gupta *et al.*, 1984; Bates *et al.*, 1987). Furthermore, progeny analyses were made only by Bates *et al.* (1987). In order to demonstrate the potential of plant somatic cell hybridization for transferring chromosomes between plant species, we describe here a series of highly asymmetric hybrid plants in an intertribal combination between *Nicotiana plumbaginifolia* and *Atropa belladonna* (Gleba *et al.*, 1988).

Wild-type *Atropa belladonna* leaf protoplasts ($2n = 2x = 72$ chromosomes) were irradiated with several doses of gamma-rays (10, 30, 50 and 100 krad) from a cobalt-60 source, and subsequently fused with leaf protoplasts of a nitrate-deficient co-factor mutant (genetic marker) of *N. plumbaginifolia* ($2n = 2x = 20$ chromosomes).

Fusion products were recovered following selection for metabolic complementation in the defective function of the recipient partner (nitrate reductase co-factor mutant). Corrected-phenotype plants for nitrate metabolism were produced at all irradiation doses used and ranged between 0.7 and 3.7%. About 20–25% of the stable gamma fusion products could be regenerated into relatively normal recipient-type regenerants, which were further analysed.

Cytological analyses showed that in all hybrids tested polyploid (tetra- to hexaploid) chromosome sets of *N. plumbaginifolia* ($n = 10$) were present along with 6–29 *A. belladonna* chromosomes or chromosome fragments. Biochemical (isoenzymes) and molecular (ribosomal DNA) analyses showed that besides the selectable marker (nitrate reductase gene), other genetic material from the donor partner was present and expressed in different lines.

For genetic analyses (progeny) the best regenerants identified under *in vitro* conditions were transferred to the greenhouse. Independently identified regenerants showed some differences in morphology: some remained at the vegetative rosette stage while others produced abnormal flower structures to various degrees. All plants tested were male-sterile, but backcrossing with the wild-type recipient partner resulted in fruit swelling and abortive seed production for one of the R_0 regenerants. This R_0 plant, which contained 35 *Nicotiana* along with seven *Atropa belladonna* chromosomes (Fig. 6.3a), was chosen for further tests in embryo rescue experiments. The F_1 progeny obtained after backcrossing this R_0 plant to the diploid and tetraploid recipient partner associated with the embryo rescue were male-sterile again, but their phenotype was more uniform and very similar to that of the recipient species. Cytological analyses of these F_1 plants revealed the presence of only two *Atropa* chromosome fragments. A second backcross of these highly asymmetric triploid plants to diploid wild type resulted in improved seed setting and seed germination. Seventy-five percent of this second backcross progeny were fertile. Cytogenetic analyses of some of these backcrossed plants showed that the *Atropa* complement

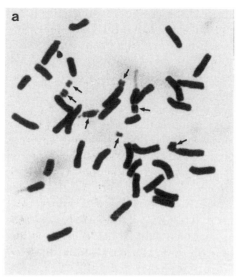

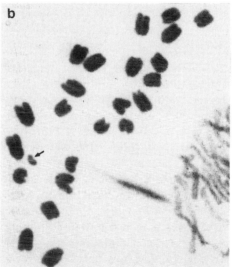

Fig. 6.3. Metaphase plates of *Nicotiana plumbaginifolia* (Np)+*Atropa belladonna* (Ab) somatic hybrids. *Atropa* chromosomes (or fragments) are indicated by arrows.
(a) R_0 regenerant N200/30 krad: 35 Np+7 Ab chromosomes.
(b) Plant obtained after two successive backcrosses of the R_0 regenerant: 23 Np+1 Ab chromosomes.

was reduced to a single chromosome fragment (Fig. 6.3b). Although we cannot show at the molecular level that the complementing gene is present on this extra-chromosome fragment (the probe is not available), we can speculate, as segregation ratios for the nitrate reductase marker of seedlings obtained after selfing of some second backcross progeny plants did not follow the 3:1 ratio.

Although we still do not control the process of chromosome elimination in somatic hybrid fusion products, it is clear that by screening a large number of fusion hybrids one might identify individuals with relatively low numbers of donor chromosomes. Backcrossing of these asymmetric lines to the recipient partner represents a very efficient way to accelerate and direct the process of chromosome elimination. Since at the present time we are analysing very promising, highly asymmetric, intergeneric hybrids between *Nicotiana* and respectively *Petunia* and *Lycopersicon*, we conclude that, at least in some combinations, plant somatic cell hybridization might be an alternative for alien gene and chromosome transfer.

Hybrid cereals

The use of species-specific DNA probes

In situ hybridization with repetitive DNA sequences represents an extremely powerful method for the identification and characterization of chromosomes. To date, many different classes of repetitive DNA have been isolated from plant genomes and characterized. This is particularly true for the Triticae (Bedbrook *et al.*, 1980; Jones & Flavell, 1982a, b; Appels & McIntyre, 1985; Rayburn & Gill, 1985, 1986; Metzlaff *et al.*, 1986; Guidet *et al.*, 1990). Of the species isolated, the most extensively studied have been the various tandemly repeated repetitive sequences found in high copy number throughout the genomes of many members of the Triticae (Bedbrook *et al.*, 1980; Gerlach & Peacock, 1980; Jones & Flavell, 1982a, b). Some of these families have been demonstrated to be species- or genome-specific and hence very useful in understanding genome structure and elucidating evolutionary relationships between different members of this tribe (Bedbrook *et al.*, 1980; Gerlach & Peacock, 1980; Jones & Flavell, 1982a, b). For instance, Jones & Flavell (1982b), using four families of highly repeated sequences isolated from *Secale cereale,* were able to obtain information on the evolutionary relationship between the various *Secale* species, as well as determining the accuracy of earlier classifications of *Secale* species on the basis of hybridization patterns, the strength of hybridization, and the number of chromosomes which showed hybridization with these repetitive sequences. Other workers have used rye-specific telomeric sequences to identify and analyse alien rye chromosomes in various wheat–rye hybrids such as triticale (Jones & Flavell, 1982a, b; Appels *et al.*, 1986) and to investigate the nature of chromosome pairing in *Secale–Aegilops* hybrids at meiosis (Hutchinson *et al.*, 1980).

The use of these types of repetitive DNA sequences, while providing a great deal of valuable information, is limited, due to the localization of these sequence families to defined loci. Intervarietal and intravarietal polymorphisms in the hybridization patterns of some of these repetitive sequences have also been noted (Jones & Flavell, 1982a, b). Dispersed repetitive sequences such as the 120 bp family (Bedbrook *et al.*, 1980), λ het 5.3 (Appels & McIntyre, 1985; Appels *et al.*, 1986) and the R173 family (Guidet *et al.*, 1990; Rogowsky *et al.*, 1990) are potentially of greater interest.

The 120 bp family was first isolated from *S. cereale* constitutive telomeric heterochromatin but was subsequently found to hybridize to the entire length of all rye chromosomes, suggesting a homogeneous distribution throughout the rye genome (Bedbrook *et al.*, 1980; Lapitan *et al.*, 1986). This family has been shown to hybridize at a number of interstitial sites on wheat chromosomes also (Bedbrook *et al.*, 1980; Rayburn & Gill, 1985; Lapitan *et al.*, 1986). The application of this sequence for the identification of very small introgressed segments of (rye) chromatin, particularly into wheat, is therefore limited due to cross-hybridization with wheat chromosome, a problem associated with all non-species-specific sequences.

Middle repetitive sequences such as λ het 5.3 and R173 isolated from *S. cereale* are known to be rye-specific and should therefore prove to be more suitable. Both sequences appear to be quite well dispersed throughout the rye genome except for an apparent absence from some telomeres and the nucleolar organizer region (Appels *et al.*, 1986; Guidet *et al.*, 1990; Fig. 6.4). Both sequences have been used successfully to identify rye chromosomes in wheat–rye hybrids (Appels & McIntyre, 1985; Guidet *et al.*, 1990; Fig. 6.4). One member of this family, G10 4', has been non-radioactively labelled with biotin and used *in situ* to identify all 14 rye chromosomes in hexaploid triticale. In Fig. 6.4, all 14 rye chromosomes present in the chromosome spread are heavily labelled along the entire length of both arms. Interestingly, it appears that R173 is present at the telomeres but absent from around the centromere regions of rye chromosomes. This family of sequences therefore appears to be well distributed throughout the rye genome, making the cloned members of this family potentially useful in the detection and identification of small fragments of rye chromosomes introgressed into alien (particularly wheat) chromosomes.

Although the use of cloned repetitive sequences has provided a great deal of information on the structure, organization and evolution of plant genomes, their use has been essentially confined to the more economically important crop species. This is a direct consequence of the considerable effort required to isolate and characterize such sequences, but their use does have a number of advantages. For instance, the species specificity of certain sequences allows, in general, a more conclusive identification of alien

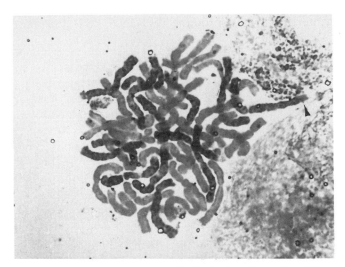

Fig. 6.4. *In situ* hybridization of a biotinylated member of the R173 family, G10 4' to chromosomes of triticale cv. Carman, carried out as described by Gustafson *et al.* (1990). All 14 rye chromosomes are heavily labelled along both arms, including the centromeres and most telomeres, whereas the wheat chromosomes remain unlabelled. The arrow indicates an unlabelled telomere.

chromatin in hybrid backgrounds due to the absence of cross-hybridization. Also, as these sequences are more recently evolved (Bedbrook *et al.,* 1980; Appels *et al.,* 1986), it is possible to elucidate evolutionary relationships (Gerlach & Peacock, 1980; Jones & Flavell, 1982b; Rayburn & Gill, 1986). Further, these sequences are often not directly associated with the constitutive heterochromatic blocks of chromatin (Jones & Flavell, 1982a; Appels *et al.,* 1986). Additional information which may be related to the base sequence of these repetitive DNAs can therefore be obtained. This approach may also provide a means for establishing molecular karyotypes for members of this tribe. It is known that certain repetitive sequences are chromosome- or region-specific, and to date, some of these sequences have already been used to identify individual chromosomes on the basis of their hybridization patterns (Gerlach & Peacock, 1980; Jones & Flavell, 1982a, b; Rayburn & Gill, 1986, 1987).

The use of total genomic DNA probes

Improved wheat cultivars can be produced through interspecific gene transfer from rye, *Agropyron, Elymus, Hordeum* and other species. This confers improved winter hardiness, sprouting resistance, resistance to *Fusarium* head blight, barley yellow, yellow dwarf virus, stem rust and leaf rust, for example.

The traditional approach, which involves the production of interspecific hybrids and disomic lines, is used. Addition lines are identified by isozymes and RFLP markers, chromosome banding, and *in situ* hybridization with localized repetitive DNA sequences. In order to evenly label all chromosome segments, dispersed repetitive DNA sequences are sought. However, in the absence of suitable dispersed sequences, it has been found that a total genomic library from the alien species can be labelled with biotin and used to detect segments of alien chromosomes introgressed into wheat.

In situ hybridization with biotin-labelled genomic DNA of rye preferentially labelled the rye chromosomes in Welsh triticale. Only 12 chromosomes were labelled, which corresponds with previous results suggesting that Welsh is a secondary triticale which carries only six pairs of rye chromosomes. Cytological preparations, probe labelling with biotin and hybridization procedures were carried out as previously reported in Le *et al.* (1989).

In Kavkaz, a wheat stock known to carry the 1Ds/1Rs interchange, only one pair of chromosome arms was seen to be labelled with this kind of probe. The signal on the 1Rs chromosome arm in Kavkaz was detectable from cells in interphase (or at least extremely early prophase) and prophase cells (Fig. 6.5). These results allow a rough screening for the presence of interchange chromosomes from interphase or early prophase cells and for a finer analysis of the labelled segment from prophase as opposed to highly condensed metaphase chromosomes.

Recombinant stocks from the 1Ds and 1Rs chromosomes have also been analysed (Dr K. Shepherd, Australia) and it has been determined that rye segments representing roughly 1/4–1/6 of the rye arm can be detected with this technique. The resolution should prove to be still finer.

Agropyron biotin-labelled genomic DNA was used to label the *Agropyron* chromosomes in disomic addition lines of *Agropyron distichum* into durum wheat. This signal was also detectable at interphase to prophase stages.

Meiotic cells of Welsh triticale were shown to contain label on only the rye chromosomes at metaphase I and anaphase. It will be possible to analyse chromosome pairing in hybrids and cytogenetic stocks to determine the expected frequency of gene introgression. Recently Anamthawat-Jonsson and co-workers (1990) have used this technique to identify parental chromosomes within *Hordeum chilense* and *Secale africanum* hybrids.

The obvious advantages to this approach, particularly where breeding objectives are concerned, are simplicity and flexibility. Many different species and genera can be analysed without the necessity to isolate species-specific sequences, relatively small segments of alien chromatin can be identified and non-radioactive methods for labelling DNA can be used. The

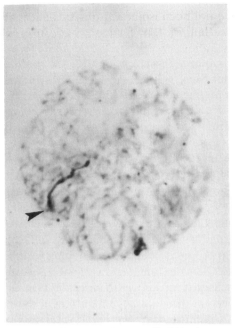

Fig. 6.5. Early prophase cell of Kavkaz wheat with 1Rs arm labelled (arrow) with biotin dUTP.

system is therefore fast and sensitive. However, since it is a heterogeneous mixture of DNA species in unknown proportion which hybridizes, the detection of very small segments of introgressed chromatin may prove more difficult compared with species-specific sequences. Rather, this may require the use of an homogeneous probe mixture enriched for specific sequences. Although the use of total genomic DNA as probes may provide the means for initial characterization of hybrids, the use of cloned repetitive sequences (i.e. species-specific dispersed sequences) will still maintain an important role in the identification and characterization of introgressed DNA from any one species into another, particularly as the sensitivity of detection for non-radioactive *in situ* hybridization of plant chromosomes improves, and as this will allow the use of less highly reiterated sequences.

Conclusions

Gene transfer in plants may involve either single genes or gene families. The introgression of unique DNA sequences may be carried out using technologies such as direct gene transfer or agroinfection. In the case of multigenic traits, particularly when genes are known only by their function or when they have not been isolated, somatic or sexual hybridization is at present the only method for transferring genetic information.

In this chapter we have described the cytogenetic and *in situ* hybridization methods currently used to identify both the chromosome set of the recipient partner and the alien genetic information introduced either as small DNA sequences (single-copy genes) or as whole chromosomes or chromosome fragments. These strategies for the identification of alien DNA are different in their application although identical in principle.

When chromosomes from the donor and recipient partner are similar and cannot be identified by classical methods or when translocations occur, *in situ* hybridization has proved to be a useful approach. Indeed, it has been possible to identify the alien chromatin in somatic hybrids or in some sexual intergeneric hybrids, using as a probe either specific DNA sequences or total genomic DNA from the donor partner.

When gene(s) of interest have been introduced as single- or low-copy sequences, the localization on chromosomes of alien DNA is more difficult than for repetitive sequences or chromosome fragments. In these cases it seems that radioactively labelled probes are preferable, as they appear to be more sensitive than non-radioactive probes; their main drawback, however, lies in the length of time required for autoradiography to display the target DNA.

It should be emphasized that improvements are expected, particularly in the methodology of detection of hybrid labelled molecules. Increased reliability and resolution of *in situ* hybridization should allow this method to be used routinely in chromosome identification and as a complementary method to RFLP analysis for gene mapping in plants.

References

Akiyoshi, D. E., Morris, R. O., Hinz, R., Mischke, B. S., Kosuge, T., Appels, R., Gustafson, J. P. & May, C. E. (1982) Structural variation in the heterochromatin of rye chromosomes in triticales. *Theoretical and Applied Genetics* 63, 235–44.

Ambros, P. F., Matzke, A. J. M. & Matzke, A. M. (1986) Localization of *Agrobacterium rhizogenes* T-DNA in plant chromosomes by *in situ* hybridization. *EMBO Journal* 5, 2073–7.

Anamthawat-Jonsson, K., Schwarzacher, T., Leitch, A. R., Bennet, M. D. & Heslop-Harrison, J. S. (1990) Discrimination between closely related Triticae species using genomic DNA as a probe. *Theoretical and Applied Genetics* 79, 721–8.

Appels, R. & McIntyre, C. L. (1985) Cereal genome organization as revealed by molecular probes. *Oxford Survey of Plant Molecular and Cell Biology* 2, 235–52.

Appels, R., Moran, L. B. & Gustafson, J. P. (1986) Rye heterochromatin. I. Studies on clusters of the major repeating sequence and identification of a new dispersed repetitive sequence element. *Canadian Journal of Genetics and Cytology* 28, 645–57.

Bates, G. W., Hasenkampf, C. A., Contolini, C. L. & Piastuch, W. C. (1987) Asymmetric hybridization in *Nicotiana* by fusion of irradiated protoplasts. *Theoretical and Applied Genetics* 74, 718–26.

Bedbrook, C. K., Jones, J., O'Dell, M., Thompson, R. D. & Flavell, R. B. (1980) A molecular description of telomeric heterochromatin in secale species. *Cell* 19, 545–60.

Chevre, A. M., Mouras, A. & Salesses, G. (1985) Caryotype de *Castanea sativa* Miller étudié après traitement à la cycloheximide et obtention de protoplastes, avec emploi d'un algorithme pour classer les chromosomes. *Comptes Rendus Académie des Sciences Paris* t. 301, Série III, no. 10, 535–8.

Clark, M. & Karp, A. (1989) Physical mapping of the B-hordein loci on barley chromosome 5 by *in situ* hybridization. *Genome* 32, 925–9.

Darvey, N. L. & Gustafson, J. P. (1975) Identification of rye chromosomes in wheat–rye addition lines and triticale by heterochromatin bands. *Crop Sciences* 15, 239–43.

De Vos, G., De Beuckeleer, M., Van Montagu, M. & Schell, J. (1981) Restriction endonuclease mapping of the octopine-tumour-inducing plasmid pTi Ach5 of *Agrobacterium tumefaciens*. *Plasmid* 6, 243–53.

Dudits, D., Fejer, O., Hadlaczky, G. Y., Koncz, C. S., Lazar, G. B. & Horvath, G. (1980) Intergeneric gene transfer mediated by plant protoplast fusion. *Molecular and General Genetics* 179, 283–8.

Eriksson, T. (1966) Partial synchronization of cell division in suspension cultures of *Haplopappus gracilis*. *Physiologia Plantarum* 20, 348–54.

Essad, S., Arnoux, J. & Maia, N. (1966) Controle de validité des caryogrammes: application au caryotype de *Lolium perrene* L. *Chromosoma* (Berlin) 20, 202–20.

Feinberg, A. P. & Vogelstein, B. (1983) A technique for radio-labelling DNA restriction endonuclease fragments to high specific activity. *Analytical Biochemistry* 132, 6–13.

Fraley, R. T., Rogers, S. G., Horsch, R. N., Eichholtz, D. A., Flick, J. S., Fink, C. L., Hoffman, N. L. & Sanders, P. R. (1985) The SEV system: a new disarmed Ti plasmid vector for plant transformation. *Biotechnology* 3, 629–35.

Gerhard, K. S., Kawasaki, E. S., Carter Bancroft, F. & Szabo, P. (1981) Localization of a unique gene by direct hybridization *in situ*. *Proceedings of the National Academy of Sciences USA* 78, 3755–9.

Gerlach, W. L. & Peacock, W. J. (1980) Chromosomal locations of highly repeated DNA sequences in wheat. *Heredity* 44, 269–76.

Gleba, Y. Y., Hinnisdaels, S., Sidorov, V. A., Parokonny, A. S., Boryshuk, N. V., Cherep, N. N., Negrutiu, I. & Jacobs, M. (1988) Intergeneric asymmetric hybrids between *Nicotiana plumbaginifolia* and *Atropa belladonna* obtained by 'gamma-fusion'. *Theoretical and Applied Genetics* 76, 760–6.

Guidet, F., Rogowsky, P., Taylor, C. & Langridge, P. Cloning and characterisation of a new rye specific repeater sequence. *Genome,* in press.

Gupta, P. P., Schieder, O. & Gupta, M. (1984) Intergeneric nuclear gene transfer between somatically and sexually incompatible plants through asymmetric protoplast fusion. *Molecular and General Genetics* 197, 30–5.

Gustafson, J. P., Butler, E. & McIntyre, C. L. (1990) Physical mapping of low-copy DNA sequence in rye (*Secale cereale* L.). *Proceedings of the National Academy of Sciences USA* 87, 1899–902.

Huang, P. L., Hahlbrock, K. & Somssich, I. E. (1988) Detection of single copy gene on plant chromosomes by *in situ* hybridization. *Molecular and General Genetics* 211, 143–7.

Huang, P. L., Hahlbrock, K. & Somssich, I. E. (1989) Chromosomal localization of parsley 4-coumarate: CoA ligase gene by *in situ* hybridization with complementary DNA. *Plant Cell Reports* 8, 59–62.

Hutchinson, J., Chapman, V. & Miller, T. E. (1980) Chromosome pairing at meiosis in hybrids between *Aegilops* and *Secale* species: a study by *in situ* hybridization using cloned DNA. *Heredity* 45, 245–54.

Hutchinson, J., Flavell, R. B. & Jones, J. (1981) Physical mapping of plant chromosomes by *in situ* hybridization. In: Setlow, J. K. & Hollaender, A. (eds), *Genetic Engineering*, vol. 3. Plenum Publishing Co., New York, pp. 207–22.

Jones, J. D. G. & Flavell, R. B. (1982a) The mapping of highly-repeated DNA families and their relationship to C-bands in chromosomes of *Secale cereale*. *Chromosoma* 86, 595–612.

Jones, J. D. G. & Flavell, R. B. (1982b) The structure, amount and chromosomal localization of defined repeated DNA sequences in species of the genus *Secale*. *Chromosoma* 86, 613–41.

Jouanneau, J. (1971) Controle par les cytokinine de la synchronisation des mitoses dans les cellules de tabac. *Experimental Cell Research* 67, 327–37.

Jouve, N., Galindo, C., Montserrat, M., Diaz, F. & Abella, B. (1989). Changes in triticale chromosome heterochromatin visualized by C-banding. *Genome* 32, 735–42.

Karp, A. (1989) Can genetic instability be controlled in plant cell cultures. *Newsletter – International Association for Plant Tissue Culture*, no. 58, 3–11.

Lange, W. (1988) Cereal cytogenetics in retrospect. What came true of some cereal cytogeneticists' pipe dreams? *Euphytica* (supplement), 7–25.

Lapitan, N. L. V., Sears, R. G., Rayburn, A. L. & Gill, B. S. (1986) Wheat–rye translocations: detection of chromosome breakpoints by *in situ* hybridization with a biotin-labelled probe. *Journal of Heredity* 77, 415–19.

Le, H. T., Armstrong, K. C. & Miki, B. (1989) Detection of rye DNA in wheat–rye hybrids and wheat translocation stocks using total genomic DNA as a probe. *Plant Molecular Biology Reporter* 7(2), 150–8.

Lelley, T., Josifek, K. & Kaltsikes, P. J. (1978) Polymorphism in the giemsa C-banding pattern of rye chromosomes. *Canadian Journal of Genetics and Cytology* 20, 307–12.

Metzlaff, M., Proebner, W., Baldauf, F., Schlegel, R. & Cullum, J. (1986) Wheat-specific repetitive DNA sequences. construction and characterization of four different genomic clones. *Theoretical and Applied Genetics* 72, 207–10.

Morel, G. (1948) Recherches sur la culture associée de parasites obligatoires et de tissus végétaux. Thèse, Paris. *Annales Epiphytes (Paris)* 14, 123–234.

Mouras, A. (1981) 'Relation entre mutations chromosomiques et pouvoir tumoral chez des clones issus d'une souche tumorale de tabac.' Thèse, Université de Bordeaux II.

Mouras, A. (1984) Marker chromosomes associated with tumour growth potential in plants. *Differentiation* 27, 88–93.

Mouras, A. & Lutz, A. (1983) Plant tumour reversal associated with the loss of marker chromosomes in tobacco cells. *Theoretical and Applied Genetics* 65, 283–8.

Mouras, A. & Negrutiu, I. (1989) Localization of the T-DNA on marker chromosomes in transformed tobacco cells by *in situ* hybridization. *Theoretical and Applied Genetics* 78, 715–20.

Mouras, A., Salesses, G. & Lutz, A. (1978) Sur l'utilisation des protoplastes en cytologie: amélioration d'une méthode récente en vue de l'identification des chromosomes mitotiques des genre *Nicotiana* et *Prunus. Caryologia* 31, 117–27.

Mouras, A., Wildenstein, C. & Salesses, G. (1986) Analysis of Karyotype and C-banding pattern of *Nicotiana plumbaginifolia* using two techniques. *Genetica* 68, 197–202.

Mouras, A., Negrutiu, I. & Dessaux, Y. (1987a) Phenotypic and genetic variations in crown-gall tumour cells of tobacco. *Theoretical and Applied Genetics* 74, 253–60.

Mouras, A., Saul, M. W., Essad, S. & Potrykus, I. (1987b) Localization by *in situ* hybridization of low copy chimaeric resistance gene introduced into plants by direct gene transfer. *Molecular and General Genetics* 207, 204–9.

Mouras, A., Negrutiu, I., Horth, M. & Jacobs, M. (1989) From repetitive DNA sequences to single-copy gene mapping in plant chromosomes by *in situ* hybridization. *Plant Physiology and Biochemistry* 27, 161–8.

Nagata, T., Okada, K. & Takebe, I. (1982) Mitotic protoplasts and their infection with tobacco mosaic virus RNA encapsulated in liposomes. *Plant Cell Reports* 1, 250–2.

Paszkowski, J., Shillito, R. D., Saul, M. W., Mandak, V., Hohn, T., Hohn, B. & Potrykus, I. (1984) Direct gene transfer to plants. *EMBO Journal* 3, 2717–22.

Pijnacker, L. P. & Ferwerda, M. A. (1984) Giemsa C-banding of potato chromosomes. *Canadian Journal of Genetics and Cytology* 26, 415–19.

Potrykus, I., Paszkowski, J., Saul, M. W., Petruska, J. & Shillito, R. D. (1985) Molecular and general genetics of a hybrid foreign gene introduced into tobacco by direct gene transfer. *Molecular and General Genetics* 199, 169–77.

Rayburn, A. L. & Gill, B. S. (1985) Use of biotin-labeled probe to map specific DNA sequences on wheat chromosomes. *Journal of Heredity* 76, 78–81.

Rayburn, A. L. & Gill, B. S. (1986) Molecular identification of the D-genome chromosomes of wheat. *Journal of Heredity* 77 253–5.

Rayburn, A. L. & Gill, B. S. (1987) Use of repeated DNA sequences as cytological markers. *American Journal of Botany* 74, 574–80.

Rigby, P. W. J., Dieckmann, M., Rhodes, C. & Berg, P. (1977) Labelling deoxyribonucleic acid to a high specific activity *in vitro* by nick-translation with DNA polymerase I. *Molecular Biology* 113, 237–51.

Rogowsky, P. M., Manning, S., Liu, J. Y. & Landgrige, P. (1990) The R173 family of rye specific repetitive DNA sequences: a structural analysis. *Genome* (in press).

Schell, J. S. (1987) Transgenic plants as tools to study the molecular organization of plant genes. *Science* 237, 1176–82.

Sharma, A. M. & Sharma, A. S. (1965) *Chromosomes: Techniques, Theory and Practice.* Butterworth, London.

Simpson, P. R., Newman, M. A. & Davies, D. R. (1988) Detection of legumin gene sequences in pea by *in situ* hybridization. *Chromosoma* 96, 454–8.

Tanksley, S. D., Young, N. D., Paterson, A. H. & Bonierbale, M. W. (1989) RFLP mapping in plant breeding: new tools for an old science. *Biotechnology* 7, 257–64.

Van Lijsebettens, M., Inzé, D., Schell, J. & van Montagu, M. (1986) Transformed cell clones as a tool to study T-DNA integration mediated by *Agrobacterium tumefaciens*. *Journal of Molecular Biology* 188, 129–34.

Chapter 7
Somatic Embryogenesis: Potential for Use in Propagation and Gene Transfer Systems

W. A. Parrott[1], S. A. Merkle[2] & E. G. Williams[3]
[1]Department of Agronomy, [2] School of Forest Resources and [3]Department of Botany, University of Georgia, Athens, GA 30602

Introduction

Since 1958 when the first plant embryos were obtained from somatic tissues of carrot (*Daucus carota*) cultured *in vitro* (Reinert, 1958; Steward, 1958), ever increasing numbers of species and tissues have been induced to form somatic embryos. In the process, a greater understanding of the phenomenon has been achieved, and the ability to exploit it has increased. Although somatic embryos are derived from somatic cells, they closely resemble their sexual counterparts and presumably result from expression of genes regulating the same developmental pathway. The evolution of this pathway must date back at least to the evolution of seed plants, and its basic features appear to be highly conserved. As a result, it is possible to apply the most basic principles to somatic embryogenesis across the range of seed-bearing plants.

A wide range of plant tissues has been used as explant sources from which to obtain somatic embryos, with tissues derived from immature embryos being especially amenable to induced embryogenesis. Responding species are as diverse as wheat, *Triticum aestivum* (Ozias-Akins & Vasil, 1982), and soybean, *Glycine max* (Lippmann & Lippmann, 1984) (Fig. 7.1). The use of zygotic embryos as an explant source is normally not a limitation, except in the case of cross-pollinated species, in which case a zygotic embryo (or seedling) represents an unknown genotype.

Cells within a zygotic embryo are believed to express the genes necessary for the embryogenic developmental programme. In simplest terms, they need only to become independent from positional constraints imposed by

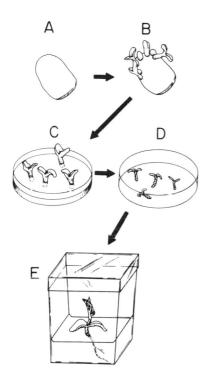

Fig. 7.1. Basic regeneration protocol for soybean (based on Lazzeri *et al.*, 1985; Parrott *et al.*, 1988). A). Immature cotyledons, with the embryonic axes removed, are exposed to 10 mg/l of α-naphthalene acetic acid (NAA) for 7–10 days, and then transferred to hormone-free basal medium. B) Somatic embryos develop around the distal periphery of the cotyledon. After 20 days on hormone-free basal medium, the somatic embryos are removed from the explant tissue and transferred to hormone-free basal medium. C) Somatic embryos are allowed to mature for at least one month on basal medium. D) Somatic embryos are desiccated for one week. E) Desiccated embryos germinate when transferred to basal medium. (Artwork by J. Adang.)

other cells within the tissue. Consequently, the disruption of such tissues alone can be sufficient in some species to permit the development of somatic embryos on hormone-free media, as has been observed with East Indies walnut (*Albizia lebbeck*) and carrot (Gharyal & Maheshwari, 1981; Smith & Krikorian, 1989). In recognition of the existence of an embryogenic state within cells prior to the formation of somatic embryos, the term pre-embryogenic determined cell (PEDC) was coined (Sharp *et al.*, 1982). Because PEDCs can divide to form an embryo directly, the process has been referred to as direct somatic embryogenesis.

Embryogenic developmental programmes are not expressed in other, more differentiated plant tissues that give rise to somatic embryos, and several cell division cycles in the presence of an auxin are normally required before the embryogenic developmental pathway is once again expressed. The term induced embryogenic determined cell (IEDC) is used to identify cells that acquire embryogenic potential in culture (Sharp *et al.*, 1982). The various cell generations that intervene between the original explant and the formation of somatic embryos are manifested as a callus. In this case, somatic embryogenesis is said to be indirect. Once formed, an IEDC and a PEDC are functionally equivalent, and the term embryogenic cell (EC) could be considered to be a better descriptor.

The terms PEDC and IEDC provide convenient ways to classify tissues, but in reality represent extremes of a continuum. Apparently, the embryogenic developmental programme is not terminated abruptly and replaced by another, but alters gradually over a number of cell cycles. The net result is that increasingly greater amounts of reprogramming are necessary to form ECs as the cells diverge in time and space from the original embryogenic condition. The observation that older cells from immature cotyledons of soybean form more abnormal somatic embryos than do younger cells (Hartweck *et al.*, 1988) fits this model of somatic embryogenesis. Such gradients in the morphology of somatic embryos have also been observed in flax (*Linum usitatissimum*) by Pretova & Williams (1986b) and in tomato (*Lycopersicon*) species by Young *et al.* (1987). In addition, explants derived from tissues associated with reproduction (e.g. nucellar or anther tissues) and hypocotyls or cotyledons of young seedlings tend to be more easily induced to an embryogenic state than do cells from mature tissues (Williams *et al.*, 1990).

When a tissue consists of ECs, the stimulation of cell division may be all that is necessary to perpetuate the embryogenic state and form somatic embryos. This could explain the formation of somatic embryos following the exposure of immature zygotic embryos to low concentrations of cytokinin in oilseed rape, *Brassica napus* ssp. *oleifera* (Pretova & Williams, 1986a), flax (Pretova & Williams, 1986b), alfalfa (*Medicago sativa*) and various species of clover (*Trifolium*) (Maheswaran & Williams, 1984, 1985, 1986a, b). The perpetuation of cycles of repetitive embryogenesis in microspore-derived somatic embryos of oilseed rape on hormone-free medium has been attributed to high concentrations of endogenous cytokinins within the ECs of the somatic embryos (Loh *et al.*, 1983). In this particular case, repetitive embryogenesis is stopped by exposure to exogenous cytokinin, presumably because exogenous cytokinins are able to activate the metabolic machinery that ordinarily maintains low concentrations of internal cytokinins.

Whenever an IEDC must be produced from a non-embryogenic cell, exposure to an auxin is necessary, and exposure to a cytokinin can be detrimental to the process (e.g. Wenck *et al.*, 1988). In fact, high concentrations of endogenous cytokinins have been associated with lack of embryogenic capacity in Napiergrass, *Pennisetum purpureum*, and orchardgrass, *Dactylis glomerata* (Rajasekaran *et al.*, 1987; Wenck *et al.*, 1988). Alternatively, in A188, an embryogenic genotype of maize (*Zea mays*), endogenous auxin concentrations in the ovule are 16–20 times lower than those in non-embryogenic genotypes (Carnes & Wright, 1988). The mode of action of auxin in the redetermination of an IEDC remains unknown, and is probably not due to a single factor. On exposure to auxin, substantial DNA methylation occurs (LoSchiavo *et al.*, 1989), which may stop or hinder the expression of existing developmental programmes within the cell. Stress may also have the same effect (Kamada *et al.*, 1989). Isolation of cells and disruption of tissue may also play a role in the redetermination of cells. This factor has been reviewed extensively by Williams & Maheswaran (1986) and by Smith & Krikorian (1989). Although factors as diverse as plasmolysis (Wetherell, 1984) and tissue necrosis (Hartweck *et al.*, 1988; Trigiano *et al.*, 1989) obviously lead to isolation of cells from the rest of a tissue, auxins themselves induce the property of friability, which is caused by rapid cell separation (Evans *et al.*, 1981). Finally, once a cell or small group of cells has been isolated, an auxin may help establish polarity. Polarity has been found to precede and accompany differentiation in cell aggregates of carrot (Brawley *et al.*, 1984; Nomura & Komamine, 1986; Gorst *et al.*, 1987; Rathore *et al.*, 1988; Timmers *et al.*, 1989) and sweet potato, *Ipomoea batatas* (Chée & Cantliffe, 1989). Application of a low-voltage field greatly increased the frequency of somatic embryos formed on callus of alfalfa (Dijak *et al.*, 1986), plausibly in response to polarity imposed by the electrical field.

Once an EC has been obtained, the continued presence of an auxin can be detrimental to normal development of somatic embryos (Halperin & Wetherell, 1964; Parrott *et al.*, 1988). The strikingly negative effect that 2,4-D can have on the quality of somatic embryos was illustrated by Kamada *et al.* (1989). Carrot somatic embryos were induced by treatment with high concentrations of sucrose, cadmium ions, sodium hypochlorite, or 2,4-D, and encapsulated in calcium alginate to produce synthetic seeds. For synthetic seeds containing a single large embryo, 30–50% of encapsulated embryos from cadmium treatments developed both a radicle and a green bud, whereas only 15% of embryos derived from 2,4-D treatment did so. The best time to withdraw auxin from the medium is not clear, but probably corresponds to the time that auxin contents drop naturally in ovules developing *in planta* (Carnes & Wright, 1988; Carman, 1989).

If the concentration of auxin is high enough, instead of proceeding to the next stage of its ontogeny, a somatic embryo may instead give rise to new somatic embryos. Such a process has been described at various times as secondary, recurrent or repetitive embryogenesis (Fig. 7.2). Depending on the species and the culture system in use, the repetitive cycle may be expressed as continuous propagation of various embryonic stages, including proembryogenic masses (PEMs) as in carrot (e.g. Halperin & Wetherell, 1965), globular embryos as in *Citrus* (e.g. Button *et al.*, 1974) and soybean (Finer, 1988; Finer & Nagasawa, 1988), or even cotyledonary stage embryos, as in alfalfa (Lupotto, 1983, 1986). Other factors that can influence the proliferation of somatic embryos include the ammonium to nitrate ratio in the medium (Smith & Krikorian, 1989) and a low pH (4.0) in the medium (Smith & Krikorian, 1990). Such cycles of repetitive embryogenesis can be broken by the removal or reduction of auxin in the culture medium, permitting the development of mature embryos. It is the ability of repetitive embryogenesis to perpetuate the embryogenic state indefinitely and produce large numbers of embryos that makes somatic embryogenesis a

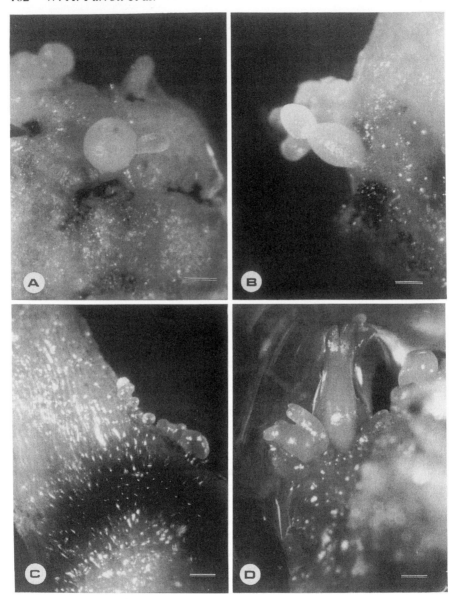

Fig. 7.2. Repetitive somatic embryogenesis. A) B) Globular somatic embryos of soybean giving rise to further globular embryos. C) Globular and D) torpedo-stage embryos that have formed along the hypocotyl of larger, cotyledonary-stage somatic embryos of alfalfa.

powerful tool capable of being exploited for diverse goals such as mass propagation and the production of transgenic plants.

Factors affecting the application of embryogenic cultures and somatic embryos

Genetic control of regeneration

Biotechnology has been expected to assist plant breeding programmes, but applications of biotechnology to plant breeding have frequently been limited by two major factors: the lack of regeneration capacity in desired tissues, cultivars or crops, and the lack for most plant species of a simple reliable method for genetic transformation with defined DNA sequences. The advent of biolistic technology has removed a major barrier for the delivery of exogenous DNA into plant cells, leaving the recovery of plants from transgenic cells as a major limitation. It may be that before biotechnology can make a significant contribution to the improvement of many crops, it will be necessary to breed for increased regeneration capacity. As an example, the argument for breeding for regeneration has been made for potato (*Solanum tuberosum*), in which the ability to form somatic embryos in cultured anthers is a heritable trait (Sonnino *et al.*, 1989), and crosses between parents with low embryogenic capacity can result in progeny with much higher regeneration capacities (Jacobsen & Sopory, 1978).

The ability to form somatic embryos is, in most cases, not merely an intrinsic property of a species. Instead, it is a property under genetic control, such that individual genotypes within a species can differ in their ability to undergo somatic embryogenesis. This phenomenon has been widely documented in several species, including red clover, *Trifolium pratense* (MacLean & Nowak, 1989), alfalfa (Mitten *et al.*, 1984; Brown & Atanassov, 1985; Meijer & Brown, 1985; Chen & Marowitch, 1987; Chen *et al.*, 1987; Bianchi *et al.*, 1988; Walton & Brown, 1988), soybean (Komatsuda & Ohyama, 1988; Parrott *et al.*, 1989), cotton (Trolinder & Xhixian, 1989), maize (Green & Phillips, 1975, Bartkowiak, 1982; Lu *et al.*, 1983a; Beckert & Qing, 1984; Duncan *et al.*, 1985; Hodges *et al.*, 1986), rice, *Oryza sativa* (Abe & Futsuhara, 1986), barley, *Hordeum vulgare* (Hanzel *et al.*, 1984; Lührs & Lörz, 1987; Ohkoshi *et al.*, 1987) and wheat (Sears & Deckard, 1982; Lazar *et al.*, 1983; Maddock *et al.*, 1983; He *et al.*, 1988).

The most complete definition of the genetic control of regeneration has been accomplished in alfalfa. Two dominant genes, named *Rn1* and *Rn2*, were initially found to be necessary for a high frequency of regeneration to occur (Reisch & Bingham, 1980). Two dominant genes were also found to be required for regeneration in other populations of alfalfa. These were named *Rn3* and *Rn4* by Wan *et al.* (1988) and *Rna* and *Rnb* by Hernández-Fernández and Christie (1989). Although given different names, these three gene pairs have not been tested for allelism, making it conceivable that they are allelic in all populations. In cucumber, two dominant genes are necessary for an intermediate frequency of somatic embryogenesis, while the presence of a third dominant gene provides a high frequency of regeneration. A single dominant gene conditions for embryogenesis from leaf tissues of orchardgrass (Gavin *et al.*, 1989). Only in red clover has regeneration capacity been attributed to a recessive gene, *rg*, while the ability to form callus prior to the initiation of somatic embryos is due to a single dominant gene, *C* (Broda, 1984).

Wheat, although a hexaploid species, makes an excellent organism in which to study the genetic control of regeneration, as a wide range of cytogenetic stocks are available. Using these stocks, a gene (or genes) on chromosome 4B has been found to be important for regeneration (Mathias & Fukui, 1986; Higgins & Mathias, 1987). In addition, a gene with major influence on regeneration capacity is located on the long arm of chromosome 2D, and genes with smaller influences have been located to the long arm of chromosome 2A and the short arm of chromosome 2B. These three genes appear to be regulated by a gene on the long arm of chromosome 2B (Kaleikau *et al.*, 1989). Finally, an example of the role cytoplasm may have on regeneration capacity is found in wheat. A rearrangement of the mitochondrial genome has been found in cultures of the cultivar Chinese Spring which have lost their embryogenic capacity. This rearrangement is like the arrangement normally found in the cultivars Talent, Captitole and Thésée, which all lack regeneration capacity (Rode *et al.*, 1988).

In most other species in which the genetic control of regeneration has been studied, it has not been possible thus far to identify individual genes, and investigators have resorted to the evaluation of regeneration capacity in quantitative terms. This is partly due to the large environmental influence on somatic embryogenesis. In a study of regeneration capacity in Brussels sprouts (*Brassica oleracea* var. *gemmifera*), genetic effects accounted for only 50% of the observed variability in somatic embryogenesis from microspores. Partial dominance was observed for regeneration capacity, and narrow-sense heritability was estimated at 0.48 (Ockendon & Sutherland, 1987).

One of the earliest attempts at a quantitative study of embryogenic capacity was conducted on red clover, in which additive genetic effects were found to be the most important determinants of the potential for somatic embryogenesis (Keyes *et al.*, 1980). This observation has held with almost all other species studied.

In an investigation of five maize inbreds differing in ability to form embryogenic callus, and F_1 and F_2 plants derived from a partial diallel, additive genetic effects explained 70% of observed variability for regeneration capacity, which was highly heritable (Tomes & Smith, 1985). Cytoplasmic effects were significant in depressing embryogenic capacity. In a separate set of crosses between A188, a genotype with a high regeneration capacity, and the widely used inbreds, B73 and Mo17 (Willman *et al.*, 1989), additive effects were more important than dominance effects in determining both the percentage of immature embryos forming somatic embryos and the number of plants regenerated per explant. Differences in regeneration from reciprocal crosses suggested that cytoplasmic or other maternal effects influence regeneration capacity. Based on analysis of genetic variances, the authors suggested that at least one gene or one group of linked genes was responsible for determining the frequency of somatic embryogenesis.

Analysis of a diallel between four genotypes of rice (Peng & Hodges, 1989) suggested that additive genetic effects (general combining ability) explained 49% of the variability observed for the number of plants regenerated, while dominance genetic effects (specific combining ability) and cytoplasmic effects accounted for 40% and 90%, respectively, of the observed variation. When frequency of explants undergoing regeneration was the parameter evaluated, additive, dominance and cytoplasmic effects accounted for 61%, 29% and 10% of the observed variation, respectively.

Diallel analysis of five winter wheat cultivars showed that additive, non-additive and cytoplasmic effects all influenced regeneration, with additive effects being the most important. Cytoplasmic effects are nevertheless sufficient to necessitate careful selection of maternal parents to ensure regeneration success (Ou *et al.*, 1989). Only in barley have dominance genetic effects been found to be more important than additive effects, as determined from the analysis of a diallel between seven genotypes (Komatsuda *et al.*, 1989). Nevertheless, broad-sense heritability of regeneration was estimated at 0.86, while narrow-sense heritability was estimated to be 0.72.

As with any trait under genetic control, it is possible to breed for the capacity to regenerate via somatic embryogenesis. The earliest success was in the breeding of tetraploid alfalfa for regeneration capacity, and resulted in Regen-S, a genotype with a high frequency of somatic embryo formation (Bingham *et al.*, 1975). The availability of Regen-S has facilitated a broad range of studies, including:

1. physiological aspects of somatic embryogenesis (Walker *et al.*, 1979; Walker & Sato, 1981; Stuart & Strickland, 1984a, b);
2. *in vitro* selection and subsequent recovery of plants resistant to the toxin produced by *Fusarium oxysporum f.* sp. *medicaginensis* (Hartman *et al.*, 1984);
3. regeneration of plants from protoplasts (Johnson *et al.*, 1981, Dijak *et al.*, 1986);
4. genetic transformation of alfalfa (D'Halluin *et al.*, 1990).

More recently, diploid alfalfa has also been bred for regeneration capacity (Ray & Bingham, 1989).

In addition to the breeding efforts in alfalfa, breeding and selection have increased the capacity for somatic embryo formation from microspores in maize (Petolino *et al.*, 1988) and rapid-cycling *Brassica campestris* (Aslam *et al.*, 1990). In maize, only 3.5% of F_1 plants from a three-way cross were embryogenic, while this percentage increased to 23.4% of embryogenic S_1 plants after the first *in vitro* selection cycle. In *Brassica*, only 46% of plants in the parental population were embryogenic. This frequency increased to include 96% of the plants in the S_3 generation.

Obtaining plants from somatic embryos

It is self-evident that recovery of plants from somatic embryos, a process referred to as conversion, is essential for the ultimate success of any system based on somatic embryogenesis. Ease with which somatic embryos convert to plants differs widely across genotypes, species and culture systems, representing the entire continuum from the readily accomplishable to the currently impossible.

It is difficult to evaluate the comparative conversion efficiencies of various species and culture systems, as this statistic is seldom reported in the literature. Many authors have limited their reporting to the successful recovery of plants, providing little or no information on conversion rates. Many studies have not been designed to provide this information. For some species, the absolute conversion rate may not be a critical factor. For example, recovery of plants from somatic embryos of carrot and alfalfa, and microspores of oilseed rape is generally not considered limiting, although this is not necessarily due to high conversion rates. Instead, since large numbers of somatic embryos are readily obtainable, a low conversion rate becomes inconsequential. On the other hand, low numbers of somatic embryos coupled with a low conversion rate make it very difficult, if not impossible, to recover plants. For example, there are no published reports of plants recovered from somatic embryos of bean, *Phaseolus vulgaris* (Martins & Sondahl, 1984), and recovery of plants from somatic embryos of red bud, *Cercis canadensis,* has been very limited (Trigiano *et al.,* 1988).

Several factors probably contribute to lack of conversion capacity in somatic embryos. Just as the formation of somatic embryos is under genetic control, there is some evidence that conversion ability is also under genetic control, e.g. in alfalfa (Seitz Kris & Bingham, 1988) and soybean (Komatsuda & Ohyama, 1988). This control is separate from that controlling embryogenesis. In triticale (*X. triticosecale*), the formation of somatic embryos from microspores is under the control of nuclear genes with additive action, while conversion of these embryos is influenced by additive, non-additive and cytoplasmic effects (Charmet & Bernard, 1984). To the degree that they can be manipulated *in vitro*, simulation of the events that occur *in planta* is probably the best way to optimize both the maturation of somatic embryos and their conversion to plantlets. Timing of growth regulator applications to match those that occur *in planta*, use of low oxygen tensions and desiccation of mature embryos have all been suggested as treatments that can increase the efficiency of plantlet production (Carman, 1988, 1989).

Auxin

Although auxin is used to induce somatic embryogenesis, continued exposure to auxin has long been known to have a detrimental effect on the development of the apical meristem of somatic embryos (Halperin & Wetherell, 1964). Previous discussion in this chapter centred on continued exposure to auxin as a means of locking somatic embryos into a given developmental phase and establishing cultures undergoing repetitive embryogenesis. Continuous exposure can, however, be detrimental to normal development once ECs have been induced and the resulting somatic embryos have completed the initial stages of development. Exposing the explant tissue to an auxin for a short period followed by transfer to hormone-free basal medium has led to the recovery of somatic embryos and plants from alfalfa (Walker & Sato, 1981), pecan, *Carya illinoensis* (Wetzstein *et al.*, 1989), and black locust, *Robinia pseudoacacia* (Merkle & Wiecko, 1989). Other examples of the transfer of young proembryos to low- or no-auxin medium for their development include carrot (Gorst *et al.*, 1987) and *Eucalyptus citriodora*. Work in soybean (Parrott *et al.*, 1988) showed that only one week's exposure of explant tissues to auxin was necessary to obtain somatic embryos, and that increasingly longer exposures resulted in decreased development of the shoot meristem in the somatic embryos subsequently formed. In certain systems, auxin effects may persist for some time after removal of exogenous auxin. Activated charcoal has been added to the culture medium to further remove auxin from tissues (Buchheim *et al.*, 1989).

Desiccation

Desiccation, another characteristic of zygotic embryogenesis *in planta*, may play a role in terminating embryogenic developmental processes and triggering the germination and seedling developmental programmes (Rosenberg & Rinne, 1986, 1987, 1988). In those instances where high sucrose concentrations in the medium have been used to achieve conversion competence in somatic embryos (e.g. Lee & Thomas, 1985; Janick, 1986; Buchheim *et al.*, 1989; Carman, 1989), osmotic desiccation is a plausible explanation for the success of the treatment.

The imposition of an outright desiccation step has assisted in the conversion of somatic embryos of some species. Initial attempts at the desiccation of somatic embryos, coupled with the successful recovery of plants, were carried out in orchardgrass, and resulted in a 4% conversion rate following 3 weeks of storage (Gray *et al.*, 1987). Twenty percent of grape (*Vitis longii*) somatic embryos converted after desiccation, as opposed to only 5% conversion of somatic embryos that had not been desiccated (Gray, 1987). A subsequent study identified one genotype of grape that produced well-developed embryos and responded well to desiccation, resulting in a 34% conversion rate. Conversion for this genotype was higher following desiccation than following treatments with benzyladenine (BA), gibberellin (GA) or abscisic acid (ABA) (Gray, 1989). Soybean embryos matured for 4 weeks on a basal medium survived desiccation and converted into plants at ten times the frequency of their non-desiccated counterparts (Hammatt & Davey, 1987; Parrott *et al.*, 1988). Similarly, a 5-day desiccation of pecan somatic embryos also raised germination rates (Wetzstein *et al.*, 1989). Finally, desiccation of interior spruce somatic embryos under high relative humidity resulted in a very high, uniform germination, and a conversion rate of 50%. In contrast, only 5% of embryos that were not desiccated converted into plants (Roberts *et al.*, 1990b).

Zygotic embryos developing *in planta* undergo a period during which they contain high concentrations of ABA. The role of ABA during embryogenesis is not completely clear, but there is evidence that ABA helps regulate both the accumulation of storage proteins and proteins that may help to protect the embryos during desiccation (Galau *et al.*, 1986, 1990; Hughes & Galau, 1989). The addition of 10^{-7} M ABA to developing somatic embryos promoted the development of mature, well-formed embryos of caraway (Ammirato, 1974) and carrot (Ammirato, 1983). In some conifers, including spruce (*Picea*) species and hybrids, exogenous ABA is critical for normal accumulation of storage reserves and subsequent viability of somatic embryos (Roberts *et al.*, 1990a). Application of ABA has also been associated with acquisition of desiccation tolerance in somatic embryos of alfalfa, such that 60% of somatic embryos exposed to ABA survived desiccation treatment (Senaratna *et al.*, 1989, 1990).

Mass propagation systems

Mass propagation of embryos

The fact that many embryogenic systems can be perpetuated via repetitive embryogenesis makes them potentially attractive for mass production of clonal plantlets. Theoretically, a culture initiated from a single explant can produce an unlimited number of embryos. This potential for virtually unlimited multiplication gives somatic embryogenesis a huge advantage over conventional vegetative propagation systems such as rooted cuttings, which are limited to the amount of material that can be harvested from the mother plant. Multiplication rates from embryogenic cultures also generally exceed those attainable with other tissue culture regeneration systems such as shoot micropropagation. A high multiplication rate, however, is only the first of several potential advantages offered by somatic embryogenesis in comparison with other vegetative propagation systems. For many species, particularly those characterized by the production of PEMs, both the proliferation of embryogenic cells and the development of individual somatic embryos can be accomplished in liquid medium, making possible the manipulation of very large numbers of propagules with a minimum of labour. For example, Drew (1980) estimated that a single litre of an embryogenic carrot suspension culture contained 1.35 million somatic embryos. If this capacity for liquid culture can be combined with automated, continuous culture technologies, there is the potential for even greater economies of scale.

Besides the features of embryogenic cultures which make them directly adaptable to large-scale production of plants, somatic embryos themselves have several desirable characteristics as propagules. Chief among these is the fact that the product of somatic embryogenesis is an embryo, in many cases very similar to the embryo in a seed of the same species, and potentially able to be exploited in the same way as a complete propagule comprising embryonic shoot and root tissues. Even more important than these physical attributes, however, is the fact that a somatic embryo carries the 'program' to make a complete plant, and is capable of doing so with very little labour input. Unlike other vegetative propagation systems, there is no requirement for separate shoot growth and rooting steps to obtain complete plantlets. Furthermore, in contrast to regeneration systems that rely on organogenesis or axillary branching, many embryogenic systems are capable of producing individual embryos, unattached to either mother tissue or other embryos. No additional labour input is required to obtain individual propagules before further manipulation. Thus, embryogenic cultures produce propagules that are not only complete, but also discrete as well. The combination of these two properties gives somatic embryos potential for direct delivery to the greenhouse or field as components of artificial seeds or in fluid drilling systems. Such applications have been the subject of a number of studies over the past five years.

Somatic embryos do differ from their sexual counterparts in that they have bypassed genetic segregation and recombination. Consequently they represent a method of clonal propagation, maintaining the genotype of the plant from which the explant tissues came, and subject only to the natural mutation rate and mutations that may be induced *in vitro*. The latter mutation rate has never been strictly quantified, but is lower than that incurred by organogenic regeneration systems (Hanna *et al.*, 1984). Since somatic embryos must express many more developmental genes in order for ontogeny to be completed successfully, they may be less tolerant of mutations and epigenetic changes than organogenic cultures (Ozias-Akins & Vasil, 1988). Hence, there is the potential for lower variability among plantlets derived via embryogenesis than among those produced via organogenesis. This may be important to plant propagators interested in maintaining fidelity among regenerants.

For at least one plantation species, oil palm (*Elaeis guineensis*), the large-scale production of somatic embryos to provide clonal material is already a reality. Since 1983, over 280 ha of experimental oil palm plantations have been established in Ivory Coast, using somatic embryo-derived plants. Recently, the first performance assessments of these plants showed that the clones provided the desired homogeneity for such characters as percentage of oil per mesocarp. Although a few abnormalities were observed in floral morphology, these apparently did not affect overall oil yield (Durand-Gasselin *et al.*, 1990).

Two features mentioned in this section which may be of special significance for commercialization purposes are the potential for somatic embryos to be grown in large volumes in continuous liquid culture and the potential for these embryos to be used as directly delivered propagules. Detailed discussions of these areas of research follow.

Scale-up potential

As noted earlier, an important advantage of many embryogenic cultures with respect to mass propagation is the ability not only to maintain the cultures in suspension, but to induce embryo development in this state as well. Over the past decade, researchers have developed several techniques to obtain large numbers of well-developed somatic embryos from suspension cultures. Although the primary aim of these early experiments was elucidation of the process involved in embryo development, the application of these techniques to mass propagation is now apparent. Fujimura & Komamine (1975) reported that high rates of synchronous embryo formation could be obtained in embryogenic carrot suspension cultures by thorough removal of auxin from the culture medium, combined with sieving the suspensions through nylon screens to obtain cell clusters of uniform size. Later this method was refined by subjecting the fractionated cell clusters obtained by sieving to density-gradient centrifugation in Ficoll solutions. This step was followed by repeated centrifugation of the heaviest fraction of cell clusters at low speed for very short periods in culture medium (Fujimura & Komamine, 1979). The density-gradient fractionation step effectively removed larger vacuolated cells from the suspension, resulting in synchronous embryo formation, with over 90% of the initial cell clusters forming embryos. The synchronous embryo populations were subsequently employed in studies designed to define the morphological and biochemical stages of embryo development (Fujimura & Komamine, 1980; Fujimura *et al.*, 1980, 1981). Nomura & Komamine (1985) extended the fractionation of embryogenic suspension cultures to single cells, which could be separated by density-gradient centrifugation in Percoll solutions. Embryogenic frequencies of cell populations obtained in this manner were as high as 30%, and the frequency could be increased to 90% by manual selection of single spherical cells.

Populations of suspension-cultured somatic embryos separated by developmental stage have also been purified without the necessity of density-gradient centrifugation. Giuliano *et al.* (1983) fractionated carrot PEMs on nylon sieves, cultured the PEMs in basal medium for 6–8 days and fractionated the developing embryos on a second set of sieves. Differential sedimentation of early embryo stages and undifferentiated cells in the liquid medium following fractionation made it possible to further purify the embryos.

While fractionation using sieves and/or density-gradient centrifugation has been sufficient to produce synchronous populations of embryos in carrot, there have not been corresponding reports of similar success with embryogenic suspensions of other species. It is likely that other systems require different physical or chemical conditions to promote synchrony. As previously discussed in the section on plantlet production from somatic embryos, Ammirato (1974) showed that 10^{-7} M ABA promoted the production of relatively synchronous populations of mature somatic embryos of caraway (*Carum carvi*), preventing precocious germination and the production of multiple embryo clusters. Later, when he reported that the same level of ABA had a similar effect on suspension-cultured carrot somatic embryos (Ammirato, 1983), he proposed that regulation of embryo maturation by ABA might be used to facilitate large-scale batch cultures of somatic embryos for applied purposes. Nadel *et al.* (1990) combined ABA and sieving on metal mesh screens to obtain synchronous populations of celery (*Apium graveolens*) somatic embryos. These authors stressed that, while the effect of ABA was significant in synchronization of the embryos, it did not improve singularization or maturation of the embryos.

The promising experimental results noted above, obtained using model systems of embryogenic cells and somatic embryos grown in liquid medium, have prompted researchers to test the application of engineering technology to these cultures. The ultimate goal of this research is the development of large-scale mechanized or automated culture systems for commercialization. Such systems have the potential to generate huge numbers of embryos with low labour inputs, decreasing the costs per propagule to the point where, depending on the crop, they may be competitive with sexual seeds. Furthermore, in combination with value-added inputs to the propagules made possible by hybridity or by genetic engineering, such large-scale clonal multiplication systems may someday be the preferred means of production for some specialized crops (see sections on economics and gene transfer).

Although the idea of applying scale-up technologies to somatic embryo production has been discussed for some time, few results from the actual testing of model systems have been reported to date. An early effort at large-scale culture of carrot cells in 20-litre carboys was reported by Backs-Hüsemann & Reinert (1970). However, few embryos were formed. Over the past 20 years, one system which has been repeatedly tested for use with plant cell cultures is the stirred-tank bioreactor (Wilson *et al.*, 1971; Martin,

1980; Kurz & Constabel, 1981), an apparatus originally designed for microbial fermentations. Most efforts to adapt this system to plant suspension cultures have had the goal of harvesting plant cell by-products (e.g. Ten Hoopen *et al.*, 1990), not propagules. Regardless of the goal, however, those employing these bioreactors with plant cells have struggled with the problem of high shear forces generated by stirring (Fowler, 1987; Leckie *et al.*, 1990). As a result, many researchers have concentrated on alternative bioreactor designs using means of agitation which produce less shear. Air-driven bioreactors are one such alternative, and have been shown to support growth of several plant cell types (Fowler, 1984). When adapted to the production of somatic embryos, air-lift bioreactors have for the most part given disappointing results. Stuart *et al.* (1987) reported that air-lift bioreactors could produce large numbers of alfalfa somatic embryos, but conversion rates of these embryos were disappointing. Walker (1989) tested relative embryo production by suspension cultures of Norway spruce (*Picea abies*) grown in air-lift bioreactors versus shaken flasks. Although conversion was not tested, somatic embryos were produced at a higher frequency in the shaken flasks.

The most efficient bioreactor designs for growth of plant cells are those allowing continuous culture, as opposed to batch culture (Styer, 1985). In a continuous-culture bioreactor, the tank is initially filled and inoculated as with a batch culture. Then, as the culture grows into the log phase, fresh medium is introduced at a low rate while the same volume of spent medium and cells is removed. Those attempting to adapt this bioreactor design to production of somatic embryos are faced with the problem of spent-medium removal without lowering the concentration of cells needed to perpetuate embryo production. Styer (1987) reported that use of a spin filter allowed removal of spent medium from the bioreactor without loss of suspension-cultured carrot cells, enabling maintenance of cell populations at the desired density. Furthermore, replacement of cell-proliferation medium with embryo-differentiation medium in this bioreactor produced a constant number of PEMs, each of which increased in cell number.

An automated system for large-scale commercial plant propagation, incorporating a bioreactor, has been developed by Plant Biotech Industries, Ltd. This system is apparently capable of handling somatic embryos, as well as other propagules such as microtubers and bulblets (Levin *et al.*, 1988; Levin & Vasil, 1989). A bioprocessor controls separation, sizing and distribution of propagules into a culture vessel, and the system is linked to an automated transplanting machine capable of transferring 8000 plantlets per hour to potting mix in greenhouse trays. Levin *et al.* (1988) claimed that compared with conventional tissue culture propagation techniques, their bioreactor-based system provided substantial savings in space, time and labour, as well as lower contamination rates, together with accurate monitoring and control of temperature, pH and gas concentrations.

Despite promising results reported in the last few years, bioreactor technology has yet to meet its apparent potential to produce somatic embryos capable of growing into plantlets on a scale that would make them economically competitive with true seedlings. Recently, Cazzulino *et al.* (1990) reported on a systematic method to improve the capability of bio-reactors to produce competent embryos, by applying a kinetic model of carrot somatic embryo development in suspension culture. The model was developed by monitoring substrate utilization, culture growth and embryo development over time in an embryogenic culture. The kinetic model will be employed to optimize bioreactor conditions for high-frequency production of mature somatic embryos capable of growing into plants.

It is probably only a matter of time until the performance of a number of embryogenic suspension culture systems will be improved to the point where they can be combined with bioreactor technology for economic, large-scale plantlet production. As mentioned earlier, however, it should be remembered that some embryogenic systems are much more amenable to suspension culture than others. For example, systems such as soybean develop in liquid medium not as true cell suspensions or PEMs, but as adherent clumps of proliferating globular embryos (Finer & Nagasawa, 1988), making procedures such as synchronization by sieving very difficult with current techniques. Clearly, if large-scale soybean somatic embryo production is to be accomplished, means of manipulating these cultures will have to be developed that are very different from those described above for carrot. Even some embryogenic systems which appear to respond well in suspension culture have not met expectations when tested for productivity of mature embryos capable of conversion to plantlets. For example, Stuart *et al.* (1987) reported that the conversion percentage of alfalfa somatic embryos grown on semisolid medium was three times higher than that of embryos grown in liquid medium in shaker flasks, and 30 times higher than that of embryos grown in bioreactors. Researchers working with alfalfa have resorted, therefore, to a modified protocol in which embryogenic cell clusters are removed from suspension before development into mature embryos. Large numbers of roughly synchronous somatic embryos are obtained by sieving embryogenic suspensions on nylon mesh and immediately placing the desired fraction of cell clusters, with the mesh, on to semisolid basal medium (McKersie *et al.*, 1989; Senaratna *et al.*, 1989, 1990).

As with alfalfa, embryogenic suspension cultures of yellow poplar (*Liriodendron tulipifera*) are very prolific with respect to embryo production, but embryos allowed to mature in liquid medium convert poorly compared with embryos developing from PEMs on semisolid medium. Therefore, as illustrated in Fig. 7.3, embryogenic suspensions of this species were fractionated by sieving and the desired fraction of PEMs was backwashed from the screen on to a disc of filter paper in a Buchner funnel. After the liquid medium was drawn off under low vacuum, the filter paper with PEMs was

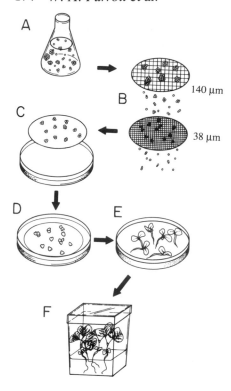

Fig. 7.3. Synchronization by sieving of yellow-poplar embryogenic suspension cultures. A) Embryogenic cells are grown in a liquid induction medium. B) Sieving the suspension through stainless steel mesh isolates PEMs, which are retained on the 38 μm sieve. C) The isolated PEMs are placed on filter paper and plated on semisolid basal medium. D) Mature embryos develop synchronously over the course of 12–14 days. E) Transfer of the mature somatic embryos to basal medium without filter paper permits germination. F) Plantlets are ready to transfer to a soil mix 6–8 weeks after transfer to plant development medium. (Artwork by J. Adang.)

placed on semisolid basal medium. This protocol of fractionation and plating, combining the high proliferation of the embryogenic suspension culture with the favourable maturation conditions of semisolid medium, resulted in production of hundreds of roughly synchronous, well-formed embryos within a few weeks (Fig. 7.4). Over 70% of the embryos obtained in this manner were capable of converting to plantlets (Merkle & Wiecko, 1990).

As illustrated by the examples above, relatively large-scale production of somatic embryos from suspension cultures is possible, even without a complete understanding of the factors controlling embryo development in liquid medium. In alfalfa and yellow poplar, problems arose from a differential ability of somatic embryos to mature correctly in liquid medium as compared with semisolid medium. The immediate solution was simply to avoid having maturation occur in liquid medium. Certainly for experimental purposes, such systems can produce adequate numbers of embryos. For mass production of propagules, however, economic application of embryogenic suspension cultures appears to depend on the ability to combine these systems with bioreactor technology and still achieve efficient production of competent embryos. If all the steps of somatic embryo proliferation, development and maturation are to be accomplished in the bioreactor, the conditions controlling these stages in liquid culture must be defined.

Delivery systems

Encapsulation of embryos

In many instances, the only major differences between somatic embryos and their zygotic counterparts arc the lack of storage reserves and the absence of a seed coat. Seed storage reserves are normally present either in the form of endosperm (megagametophyte in the case of conifer embryos) or storage cotyledons. For somatic embryos, endosperm and megagametophyte are

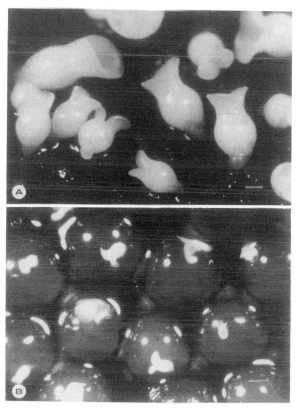

Fig. 7.4. Synchronized somatic embryos and their use as synthetic seeds. A) Mature somatic embryos of yellow poplar two weeks after sieving and plating on filter paper. B) Calcium alginate-encapsulated somatic embryos of yellow poplar. (Bar = 500 μm.)

absent, and, even in species where reserves are stored in the cotyledons, these organs do not generally achieve normal storage capacity in culture. Because of these missing seed components, additional protection and nutrition must be supplied in order for somatic embryos to be used as directly delivered propagules. Systems are needed which are capable of protecting the embryos until they are delivered to the desired growing locations, keeping them viable until conditions are favourable for plantlet survival, and supplying them with the nutrients and other compounds needed to promote early growth. The similarity of somatic embryos to zygotic embryos suggests a number of ways in which delivery systems may be developed to handle them. Basically three delivery methods have been tested over the past eight years: encapsulation to produce artificial seeds, fluid drilling, and use of desiccated, uncoated somatic embryos. Artificial seeds tested to date can be further classified into two subgroups: desiccated somatic embryos encapsulated in a water-soluble resin and hydrated somatic embryos encapsulated in hydrated gels.

The concept of a synthetic seed made by encapsulating individual somatic embryos was first presented by Murashige (1978), but the required properties of a synthetic seed coat were first clearly defined by Redenbaugh *et al.* (1986, 1987a). According to these authors, the encapsulation material must both protect the embryo, being sufficiently durable to withstand handling during transport and planting, and at the same time allow for germination and conversion. It should also be able to hold and deliver nutrients, other chemical factors and even micro-organisms to promote germination and early plant growth. In addition, if a synthetic seed is to be used by current growers, it must be sowable using existing greenhouse and farm machinery.

Encapsulation of somatic embryos was first reported by Kitto & Janick (1985a, b), who applied synthetic seed coats to clumps of carrot somatic embryos by mixing equal volumes of embryo suspension and a 5% solution of polyethylene oxide (Polyox WSR N-750), a water-soluble resin, which subsequently dried to form polyembryonic wafers. Embryo 'hardening' treatments, including treatments with 12% sucrose or 10^{-6} M ABA, chilling and high inoculum density, increased survival of encapsulated and desiccated somatic embryos to as high as 58%. Kim & Janick (1989) also reported survival for up to 9 days of desiccated somatic embryo of celery encapsulated in Polyox, while non-encapsulated embryos did not survive beyond 4 days of desiccation. The authors hypothesized that survival enhancement by Polyox encapsulation may have been due to slowed drying brought about by high relative humidity inside the wafers. Although Polyox-encapsulated, desiccated somatic embryos were demonstrated to produce plantlets, conversion frequencies were not reported for either the carrot or celery embryos.

The first work on development of artificial seeds using individually encapsulated somatic embryos was reported by Redenbaugh *et al.* (1984,

1986, 1987a, b), who encapsulated somatic embryos of alfalfa, celery and cauliflower (*Brassica oleracea*) by mixing them with sodium alginate and dropping them into a solution of calcium chloride to form calcium alginate beads via an ion exchange reaction (Fig. 7.4). Other hydrogels were tested as possible encapsulating agents, but were not found to be as useful. Redenbaugh *et al.* (1986) found that the encapsulated embryos germinated and formed seedling-quality plants *in vitro* at the rate of 29% for randomly picked alfalfa somatic embryos and 55% for celery somatic embryos that had been selected for high quality. Encapsulated alfalfa and celery embryos sown in sand or peat plugs converted at 7% and 10%, respectively. Various problems with encapsulation in alginate were identified, including rapid leaching of water-soluble nutrients out of the capsules, obstruction of root and shoot emergence, rapid drying of capsules when exposed to air and tackiness of the capsules, which caused difficulty with handling and machine planting (Redenbaugh *et al.*, 1987a). Some of these problems were resolved. For example, root and shoot penetration of the capsule was facilitated by controlling the alginate concentration and complexing time, and a coating compound prevented the capsules from drying too rapidly and lowered the tackiness of the capsules to the point where they could be machine planted. The authors emphasized, however, in this and later studies (Redenbaugh *et al.*, 1987b; Fujii *et al.*, 1989; Redenbaugh & Ruzin, 1989; Redenbaugh, 1990), that the success of an encapsulation system is determined not so much by the ability to successfully encapsulate the embryos, but by the quality of the embryos themselves, since, with or without encapsulation, conversion to plantlets is controlled by this variable. Lutz *et al.* (1985), who tested encapsulation of individual carrot somatic embryos in a gel matrix, also concluded that embryo quality was the limiting factor in developing a delivery system, although they did not report conversion frequencies. Alginate encapsulation has also been tested with loblolly pine (*Pinus taeda*) somatic embryos, which survived encapsulation and storage at 4°C for 4 months, but did not convert to plantlets (Gupta & Durzan, 1987).

In recent work, Fujii *et al.* (1989) employed selected embryos with *in vitro* conversion frequencies of 70–90% to test the effects of embryo maturation treatments on conversion frequency of alfalfa artificial seeds on potting mix. Embryo maturation with ABA at 1–5 µM resulted in greenhouse conversion rates of up to 64% when a humidity-tent watering system was used to prevent soil surface drying. Attempts to replace ABA with 8% mannitol during maturation did not result in similarly high conversion rates.

Fluid drilling

Another potential delivery system for somatic embryos is fluid drilling, in which embryos are suspended in a viscous carrier gel which is extruded into the soil. Originally, this technique was applied as a means of sowing pregerminated seeds in order to improve seedling emergence and the uniformity of the crop stand (Currah *et al.*, 1974). The primary problem with application of fluid drilling to carrot somatic embryos was found to be that embryos at a developmental stage early enough to bulk-handle for fluid drilling would not continue development and convert to autotrophic plantlets without the addition of sucrose to the gel. The sucrose level necessary to permit conversion, however, would also be expected to promote rapid growth of contaminating micro-organisms in this non-aseptic system (Drew, 1979; Lutz *et al.*, 1985). Baker (1985) conditioned carrot somatic embryos with sucrose and ABA prior to sowing them into gel supplemented with growth regulators, sucrose and nutrients, on top of soil mix in the greenhouse. Embryo survival, however, was only 4% and no embryos converted to plantlets. Schultheis *et al.* (1986) tested different gels supplemented with nutrients, vitamins and sucrose for their effect on survival and growth of sweet potato somatic embryos. While embryos did not survive beyond 6 days in some gels, one product (Natrosol 250HHR) allowed embryo growth. Schultheis & Cantliffe (1988) subsequently found that 20–25% of sweet potato somatic embryos formed plants when placed in the same gel containing MS medium supplemented with sucrose, maltose or glucose.

One concept for the generation of artificial seeds from somatic embryos does not rely on any coating or matrix to maintain embryo viability during storage or delivery. Instead, it makes use of the fact that zygotic embryos typically cease growth, lose water and become dormant or quiescent until a signal for germination is received. As discussed earlier, this natural desiccation pattern can be used to greatly enhance conversion of somatic embryos of a number of species. In addition to promoting conversion, dehydration is an alternative method for inducing somatic embryo quiescence for storage and handling (Gray, 1987). Gray (1989) also emphasized the potential role of uncoated dehydrated somatic embryos in germplasm conservation, where the primary goal is long-term storage of relatively small numbers of propagules. However, as development of propagules for commercial planting was not an objective of his study, he did not test conversion of his dehydrated grape somatic embryos under greenhouse conditions or discuss the encapsulation that presumably would be necessary for large-scale sowing.

Economics of mass propagation via somatic embryogenesis

There is general agreement that somatic embryogenesis and in particular its application as artificial seeds offer the best potential of any tissue culture regeneration system for economic mass propagation of many crop species. Most plants produced by current organogenic *in vitro* methods cannot profitably be sold for less than about 25 cents per plant, and 40–60% of this production cost is accounted for by labour (Sluis & Walker, 1985). It is also generally accepted, however, that much higher production efficiency must be achieved before somatic embryos can compete with seeds of most species on a commercial basis. For example, Redenbaugh *et al.* (1987a) estimated the cost of alfalfa artificial seeds to be 0.026 cents each, including labour, materials and overhead, compared with 0.00066 cents per true alfalfa seed at that time. Nevertheless, if artificial seeds with modest unit costs can be developed for crops such as hybrid vegetables and flowers, they may be competitive with the relatively expensive natural seeds of these plants. Furthermore, value-added components such as clonal uniformity and epistatic interactions that are not normally heritable through sexual propagation may justify substantially higher costs for somatic embryos than for natural seeds of the same species (Redenbaugh *et al.,* 1987a).

Use in gene transfer systems

Prerequisites for a gene transfer system

Several systems have been devised that are capable of introducing foreign DNA into plant cells. These include microinjection of protoplasts (Crossway *et al.,* 1986; Mathias, 1987) and electroporation of DNA into protoplasts (Horn *et al.,* 1988; Toriyama *et al.,* 1988; Chupeau *et al.,* 1989; Chapter 4), two processes that are limited by the range of species for which plants can be efficiently regenerated from protoplasts. The domestication of the crown-gall pathogen, *Agrobacterium tumefaciens*, coupled with regeneration from leaf tissues (Horsch *et al.,* 1985), has made possible the transformation of those crops that are susceptible to *Agrobacterium* and regenerate via organogenesis, especially solanaceous or cruciferous species (Chapter 3). In such a system, *Agrobacterium* inserts into individual plant cells a gene that confers resistance to an antibiotic or herbicide, along with a gene of economic or academic importance. When the corresponding antibiotic or herbicide is incorporated into a medium that would otherwise permit callus formation, only those cells that have acquired the resistance gene from *Agrobacterium* are able to divide, effectively sorting non-transformed from transformed cells. Plants regenerated from transformed cells are themselves transgenic. Many plants have been transformed using this method, but most agriculturally important plants are not included in the list (Gasser & Fraley, 1989; Table 3.2).

Implicit in any gene transfer system is the requirement that, regardless of the methods by which DNA is inserted into a plant cell, the recipient cell must be totipotent. In *Agrobacterium*-mediated transformation systems, not only must the recipient cell be totipotent, but it must also be susceptible to *Agrobacterium*, a trait that varies among species and among tissues within a plant (Matthysse & Gurlitz, 1982). Lack of totipotency associated with protoplasts of several species is perhaps the factor most limiting to their use in transformation via microinjection or electroporation. For this reason, much attention has been given to the potential for isolation and culture of protoplasts from embryogenic cells. Theoretically, since the cells from which the protoplasts were isolated are known to be regenerable, the protoplasts would be expected to retain this property and yield cultures capable of forming whole plants (Shillito et al., 1989). Thus, if embryogenic protoplasts are used in gene transfer protocols such as electroporation, microinjection or polyethylene glycol-mediated DNA uptake, there should be a high likelihood of recovering transgenic plants. In three groups of plants, embryogenic cultures have proven to be especially valuable in providing a source of totipotent protoplasts. These are the graminaceous species, citrus species, and forest trees, especially coniferous species.

Among graminaceous species, neither mesophyll-derived protoplasts nor protoplasts derived from non-morphogenic cell suspensions were found to be capable of regenerating whole plants (Vasil, 1985). However, when Vasil & Vasil (1980) isolated and cultured protoplasts from embryogenic cultures of pearl millet (*Pennisetum glaucum*), they gave rise to cell clusters from which embryos and plantlets were regenerated. Since this report, embryogenic suspension cultures have been used as sources of regenerable protoplasts of guinea-grass (*Panicum maximum*; Lu et al., 1983a), Napier-grass (Vasil et al., 1983), rice (Fujimura et al., 1985; Abdullah et al., 1986; Toriyama et al., 1986; Yamada et al., 1986; Kyozuka et al., 1987), sugarcane (*Saccharum officinarum*; Srinivasan & Vasil, 1986), perennial ryegrass (*Lolium perenne*; Dalton, 1988), tall fescue (*Festuca arundinacea*; Dalton, 1988) and orchardgrass (Horn et al., 1988). Protoplasts derived from embryogenic cultures of maize were used to regenerate somatic embryos (Kamo et al., 1988; Vasil & Vasil, 1987) and plantlets which turned out to be sterile (Rhodes et al., 1988). More recently, Shillito et al. (1989) and Prioli & Sondahl (1989) reported regeneration of fertile plants of élite inbred maize lines from protoplasts derived from embryogenic suspensions. Already, embryogenic suspension-derived protoplasts have proven useful for production of transgenic plants of graminaceous species. Transgenic (but sterile) maize plants were obtained following electroporation of embryogenic suspension-derived protoplasts (Rhodes et al., 1988) and protoplasts from embryo-derived callus of rice (Shimamoto et al., 1989). In these two examples, the tissues used to obtain protoplasts could be classified as consisting of IEDCs and PEDCs, respectively. It is clear that the embryogenic state of such cells facilitates the recovery of somatic embryos from cell colonies derived from protoplasts.

Among citrus species, Vardi et al. (1982) first reported regeneration of plantlets from protoplasts isolated from embryogenic nucellar callus of orange, mandarin and grapefruit cultivars. Since this report, plant regeneration has been achieved from embryogenic suspension-derived protoplasts of a number of citrus species and cultivars (Vardi & Galun, 1988). Although we are not aware of reports of the use of these protoplast cultures for regeneration of transgenic plants, they have made possible the production of interspecific and even intergeneric somatic hybrids via protoplast fusion (Ohgawara *et al.*, 1985; Grosser *et al.*, 1988a, b, 1989; Kobayashi & Ohgawara, 1988; Kobayashi *et al.*, 1988).

As for graminaceous species, the key to successful regeneration from protoplasts of coniferous species proved to be the isolation of protoplasts from embryogenic callus and suspension cultures. Given that the first reports of somatic embryogenesis in conifers did not appear until 1985 (Hakman *et al.*, 1985; Nagmani & Bonga, 1985), the application of embryogenic cultures to protoplast studies has had a large impact in a short period of time. To date, regeneration of somatic embryos from embryogenic suspension-derived protoplasts has been reported for loblolly pine (*Pinus taeda*; Gupta & Durzan, 1987), white spruce (*Picea glauca*; Attree *et al.*, 1987, 1989a; Bekkaoui *et al.*, 1987), Douglas-fir (*Pseudotsuga menziesii*; Gupta *et al.*, 1988), European silver fir (*Abies alba*; Lang & Kohlenbach, 1989) and black spruce (*Picea mariana*; Tautorus *et al.*, 1990). In addition, plantlet production from protoplast-derived somatic embryos has been reported for white spruce (Attree *et al.*, 1989b) and hybrid larch (*Larix×eurolepis*; Klimaszewska, 1989).

Among hardwood forest tree species, another group for which successful protoplast culture has lagged, embryogenic suspension cultures have also been shown to be a valuable source of highly regenerable protoplasts. Two hardwood species for which embryogenic suspensions have been used as sources of protoplasts capable of yielding somatic embryos and plantlets are sandalwood (*Santalum album*; Rao & Ozias-Akins, 1985) and yellow-poplar (Merkle & Sommer, 1987).

Despite the success with which embryogenic suspension cultures have been applied to protoplast culture in forest tree species, there have been no reports to date of their use for stable integration of foreign DNA into somatic embryos or plantlets. There have, however, been reports of transient expression of DNA electroporated into embryogenic suspension-derived protoplasts of loblolly pine (Gupta *et al.*, 1988), Douglas fir (Gupta *et al.*, 1988) and white spruce (Bekkaoui *et al.*, 1987). Those species whose protoplasts are totipotent are also amenable to transformation technologies less tedious than regeneration from protoplasts.

Gene transfer in indirect somatic embryogenesis systems

Regeneration via indirect somatic embryogenesis, characterized by the presence of a callus phase prior to the formation of somatic embryos, is as amenable to *Agrobacterium*-mediated transformation as an organogenic system. As described previously, transformed cells are selectively multiplied to form a callus, but an embryo, not a shoot, is ultimately formed by the callus. Plants that have been transformed via indirect somatic embryogenesis include alfalfa (Deak *et al.*, 1986; Shahin *et al.*, 1986; Chabaud *et al.*, 1988; D'Halluin *et al.*, 1990); cotton, *Gossypium hirsutum* (Firoozabady *et al.*, 1987; Umbeck *et al.*, 1987); carrot (Thomas *et al.*, 1989); eggplant, *Solanum melongena* (Filippone & Lurquin, 1989); and sunflower, *Helianthus annuus* (Everett *et al.*, 1987).

A related technique involves the use of *Agrobacterium rhizogenes*, the causal agent of hairy root disease, instead of *A. tumefaciens*. In this instance, the callus phase is replaced by proliferation of roots from transformed cells, and somatic embryos develop directly from transformed root tissues. This technique has been successfully used with both cucumber, *Cucumis sativus* (Trulson *et al.*, 1986), and alfalfa (Spano *et al.*, 1987; Sukhapinda *et al.*, 1987). There is a drawback in that growth regulators produced by the *A. rhizogenes*-derived genes responsible for rhizogenesis can alter the morphology of regenerated plants (see Chapter 3).

A compelling advantage of adapting *Agrobacterium*-mediated transformation to indirect embryogenic systems is that, once obtained, transformed calli or roots can be propagated continuously to produce large numbers of embryos. For this method to be effective, however, not only must a plant species be amenable to *Agrobacterium*, but it must also have the ability to form somatic embryos from callus. At present, only a limited number of species fulfil both of these requirements. As discussed previously in this chapter, even within a species, genotypes differ in their ability to form somatic embryos. Consequently, preselection for an embryogenic genotype may be necessary for the efficient recovery of transformants.

Gene transfer in direct embryogenic systems

A large number of species are amenable to regeneration via direct somatic embryogenesis, especially from explants derived from immature zygotic embryos. Some species produce somatic embryos by this technique alone. The combination of transformation with direct embryogenesis therefore represents the greatest potential for transformation of a number of important species.

In principle, the recovery of transgenic plants from a species that regenerates via direct somatic embryogenesis should be possible as long as three conditions are met. Firstly, somatic embryos must have a unicellular origin. Otherwise, embryos originating from multiple cells will be chimeric in nature, with transformed and non-transformed sectors arising from transformed and non-transformed cells within the original cell mass. Secondly, large numbers of single cells must regenerate; and, thirdly, the transformation system must transform large numbers of cells. The last two conditions are necessary to ensure a reasonable probability that a given cell will be transformed and will be recovered as a somatic embryo. For example, if in a given explant only 1 in 10^5 cells become embryogenic, and the available transformation technique transforms only 1 in 10^6 cells, the probability of transforming a cell that will become embryogenic is the product of the two, or 1 in 10^{11}.

In practice, the two critical properties of regeneration (from single cells at high frequency) do not commonly occur together. Nevertheless, an example of such a system is found in oilseed rape, individual microspores of which can be induced to become embryogenic. Microspores and pro-embryos derived from microspores were exposed to *Agrobacterium*, and transgenic plants subsequently recovered (Pechan, 1989). Embryos were originally believed not to have the necessary cell wall sites for infective attachment of *Agrobacterium* (Lippincott & Lippincott, 1978; Matthysse & Gurlitz, 1982; Sequeira, 1984), but the successful transformation of embryogenic cultures shows that this is not an absolute limitation.

Repetitive embryogenesis and plant transformation

An obvious limitation to plant transformation via direct somatic embryo-genesis is that somatic embryos arising directly from an explant can have multicellular origins (reviewed by Williams & Maheswaran, 1986), while available gene transfer techniques transform only individual cells. Further-more, the lack of a callus phase precludes the opportunity to preferentially propagate transformed cells before formation of somatic embryos. The result is the development of chimeric somatic embryos comprising trans-formed and non-transformed sectors.

The process of repetitive somatic embryogenesis circumvents the prob-lem of chimeric embryos, by allowing recovery of completely transformed secondary embryos from transformed sectors within a primary somatic embryo. Repetitive cycles of direct embryogenesis effectively substitute for the callus phase found in indirect embryogenic systems, and make repetitive embryogenesis a powerful method by which to obtain a wide range of transgenic plants.

As embryos formed during repetitive embryogenesis can originate from single cells (e.g. Haccius, 1977; Polito *et al.*, 1989), the number of trans-formed cells in the original embryo need not be high. This principle was first used for *Agrobacterium*-mediated transformation of embryogenic suspen-sions of carrot (Scott & Draper, 1987), and was later applied to embryogenic cultures of English walnut, *Juglans regia* (McGranahan *et al.*, 1988, 1990). Even if chimeric embryos are still recovered from the first cycle of repetitive embryogenesis, continued cycling in the presence of a selective agent eventually results in embryos consisting entirely of transformed cells. Since the transformation of walnut was first reported, this technique has also been used to transform secondary embryogenic cultures of oilseed rape originally derived from microspores (Swanson & Erickson, 1989).

Microprojectile transformation of repetitively embryogenic systems

As described above, *Agrobacterium*-mediated transformation combined with repetitive direct embryogenesis may be an efficient method by which to obtain transgenic plants, as long as the species or genotype in question is susceptible to *Agrobacterium*. This requirement, however, immediately excludes most monocotyledonous and some dicotyledonous plants of

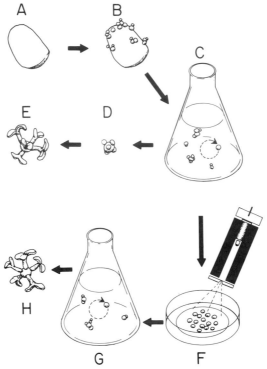

Fig. 7.5. Embryogenic suspensions of soybean, and their use in genetic transformation (based on Finer & Nagasawa, 1988; McMullen & Finer, 1990). A) Immature cotyledons, with the embryonic axes removed, are exposed to 40 mg/l of 2,4-dichlorophenoxyacetic acid (2,4-D) for 30 days. B) Somatic embryos form but their development is arrested at the globular stage. Globular-stage embryos give rise to secondary somatic embryos. C) Somatic embryos are placed in liquid culture, where the globular-stage embryos are propagated indefinitely. D) The repetitive cycle is broken by removing the embryogenic clusters from liquid culture and placing them on hormone-free basal medium. E) Globular embryos develop to the cotyledonary stage, at which time they can be separated, matured, desiccated and germinated to give plantlets. F) Embryogenic clusters growing in suspension can be subjected to microprojectile bombardment. G) Following bombardment, the embryogenic clumps are returned to liquid culture and grown in the presence of a selective agent. H) Recovery of transgenic embryos. (Artwork by J. Adang.)

agricultural importance (Chapter 3). Limitations imposed on plant transformation by the host range of *Agrobacterium* are being overcome by use of microprojectile bombardment, also known as particle-gun or gene-gun systems, to propel DNA-coated heavy metal particles into plant cells (Klein *et al.*, 1987; Sanford, 1988; Chapter 5). When applied to indirect or repetitively embryogenic systems, the technique has the advantage that, once transformed embryogenic cell lines are obtained, the capacity to produce somatic embryos is virtually unlimited.

The first to report the use of microprojectile bombardment for the transformation of an embryogenic suspension were Finer & McMullen (1990). Proliferation of a cotton embryogenic suspension after bombardment resulted in the production of transformed embryos, from which plants were subsequently obtained. McMullen & Finer (1990) later extended this technique to include the transformation of embryogenic suspensions of soybean (Fig. 7.5). The transformation of soybean is especially significant, as this species has been extremely difficult to transform using *Agrobacterium*. Although transformation of soybean following bombardment of meristematic tissues has also been reported (Christou *et al.*, 1989), this technique produces chimeric plants and lacks the potential for mass propagation associated with embryogenic systems. More recently, the bombardment of embryogenic cultures has been used to recover fertile transgenic maize plants (Fromm *et al.*, 1990; Gordon-Kamm *et al.*, 1990), using herbicide resistance as a selectable marker. Transgenic plants of the forest tree, yellow poplar, have also been obtained following microprojectile bombardment of PEMs isolated from embryogenic suspensions (cover photograph; H. D. Wilde, S. A. Merkle & R. B. Meagher, submitted).

Conclusions

Although present systems for commercial micropropagation and experimental transformation are based almost exclusively on shoot organogenesis, there are definite indications that somatic embryogenesis will contribute improved methods for future biotechnological applications. The process of somatic embryogenesis offers a means for propagating essentially limitless numbers of discrete propagules, functionally similar to seed embryos, in large-scale or continuous cultures. At present, therefore, it represents the greatest potential for scaling up volume and reducing the labour costs of mass propagation. The most immediate problems are to define the conditions required for normal development and maturation of embryos of a variety of species in liquid culture, and to develop more reliable methods for direct delivery of propagules to the greenhouse or field.

Looking further ahead, somatic embryogenesis also offers potential for efficient production of transgenic populations in a range of plant species.

Through somatic embryogenesis it should be possible to maximize the numbers of transformed cells from which regenerants can ultimately be produced, and by repetitive embryogenesis each independent transformation event should be recoverable as one to many whole, non-chimeric plantlets. When considering transformation with defined foreign genes as an option in plant improvement, it is still important to consider numbers. At present we are limited to DNA delivery systems in which the integration of foreign DNA is random within the genome, the number of integrated copies of the gene is not controlled and rearrangements during integration are relatively frequent. For these reasons there may be substantial variation in expression of introduced genes among individuals within populations of transformants. With present technology, therefore, it is important to produce numbers of transgenic plants from which desirable types can be selected. This situation will prevail at least until the development of reliable site-specific methods for integration of foreign genes.

Over the last five years, there has been conspicuous progress towards better control of somatic embryogenesis in a wide range of plants, including forest trees and important crops such as maize and soybean. Already the power of biolistic technology combined with repetitive embryogenesis has been demonstrated in several commercially important species. Progress towards commercial applications, although slower than originally hoped, is clearly positive. Where propagation costs are important, where large numbers of regenerants are required, and especially where these are transgenic, somatic embryogenesis will, in all probability, provide the basis for the technology of choice.

References

Abdullah, R., Cocking, E. C. & Thompson, J. A. (1986) Efficient plant regeneration from rice protoplasts through somatic embryogenesis. *Bio/Technology* 4, 1087–90.

Abe, T. & Futsuhara, Y. (1986) Genotypic variability for callus formation and plant regeneration in rice (*Oryza sativa* L.). *Theoretical and Applied Genetics* 72, 3–10.

Ammirato, P. V. (1974) The effects of abscisic acid on the development of somatic embryos from cells of caraway (*Carum carvi* L.). *Botanical Gazette* 135, 328–37.

Ammirato, P. V. (1983) The regulation of somatic embryo development in plant cell cultures: suspension culture techniques and hormone requirements. *Bio/Technology* 1, 68–74.

Aslam, F. N., MacDonald, M. V., Loudon, P. & Ingram, D. S. (1990) Rapid-cycling *Brassica* species: inbreeding and selection of *B. campestris* for anther culture ability. *Annals of Botany* 65, 557–66.

Attree, S. M., Bekkaoui, F., Dunstan, D. I. & Fowke, L. C. (1987) Regeneration of somatic embryos from protoplasts isolated from an embryogenic suspension of white spruce (*Picea glauca*). *Plant Cell Reports* 6, 480–3.

Attree, S. M., Dunstan, D. I. & Fowke, L. C. (1989a) Initiation of embryogenic callus and suspension cultures, and improved embryo regeneration from protoplasts of white spruce (*Picea glauca*). *Canadian Journal of Botany* 67, 1790–5.

Attree, S. M., Dunstan, D. I. & Fowke, L. C. (1989b) Plantlet regeneration from embryogenic protoplasts of white spruce (*Picea glauca*). *Bio/Technology* 7, 1060–2.

Backs-Hüsemann, D. & Reinert, J. (1970) Embryobildung durch isolierte Einzelzellen aus Gewebekulturen von *Daucus carota*. *Protoplasma* 70, 49–60.

Baker, C. M. (1985) 'Synchronization and fluid sowing of carrot, *Daucus carota*, somatic embryos.' MS thesis, University of Florida, Gainesville, Florida.

Bartkowlak, E. (1982) Tissue culture of maize. 3. Plantlet regeneration from scutellar callus. *Genetica Polonica* 23, 207–14.

Beckert, M. & Qing, C. M. (1984) Results of a diallel trial and a breeding experiment for *in vitro* aptitude in maize. *Theoretical and Applied Genetics* 68, 247–51.

Bekkaoui, F., Saxena, P. K., Attree, S. M., Fowke, L. C. & Dunstan, D. I. (1987) The isolation and culture of protoplasts form an embryogenic suspension culture of *Picea glauca* (Moench) Voss. *Plant Cell Reports* 6, 476–9.

Bianchi, S., Flament, P. & Dattee, Y. (1988) Somatic embryogenesis and organogenesis in alfalfa; genotypic variation in regeneration ability. *Agronomie* 8, 121–6.

Bingham, E. T., Hurley, L. V., Kaatz, D. M. & Saunders, J. W. (1975) Breeding alfalfa which regenerates from callus tissue culture. *Crop Science* 15, 719–21.

Brawley, S. H., Wetherall, D. F. & Robinson, K. R. (1984) Electrical polarity in embryos of wild carrot precedes cotyledon differentiation. *Proceedings of the National Academy of Sciences USA* 81, 6064–7.

Broda, Z. (1984) Wegetatywna propagacja koniczyny czerwonej (*Trifolium pratense* L.) proprzez kultury tkankowe ze szczególnym uwzglednieniem genetycznego uwarunkowania zdolnosci do regeneracji z kalusa. *Roczniki Akademii Rolniczej w Poznaniu. Rozprawy naukowe* 140, 5–41.

Brown, D. C. W. & Atanassov, A. (1985) Role of genetic background in somatic embryogenesis in *Medicago*. *Plant Cell, Tissue and Organ Culture* 4, 111–22.

Buchheim, J. A., Colburn, S. M. & Ranch, J. P. (1989) Maturation of soybean somatic embryos and the transition to plantlet growth. *Plant Physiology* 89, 768–75.

Button, J., Kochba, J. & Bornman, C. H. (1974) Fine structure of and embryoid development from embryogenic ovular callus of 'Shamouti' orange (*Citrus sinensis* Osb.). *Journal of Experimental Botany* 25, 446–57.

Carman, J. G. (1988) Improved somatic embryogenesis in wheat by partial simulation of the in-ovulo oxygen, growth-regulator and desiccation environments. *Planta* 175, 417–24.

Carman, J. G. (1989) The *in ovulo* environment and its relevance to cloning wheat via somatic embryogenesis. *In Vitro Cell and Developmental Biology* 25, 1155–62.

Carnes, M. G. & Wright, M. S. (1988) Endogenous hormone levels of immature corn kernels of A188, Missouri-17, and Dekalb XL-12. *Plant Science* 57, 195–203.

Cazzulino, D. L., Pedersen, H., Chin, C. K. & Styer, D. (1990) Kinetics of carrot somatic embryo development in suspension culture. *Biotechnology and Bioengineering* 35, 781–6.

Chabaud, M., Passiatore, J. E., Cannon, F. & Buchanan-Wollaston, V. (1988) Parameters affecting the frequency of kanamycin resistant alfalfa obtained by *Agrobacterium tumefaciens* mediated transformation. *Plant Cell Reports* 7, 512–16.

Charmet, G. & Bernard, S. (1984) Diallel analysis of androgenetic plant production in hexaploid Triticale (*X. triticosecale,* Wittmack). *Theoretical and Applied Genetics* 69, 55–61.

Chée, R. P. & Cantliffe, D. J. (1989) Embryo development from discrete cell aggregates in *Ipomoea batatas* (L.) Lam. in response to structural polarity. *In Vitro Cell and Developmental Biology* 25, 757–60.

Chen, T. H. H. & Marowitch, J. (1987) Screening of *Medicago falcata* germplasm for *in vitro* regeneration. *Journal of Plant Physiology* 128, 271–7.

Chen, T. H. H., Marowitch, J. & Thompson, B. G. (1987) Genotypic effects on somatic embryogenesis and plant regeneration from callus cultures of alfalfa. *Plant Cell, Tissue and Organ Culture* 8, 73–81.

Christou, P., Swain, W. F., Yang, N. S. & McCabe, D. E.(1989) Inheritance and expression of foreign genes in transgenic soybean plants. *Proceedings of the National Academy of Sciences USA* 86, 7500–4.

Chupeau, M.-C., Bellini, C., Guerche, P., Maisonneuve, G., Vastra, G. & Chupeau, Y. (1989) Transgenic plants of lettuce (*Lactuca sativa*) obtained through electroporation of protoplasts. *Bio/Technology* 7, 503–8.

Crossway, A., Hauptli, H., Houck, C. M., Irvine, J. M., Oakes, J. V. & Perani, L. A. (1986) Micromanipulation techniques in plant biotechnology. *BioTechniques* 4, 320–34.

Currah, I. E., Gray, D. & Thomas, T. H. (1974) The sowing of germinating vegetable seeds using a fluid drill. *Annals of Applied Biology* 76, 311–18.

Dalton, S. J. (1988) Plant regeneration from cell suspension protoplasts of *Festuca arundinacea* Schreb. (tall fescue) and *Lolium perenne* L. (perennial ryegrass). *Journal of Plant Physiology* 132, 170–5.

Deak, M., Kiss, G. B., Koncz, C. & Dudits, D. (1986) Transformation of *Medicago* by *Agrobacterium* mediated gene transfer. *Plant Cell Reports* 5, 97–100.

D'Halluin, K., Botterman, J. & De Greef, W. (1990) Engineering of herbicide-resistant alfalfa and evaluation under field conditions. *Crop Science* 30, 866–71.

Dijak, M., Smith, D. L., Wilson, T. J. & Brown, D. C. W. (1986) Stimulation of direct embryogenesis from mesophyll protoplasts of *Medicago sativa*. *Plant Cell Reports* 5, 468–70.

Drew, R. L. K. (1979) The development of carrot (*Daucus carota* L.) embryoids (derived from cell suspension culture) into plantlets on a sugar-free basal medium. *Horticultural Research* 19, 79–84.

Drew, R. L. K. (1980) A cheap, simple apparatus for growing large batches of plant tissue in submerged liquid culture. *Plant Science Letters* 17, 227–36.

Duncan, D. R., Williams, M. E., Zehr, B. E. & Widholm, J. M. (1985) The production of callus capable of plant regeneration from immature embryos of numerous *Zea mays* genotypes. *Planta* 165, 322–32.

Durand-Gasselin, T., Le Guen, V., Konan, K. & Duval, Y.(1990) Plantations en Côte-d'Ivoire de palmiers à huile (*Elaeis guineensis* Jacq.), obtenues par culture *in vitro.* Premiers résultats. *Oléagineux* 45, 1–11.

Evans, D. A., Sharp, W. R. & Flick, C. E. (1981) Growth and behaviour of cell cultures: embryogenesis and organogenesis. In: Thorpe, T. A. (ed.), *Plant Tissue Culture: Methods and Applications in Agriculture. Proceedings of UNESCO Symposium, Sao Paulo, Brazil.* Academic Press, New York, pp. 45–113.

Everett, N. P., Robinson, K. E. P. & Mascarenhas, D. (1987) Genetic engineering of sunflower (*Helianthus annuus* L.). *Bio/Technology* 5, 1201–4.

Filippone, E. & Lurquin, P. F. (1989) Stable transformation of eggplant (*Solanum melongena* L.) by cocultivation of tissues with *Agrobacterium tumefaciens* carrying a binary plasmid vector. *Plant Cell Reports* 8, 370–3.

Finer, J. J. (1988) Apical proliferation of embryogenic tissue of soybean [*Glycine max* (L.) Merrill]. *Plant Cell Reports* 7, 238–41.

Finer, J. J. & McMullen, M. D. (1990) Transformation of cotton (*Gossypium hirsutum* L.) via particle bombardment. *Plant Cell Reports* 8, 586–9.

Finer, J. J. & Nagasawa, A. (1988) Development of an embryogenic suspension culture of soybean (*Glycine max* Merrill.). *Plant Cell, Tissue and Organ Culture* 15, 125–36.

Firoozabady, E,. DeBoer, D. L., Merlo, D. J., Halk, E. L., Amerson, L. N., Rashka, K. E. & Murray, E. E. (1987) Transformation of cotton (*Gossypium hirsutum* L.) by *Agrobacterium tumefaciens* and regeneration of transgenic plants. *Plant Molecular Biology* 10, 105–16.

Fowler, M. W. (1984) Large-scale cultures of cells in suspension. In: Vasil, I. K. (ed.), *Cell Culture and Somatic Cell Genetics of Plants,* Vol. 1. Academic Press, New York, pp. 167–74.

Fowler, M. W. (1987) Process systems and approaches for large scale plant cell culture. In: Green, C. E., Somers, D. A., Hackett, W. P. & Biesboer, D. D., (eds), *Plant Tissue and Cell Culture* Alan R. Liss, Inc. New York, pp. 459–71.

Fromm, M. E., Morrish, F., Armstrong, C., Williams, R., Thomas, J. & Klein, T. M. (1990) Inheritance and expression of chimeric genes in the progeny of transgenic maize plants. *Bio/Technology* 8, 833–9.

Fujii, J. A. A., Slade, D. & Redenbaugh, K. (1989) Maturation and greenhouse planting of alfalfa artificial seeds. *In Vitro Cell and Developmental Biology* 25, 1179–82.

Fujimura, T. & Komamine, A. (1975) Effects of various growth regulators on the embryogenesis in a carrot cell suspension culture. *Plant Science Letters* 5, 359–64.

Fujimura, T. & Komamine, A. (1979) Synchronization of somatic embryogenesis in a carrot cell suspension culture. *Plant Physiology* 64, 162–4.

Fujimura, T. & Komamine, A. (1980) The serial observation of embryogenesis in a carrot cell suspension culture. *New Phytologist* 86, 213–18.

Fujimura, T., Komamine, A. & Matsumoto, H. (1980) Aspects of DNA, RNA and protein synthesis during somatic embryogenesis in a carrot cell suspension culture. *Physiologia Plantarum* 49, 255–60.

Fujimura, T., Komamine, A. & Matsumoto, H. (1981) Changes in chromosomal proteins during early stages of synchronized embryogenesis in a carrot cell suspension culture. *Zeitschrift für Pflanzenphysiologie* 102, 293–8.

Fujimura, T., Sakurai, M., Agaki, H., Negishi, T. & Hirose, A. (1985) Regeneration of rice plants from protoplasts. *Plant Tissue Culture Letters* 2, 74–5.

Galau, G. A., Hughes, D. W. & Dure, L., III (1986) Abscisic acid induction of cloned cotton late embryogenesis-abundant (Lea) mRNAs. *Plant Molecular Biology* 7, 155–70.

Galau, G. A., Jakobsen, K. S. & Hughes, D. W. (1990) The controls of later dicot embryogenesis. *Physiologia Plantarum* 81, 280–8.

Gasser, C. S. & Fraley, R. T. (1989) Genetically engineering plants for crop improvement. *Science* 244, 1293–9.

Gavin, A. L., Conger, B. V. & Trigiano, R. N. (1989) Sexual transmission of somatic embryogenesis in *Dactylis glomerata*. *Plant Breeding* 103, 251–4.

Gharyal, P. K. & Maheshwari, S. C. (1981) *In vitro* differentiation of somatic embryoids in a leguminous tree – *Albizzia lebbeck* L. *Naturwissenschaften* 68, 379–80.

Giuliano, G., Rosellini, D. & Terzi, M. (1983) A new method for purification of the different stages of carrot embryoids. *Plant Cell Reports* 2, 216–18.

Gordon-Kamm, W. J., Spencer, T. M., Mangano, M. L., Adams, T. R., Daines, R. J., Start, W. G., O'Brien, J. V., Chambers, S. A., Adams, Jr, W. R., Willetts, N. G., Rice, T. B., Mackey, C. J., Krueger, R. W., Kausch, A. P. & Lemaux, P. G. (1990) Transformation of maize cells and regeneration of fertile transgenic plants. *Plant Cell* 2, 603–18.

Gorst, J., Overall, R. L. & Wernicke, W. (1987) Ionic currents traversing cell clusters from carrot suspension cultures reveal perpetuation of morphogenetic potential as distinct from induction of embryogenesis. *Cell Differentiation* 21, 101–9.

Gray, D. J. (1987) Quiescence in monocotyledonous and dicotyledonous somatic embryos induced by dehydration. *HortScience* 22, 810–14.

Gray, D. J. (1989) Effects of dehydration and exogenous growth regulators on dormancy, quiescence and germination of grape somatic embryos. *In Vitro Cell and Developmental Biology* 25, 1173–78.

Gray, D. J., Conger, B. V. & Songstad, D. D. (1987) Desiccated quiescent somatic embryos of orchardgrass for use as synthetic seeds. *In Vitro Cell and Developmental Biology* 25, 1173–8.

Green, C. E. & Phillips, R. L. (1975) Plant regeneration from tissue cultures of maize. *Crop Science* 15, 417–21.

Grosser, J. W., Gmitter, F. G. Jr, & Chandler, J. L. (1988a) Intergeneric somatic hybrid plants of *Citrus sinensis* cv. Hamlin and *Poncinus trifoliata* cv. Flying Dragon. *Plant Cell Reports* 7, 5–8.

Grosser, J. W., Gmitter, F. G., Jr. & Chandler, J. L. (1988b) Intergeneric somatic hybrid plants from sexually incompatible woody species: *Citrus sinensis* and *Severina disticha*. *Theoretical and Applied Genetics* 75, 397–401.

Grosser, J. W., Moore, G. A. & Gmitter, F. G. Jr. (1989) Interspecific somatic hybrid plants from the fusion of 'Key' lime (*Citrus aurantifolla*) with 'Valencia' sweet orange (*Citrus sinensis*) protoplasts. *Scientia Horticulturae* 39, 23–9.

Gupta, P. K. & Durzan, D. J. (1987) Biotechnology of somatic polyembryogenesis and plantlet regeneration on loblolly pine. *Bio/Technology* 5, 147–51.

Gupta, P. K., Dandekar, A. M. & Durzan, D. J. (1988) Somatic embryo formation and transient expression of a luciferase gene in Douglas fir and loblolly pine protoplasts. *Plant Science* 58, 85–92.

Haccius, B. (1977) Question of unicellular origin of non-zygotic embryos in callus cultures. *Phytomorphology* 28, 74–81.

Hakman, I., Fowke, L. C., von Arnold, S. & Eriksson, T. (1985) The development of somatic embryos in tissue cultures initiated from immature embryos of *Picea abies* (Norway spruce). *Plant Science* 38, 53–9.

Halperin, W. & Wetherell, D. F. (1964) Adventive embryony in tissue cultures of the wild carrot, *Daucus carota. American Journal of Botany* 51, 274–83.

Halperin, W. & Wetherell, D. F. (1965) Ontogeny of adventive embryos in wild carrot. *Science* 147, 756–8.

Hammatt, N. & Davey, M. R. (1987) Somatic embryogenesis and plant regeneration from cultured zygotic embryos of soybean (*Glycine max* L.) *Journal of Plant Physiology* 128, 219–26.

Hanna, W. W., Lu, C. & Vasil, I. K. (1984) Uniformity of plants regenerated from somatic embryos of *Panicum maximum* Jacq. (Guinea grass). *Theoretical and Applied Genetics* 67, 155–9.

Hanzel, J. J., Miller, J. P., Brinkman, M. A. & Endos, E. (1984) Genotype and media effects on callus formation and regeneration in barley. *Crop Science* 25, 27–31.

Hartman, C. L., McCoy, T. J. & Knous, T. R. (1984) Selection of alfalfa (*Medicago sativa*) cell lines and regeneration of plants resistant to the toxin(s) produced by *Fusarium oxysporum* f. sp. *medicaginensis. Plant Science Letters* 34, 183–94.

Hartweck, L. M., Lazzeri, P. A., Cui, D., Collins, G. B. & Williams, E. G. (1988) Auxin-orientation effects on somatic embryogenesis from immature soybean cotyledons. *In Vitro Cell and Developmental Biology* 24, 821–8.

He, D. G., Yang, Y. M. & Scott, K. J. (1988) A comparison of scutellum callus and epiblast callus induction in wheat: the effect of genotype, embryo age and medium. *Plant Science* 57, 225–33.

Hernández-Fernández, M. M. & Christie, B. R. (1989) Inheritance of somatic embryogenesis in alfalfa (*Medicago sativa* L.). *Genome* 32, 318–21.

Higgins, P. & Mathias, R. J. (1987) The effect of the 4B chromosomes of hexaploid wheat on the growth and regeneration of callus cultures. *Theoretical and Applied Genetics* 74, 439–44.

Hodges, T. K., Kamo, K. K., Imbrie, C. W. & Becwar, M. R. (1986) Genotype specificity of somatic embryogenesis and regeneration in maize. *Bio/Technology* 4, 218–23.

Horn, M. E., Shillito, R. D., Conger, B. V. & Harms, C. T. (1988) Transgenic plants of orchardgrass (*Dactylis glomerata* L.) from protoplasts. *Plant Cell Reports* 7, 469–72.

Horsch, R. B., Fry, J. E., Hoffman, N. L., Eichholtz, D., Rogers, S. G. & Fraley, R. T. (1985) A simple and general method for transferring genes into plants. *Science* 227, 1229–31.

Hughes, D. W. & Galau, G. A. (1989) Temporally nodular gene expression during cotyledon development. *Genes and Development* 3, 358–69.

Jacobsen, E. & Sopory, S. K. (1978) The influence and possible recombination of genotypes on the production of microspore embryoids in anther cultures of *Solanum tuberosum* and dihaploid hybrids. *Theoretical and Applied Genetics* 52, 119–23.

Janick, J. (1986) Embryogenesis: the technology of obtaining useful products from the culture of asexual embryos. In: Crocomo, O. J., Sharp, W. R., Evans, D. A., Bravo, J. E., Tavares, F. C. A. & Paddock, E. F. (eds), *Biotechnology of Plants and Microorganisms*. Ohio State University Press, Columbus, pp. 97–117.

Johnson, L. B., Stuteville, D. L., Higgins, R. K. & Skinner, D. Z. (1981) Regeneration of alfalfa plants from protoplasts of selected Regen S clones. *Plant Science Letters* 20, 297–304.

Kaleikau, E. K., Sears, R. G. & Gill, B. S. (1989) Control of tissue culture response in wheat (*Triticum aestivum* L.). *Theoretical and Applied Genetics* 78, 783–7.

Kamada, H., Kobayashi, K., Kiyosue, T. & Harada, H. (1989) Stress induced somatic embryogenesis in carrot and its application to synthetic seed production. *In Vitro Cell and Developmental Biology* 25, 1163–6.

Karno, K. K., Chang, K. L., Lynn, M. E. & Hodges, T. K. (1987) Embryogenic callus formation from maize protoplasts. *Planta* 172, 245–51.

Keyes, G. J., Collins, G. B. & Taylor, N. L. (1980) Genetic variation in tissue cultures of red clover. *Theoretical and Applied Genetics* 58, 265–71.

Kim, Y. H. & Janick, J. (1989) ABA and Polyox-encapsulation or high humidity increases survival of desiccated somatic embryos of celery. *HortScience* 24, 674–6.

Kitto, S. L. & Janick, J. (1985a) Production of synthetic seeds by encapsulating asexual embryos of carrot. *Journal of the American Society for Horticultural Science* 110, 277–82.

Kitto, S. L. & Janick, J. (1985b) Hardening treatments increase survival of synthetically-coated asexual embryos of carrot. *Journal of the American Society for Horticultural Science* 110, 283–6.

Klein, T. M., Wolf, E. D., Wu, R. & Sanford, J. C. (1987) High-velocity microprojectiles for delivering nucleic acids into living cells. *Nature* 327, 70–3.

Klimaszewska, K. (1989) Recovery of somatic embryos and plantlets from protoplast cultures of *Larix×eurolepis*. *Plant Cell Reports* 8, 440–4.

Kobayashi, S. & Ohgawara, T. (1988) Production of somatic hybrid plants through protoplast fusion in citrus. *Journal of Agriculture Review Quarterly* 22, 181–8.

Kobayashi, S., Ohgawara, T., Ohgawara, E., Oiyama, I. & Ishii, I. (1988) A somatic hybrid plant obtained by protoplast fusion between navel orange (*Citrus sinensis*) and satsuma mandarin. *Plant Cell, Tissue and Organ Culture* 14, 63–9.

Komatsuda, T. & Ohyama, K. (1988) Genotypes of high competence for somatic embryogenesis and plant regeneration in soybean *Glycine max*. *Theoretical and Applied Genetics* 75, 695–700.

Komatsuda, T., Enomoto, S. & Nakajima, K. (1989) Genetics of callus proliferation and shoot differentiation in barley. *Journal of Heredity* 80, 345–50.

Kurz, W. G. W. & Constabel, F. (1981) Continuous culture of plant cells. In: Calcott, P. H. (ed.), *Continuous Cultures of Cells,* Vol. 2. CRC Press, Boca Raton, Fl, pp. 141–57.

Kyozuka, J., Hayashi, Y. & Shimamoto, K. (1987) High frequency plant regeneration from rice protoplasts by novel nurse culture methods. *Molecular and General Genetics* 206, 408–13.

Lang, H. & Kohlenbach, H. W. (1989) Cell differentiation in protoplast cultures from embryogenic callus of *Abies alba* L. *Plant Cell Reports* 8, 120–3.

Lazar, M. D., Collins, G. B. & Vian, W. E. (1983) Genetic and environmental effects of the growth and differentiation of wheat somatic cell cultures. *Journal of Heredity* 74, 353–7.

Lazzeri, P. A., Hildebrand, D. F. & Collins, G. B. (1985) A procedure for plant regeneration from immature cotyledon tissue of soybean. *Plant Molecular Biology Reporter* 3, 160–7.

Leckle, F., Scragg, A. H. & Cliffe, K. C. (1990) The effect of continuous high shear stress on plant cell suspension cultures. In: Nijkamp, H. J. J., Van Der Plas, L. W. H. & Van Aartrijk, J. (eds). *Progress in Plant Cellular and Molecular Biology.* Kluwer Academic Publishers, Dordrecht, pp. 689–93.

Lee, C. W. & Thomas, J. C. (1985) Jojoba embryo culture and oil production. *HortScience* 20, 762–4.

Levin, R. & Vasil, I. K. (1989) An integrated and automated tissue culture system for mass propagation of plants. *In Vitro Cell and Developmental Biology* 25, 21A.

Levin, R., Gaba, V., Tal, B., Hirsch, S., DeNola, D. & Vasil, I. K. (1988) Automated plant tissue culture for mass propagation. *Bio/Technology* 6, 1035–40.

Lippincott, J. A. & Lippincott, B. B. (1978) Cells of crown-gall tumors and embryonic plant tissues lack *Agrobacterium* adherence sites. *Science* 199, 1075–7.

Lippmann, B. & Lippmann, G. (1984) Induction of somatic embryos in cotyledonary tissue of soybean, *Glycine max* L. Merr. *Plant Cell Reports* 3, 215–18.

Loh, C.-S., Ingram, D. S. & Hanke, D. E. (1983) Cytokinins and the regeneration of plantlets from secondary embryoids of winter oilseed rape, *Brassica napus* ssp. *oleifera. New Phytologist* 95, 349–58.

LoSchiavo, F., Pitto, L., Giuliano, G., Torti, G., Nuti-Ronchi, V., Marazziti, D., Vergara, R., Orselli, S. & Terzi, M. (1989) DNA methylation of embryogenic carrot cell cultures and its variations as caused by mutation, differentiation, hormones and hypomethylating drugs. *Theoretical and Applied Genetics* 77, 325–31.

Lu, C., Vasil, V. & Vasil, I. K. (1983a) Improved efficiency of somatic embryogenesis and plant regeneration in tissue cultures of maize (*Zea mays* L.). *Theoretical and Applied Genetics* 66, 285–9.

Lu, C., Vasil, V. & Vasil, I. K. (1983b) Isolation and culture of protoplasts of *Panicum maximum* Jacq. (Guinea grass), Somatic embryogenesis and plantlet formation. *Zeitschrift für Pflanzenphysiologie* 104, 311–18.

Lührs, R. & Lörz, H. (1987) Plant regeneration in vitro from embryogenic cultures of spring- and winter-type barley (*Hordeum vulgare* L.) varieties. *Theoretical and Applied Genetics* 75, 16–25.

Lupotto, E. (1983) Propagation of an embryogenic culture of *Medicago sativa* L. *Zeitschrift für Pflanzenphysiologie* 111, 95–104.

Lupotto, E. (1986) The use of single somatic embryo culture in propagating and regenerating lucerne (*Medicago sativa* L.). *Annals of Botany* 57, 19–24.

Lutz, J. D., Wong, J. R., Rowe, J., Tricoli, D. M. & Lawrence, R. H. (1985) Somatic embryogenesis for mass cloning of crop plants. In: Henke, R. R., Constantin, M. J. & Hollaender, A. (eds), *Tissue Culture in Forestry and Agriculture.* Plenum Press, New York, pp. 105–16.

McGranahan, G. H., Leslie, C. A., Uratsu, S., Martin, L. A. & Dandekar, A. M. (1988) *Agrobacterium*-mediated transformation of walnut somatic embryos and regeneration of transgenic plants. *Bio/Technology* 6, 800–4.

McGranahan, G. H., Leslie, C. A., Uratsu, S. L. & Dandekar, A. M. (1990) Improved efficiency of the walnut somatic embryo gene transfer system. *Plant Cell Reports* 8, 512–16.

McKersie, B. D., Senaratna, T., Bowley, S. R., Brown, D. C. W., Krochko, J. E. & Bewley, J. D. (1989) Application of artificial seed technology in the production of hybrid alfalfa (*Medicago sativa* L.). *In Vitro Cell and Developmental Biology* 25, 1183–8.

MacLean, N. L. & Nowak, J. (1989) Plant regeneration from hypocotyl and petiole callus of *Trifolium pratense* L. *Plant Cell Reports* 8, 395–8.

McMullen, M. D. & Finer, J. J. (1990) Stable transformation of cotton and soybean embryogenic cultures via microprojectile bombardment. *Journal of Cellular Biochemistry* Supplement 14E, 285 (Abstract).

Maddock, S. E., Lancaster, V. A., Risiott, R. & Franklin, J. (1983) Plant regeneration from cultured immature embryos and inflorescences of 25 cultivars of wheat (*Triticum aestivum*). *Journal of Experimental Botany* 34, 915–26.

Maheswaran, G. & Williams, E. G. (1984) Direct somatic embryoid formation on immature embryos of *Trifolium repens, T. pratense* and *Medicago sativa*, and rapid clonal propagation of *T. repens. Annals of Botany* 54, 201–11.

Maheswaran, G. & Williams, E. G. (1985) Origin and development of somatic embryoids formed directly on immature embryos of *Trifolium repens in vitro. Annals of Botany* 56, 619–30.

Maheswaran, G. & Williams, E. G. (1986a) Direct secondary somatic embryogenesis form immature sexual embryos of *Trifolium repens* cultured *in vitro. Annals of Botany* 57, 109–17.

Maheswaran, G. & Williams, E. G. (1986b) Clonal propagation of *Trifolium pratense, Trifolium resupinatum* and *Trifolium subterraneum* by direct somatic embryogenesis on cultured immature embryos. *Plant Cell Reports* 3, 165–8.

Martin, S. M. (1980) Mass culture systems for plant cell suspensions. In: Staba, E. J. (ed.), *Plant Tissue Culture as a Source of Biochemicals.* CRC Press, Boca Raton, Fl., pp. 149–66.

Martins, I. S. & Sondahl, M. R. (1984) Early stages of somatic embryo differentiation from callus cells of bean (*Phaseolus vulgaris* L.) grown in liquid medium. *Journal of Plant Physiology* 117, 97–103.

Mathias, R. J. (1987) Plant microinjection techniques. *Genetic Engineering* 9, 199–227.

Mathias, R. J. & Fukui, K. (1986) The effect of specific chromosome and cytoplasm substitutions on the tissue culture response of wheat (*Triticum aestivum*) callus. *Theoretical and Applied Genetics* 71, 797–800.

Matthysse, A. G. & Gurlitz, R. H. G. (1982) Plant cell range for attachment of *Agrobacterium tumefaciens* to tissue culture cells. *Physiological Plant Pathology* 21, 381–7.

Meijer, E. G. M. & Brown, D. C. W. (1985) Screening of diploid *Medicago sativa* germplasm for somatic embryogenesis. *Plant Cell Reports* 4, 285–8.

Merkle, S. A. & Sommer, H. E. (1987) Regeneration of *Liriodendron tulipifera* (family Magnoliaceae) from protoplast culture. *American Journal of Botany* 74, 1317–21.

Merkle, S. A. & Wiecko, A. T. (1989) Regeneration of *Robinia pseudoacacia* via somatic embryogenesis. *Canadian Journal of Forestry Research* 19, 285–8.

Merkle, S. A. & Wiecko, A. T. (1990) Somatic embryogenesis in three magnolia species. *Journal of the American Society for Horticultural Science* 115, 858–60.

Mitten, D. H., Sato, S. J. & Skokut, T. A. (1984) *In vitro* regenerative potential of alfalfa germplasm sources. *Crop Science* 24, 943–5.

Murashige, T. (1978) The impact of plant tissue culture on agriculture. In: Thorpe, T. (ed.), *Frontiers of Plant Tissue Culture 1978*. International Association for Plant Tissue Culture, University of Calgary, Calgary, Alberta, Canada, pp. 15–26.

Nadel, B. L., Altman, A. & Ziv, M. (1990) Regulation of somatic embryogenesis in celery cell suspensions. 2. Early detection of embryogenic potential and the induction of synchronized cell cultures. *Plant Cell, Tissue and Organ Culture* 20, 119–24.

Nagmani, R. & Bonga, J. M. (1985) Embryogenesis in subcultured callus of *Larix decidua*. *Canadian Journal of Forest Research* 15, 1088–91.

Nomura, K. & Komamine, A. (1985) Identification and isolation of single cells that produce somatic embryos at a high frequency in a carrot suspension culture. *Plant Physiology* 79, 988–91.

Nomura, K. & Komamine, A. (1986) Molecular mechanisms of somatic embryogenesis. *Oxford Surveys of Plant Molecular and Cellular Biology* 3, 456–66.

Ockendon, D. J. & Sutherland, R. A. (1987) Genetic and non-genetic factors affecting anther culture of Brussels sprouts (*Brassica oleracea* var. *gemmifera*). *Theoretical and Applied Genetics* 74, 566–70.

Ohgawara, T., Kobayashi, S., Ohgawara, E., Uchimiya, H. & Ishii, S. (1985) Somatic hybrid plants obtained by protoplast fusion between *Citrus sinensis* and *Poncirus trifoliata*. *Theoretical and Applied Genetics* 71, 1–4.

Ohkoshi, S., Komatsuda, T., Enomoto, S., Taniguchi, M. & Ohyama, K. (1987) Use of tissue culture for barley improvement. 1. Differences in callus formation and plant regeneration from immature embryos of 179 varieties. *Japanese Journal Breeding* 37 (Suppl 1), 42–3.

Ou, G., Wang, W. C. & Nguyen, H. T. (1989) Inheritance of somatic embryogenesis and organ regeneration from immature embryo cultures of winter wheat. *Theoretical and Applied Genetics* 78, 137–42.

Ozias-Akins, P. & Vasil, I. K. (1982) Plant regeneration from cultured immature embryos and inflorescences of *Triticum aestivum* L. (wheat): evidence for somatic embryogenesis. *Protoplasma* 110, 95–105.

Ozias-Akins, P. & Vasil, I. K. (1988) In vitro regeneration and genetic manipulation of grasses. *Physiologia Plantarum* 73, 565–9.

Parrott, W. A., Dryden, G., Vogt, S., Hildebrand, D. F., Collins, G. B. & Williams, E. G. (1988) Optimization of somatic embryogenesis and embryo germination in soybean. *In Vitro Cell and Developmental Biology* 24, 817–20.

Parrott, W. A., Williams, E. G., Hildebrand, D. F. & Collins, G. B. (1989) Effect of genotype on somatic embryogenesis from immature cotyledons of soybean. *Plant Cell, Tissue and Organ Culture* 16, 15–21.

Pechan, P. M. (1989) Successful cocultivation of *Brassica napus* microspores and proembryos with *Agrobacterium*. *Plant Cell Reports* 8, 387–90.

Peng, J. & Hodges, T. K. (1989) Genetic analysis of plant regeneration in rice (*Oryza sativa* L.). *In Vitro Cell and Developmental Biology* 25, 91–4.

Petolino, J. F., Jones, A. M. & Thompson, S. A. (1988) Selection for increased anther culture response in maize. *Theoretical and Applied Genetics* 76, 157–9.

Polito, V. S., McGranahan, G., Pinney, K. & Leslie, C. (1989) Origin of somatic embryos from repetitively embryogenic cultures of walnut (*Juglans regia* L.): implications for *Agrobacterium*-mediated transformation. *Plant Cell Reports* 8, 219–21.

Pretova, A. & Williams, E. G. (1986a) Zygotic embryo cloning in oilseed rape (*Brassica napus* L.). *Plant Science* 47, 195–8.

Pretova, A. & Williams, E. G. (1986b) Direct somatic embryogenesis from immature embryos of flax (*Linum usitatissimum* L.). *Journal of Plant Physiology* 126, 155–61.

Prioli, L. M. & Sondahl, M. R. (1989) Plant regeneration and recovery of fertile plants from protoplasts of maize (*Zea mays* L.). *Bio/Technology* 7, 589–94.

Rajasekaran, K., Hein, M. B., Davis, G. C., Carnes, M. G. & Vasil, I. K. (1987) Endogenous growth regulators in leaves and tissue cultures of *Pennisetum purpureum* Schum. *Journal of Plant Physiology* 130, 13–25.

Rao, P. S. & Ozlas-Akins, P. (1985) Plant regeneration through somatic embryogenesis in protoplast cultures of sandalwood (*Santalum album* L.). *Protoplasma* 124, 80–6.

Rathore, K. S., Hodges, T. K. & Robinson, K. R. (1988) Ionic basis of currents in somatic embryos of *Daucas carota*. *Planta* 175, 280–9.

Ray, I. M. & Bingham, E. T. (1989) Breeding diploid alfalfa for regeneration from tissue culture. *Crop Science* 29, 1545–8.

Redenbaugh, K. (1990) Application of artificial seed to tropical crops. *HortScience* 25, 251–5.

Redenbaugh, K. & Ruzin, S. E. (1989) Artificial seed production and forestry. In: Dhawan, V. (ed.), *Applications of Biotechnology in Forestry and Horticulture*. Plenum Press, New York, pp. 57–71.

Redenbaugh, K., Nichol, J., Kossler, M. & Paasch, B. (1984) Encapsulation of somatic embryos for artificial seed production. In Vitro Cell and Developmental Biology 20, 256–7.

Redenbaugh, K., Paasch, B., Nichol, J., Kossler, M., Viss, P. & Walker, K. (1986) Somatic seeds: encapsulation of asexual plant embryos. *Bio/Technology* 4, 797–801.

Redenbaugh, K., Viss, P., Slade, D. & Fujii, J. A. (1987a) Scale-up: artificial seeds. In: Green, C. E., Somers, D. A., Hackett, W. P. & Biesboer, D. D. (eds), *Plant Tissue and Cell Culture*. Alan R. Liss, Inc., New York, pp. 473–93.

Redenbaugh, K., Slade, D., Viss, P. & Fujii J. A. (1987b) Encapsulation of somatic embryos in synthetic seed coats. *HortScience* 22, 803–9.

Reinert, J. (1958) Morphogenese und ihre Kontrolle an Gewebekulturen aus Carotten. *Naturwissenshaften* 45, 344–5.

Reisch, B. & Bingham, E. T. (1980) The genetic control of bud formation from callus cultures of diploid alfalfa. *Plant Science Letters* 20, 71–7.

Rhodes, C. A., Pierce, D. A., Mettler, I. J., Mascarenhas, D. & Detmer, J. J. (1988) Genetically transformed maize plants from protoplasts. *Science* 240, 204–8.

Roberts, D. R., Flinn, B. S., Webb, D. T., Webster, F. B. & Sutton, B. C. S. (1990a) Abscisic acid and indole-3-butyric acid regulation of maturation and accumulation of storage proteins in somatic embryos of interior spruce. *Physiologia Plantarum* 78, 355–60.

Roberts, D. R., Sutton, B. C. S. & Flinn, B. S. (1990b) Synchronous and high frequency germination of interior spruce somatic embryos following partial drying at high relative humidity. *Canadian Journal of Botany* 68, 1086–90.

Rode, A., Hartman, C., De Buyser, J. & Henry, Y. (1988) Evidence for a direct relationship between mitochondrial genome organization and regeneration ability in hexaploid wheat somatic tissue cultures. *Current Genetics* 14, 387–94.

Rosenberg, L. A. & Rinne, R. W. (1986) Moisture loss as a prerequisite for seedling growth in soybean seeds (*Glycine max* L. Merr.). *Journal of Experimental Botany* 37, 1663–74.

Rosenberg, L. A. & Rinne, R. W. (1987) Changes in seed constituents during germination and seedling growth of precociously matured soybean seeds (*Glycine max*). *Annals of Botany* 60, 705–12.

Rosenberg, L. A. & Rinne, R. W. (1988) Protein synthesis during natural and precocious soybean seed (*Glycine max* [L.] Merr.) maturation. *Plant Physiology* 87, 474–8.

Sanford, J. C. (1988) The biolistic transformation. *Trends in Biotechnology* 6, 299–302.

Schultheis, J. R. & Cantliffe, D. J. (1988) Plant formation of *Ipomea batatas* Poir. from somatic embryos in gel carriers with additives. *HortScience* 23, 812.

Schultheis, J. R., Chee, R. & Cantliffe, D. J. (1986) Effect of growth regulators and gel carriers on growth and development of sweet potato (*Ipomea batatas*) somatic embryos. *HortScience* 21, 762.

Scott, R. J. & Draper, J. (1987) Transformation of carrot tissues derived from proembryogenic suspension cells: a useful model system for gene expression studies in plants. *Plant Molecular Biology* 8, 265–74.

Sears, R. G. & Deckard, E. L. (1982) Tissue culture variability in wheat: callus induction and plant regeneration. *Crop Science* 22, 546–50.

Seitz Kris, M. H. & Bingham, E. T. (1988) Interactions of highly regenerative genotypes of alfalfa (*Medicago sativa*) and tissue culture protocols. *In Vitro Cell and Developmental Biology* 24, 1047 52.

Senaratna, T., McKersie, B. D. & Bowley, S. R. (1989) Desiccation tolerance of alfalfa (*Medicago sativa* L.) somatic embryos: influence of abscisic acid, stress pretreatments and drying rates. *Plant Science* 65, 253–9.

Senaratna, T., McKersie, B. D. & Bowley, S. R. (1990) Artificial seeds of alfalfa (*Medicago sativa* L.): induction of desiccation tolerance in somatic embryos. *In Vitro Cell and Developmental Biology* 26, 85–90.

Sequeira, L. (1984) Plant-bacterial interactions. In: Linskens, H. F. & Heslop-Harrison, J. (eds), Cellular interactions. *Encyclopedia of Plant Physiology, New Series*, vol. 17. Springer-Verlag, New York, pp. 187–209.

Shahin, E. A., Spielman, A., Sukhapinda, K., Simpson, R. & Yashar, M. (1986) Transformation of cultivated alfalfa using disarmed *Agrobacterium tumefaciens*. *Crop Science* 26, 1235–9.

Sharp, W. R., Evans, D. A. & Sondahl, M. R. (1982) Application of somatic embryogenesis to crop improvement. In: Fujiwara, A. (ed.), *Plant Tissue Culture 1983. Proceedings of the Fifth International Congress of Plant Tissue Culture*. Japanese Association for Plant Tissue Culture, Tokyo, pp. 759–62.

Shillito, R. D., Carswell, G. K., Johnson, C. M., DiMaio, J. J. & Harms, C. T. (1989) Regeneration of fertile plants from protoplasts of elite inbred maize. *Bio/Technology* 7, 581–7.

Shimamoto, K., Terada, R., Izawa, T. & Fujimoto, H. (1989) Fertile transgenic rice plants regenerated from transformed protoplasts. *Nature* 338, 274–6.

Sluis, C. J. & Walker, K. A. (1985) Commercialization of plant tissue culture. *International Association for Plant Tissue Culture Newsletter* 47, 2–12.

Smith, D. L. & Krikorian, A. D. (1989) Release of somatic embryogenic potential from excised zygotic embryos of carrot and maintenance of proembryonic cultures in hormone-free medium. *American Journal of Botany* 76, 1832–43.

Smith, D. L. & Krikorian, A. D. (1990) Somatic proembryo production from excised, wounded zygotic carrot embryos on hormone-free medium: evaluation of the effects of pH, ethylene and activated charcoal. *Plant Cell Reports* 9, 34–7.

Sonnino, A., Tanaka, S., Iwanaga, M. & Schilde-Rentschler, L. (1989) Genetic control of embryo formation in anther culture of diploid potatoes. *Plant Cell Reports* 8, 105–7.

Spano, L., Mariotti, D., Pezzotti, M., Damiani, F. & Arcioni, S. (1987) Hairy root transformation in alfalfa. *Theoretical and Applied Genetics* 73, 523–30.

Srinivasan, C. & Vasil, I. K. (1986) Plant regeneration from protoplasts of sugarcane (*Saccharum officinarum* L.). *Journal of Plant Physiology* 126, 41–8.

Steward, F. C. (1958) Growth and development of cultivated cells. III. Interpretations of the growth from free cell to carrot plant. *American Journal of Botany* 45, 709–13.

Stuart, D. A. & Strickland, S. G. (1984a) Somatic embryogenesis from cell cultures of *Medicago sativa*. 1. The role of amino acid additions to the regeneration medium. *Plant Science Letters* 34, 165–74.

Stuart, D. A. & Strickland, S. G. (1984b) Somatic embryogenesis from cell cultures of *Medicago sativa*. 2. The interaction of amino acids with ammonium. *Plant Science Letters* 34, 175–82.

Stuart, D. A., Strickland, S. G. & Walker, K. A. (1987) Bioreactor production of alfalfa somatic embryos. *HortScience* 22, 800–3.

Styer, D. J. (1985) Bioreactor technology for plant propagation. In: Henke, R. R., Hughes, K. W., Constantin, M. J. & Hollaender, A. (eds), *Tissue Culture in Forestry and Agriculture*. Plenum Press, New York. pp. 117–30.

Sukhapinda, K., Spivey, R. & Shahin, E. A. (1987) Ri-plasmid as a helper for introducing vector DNA into alfalfa plants. *Plant Molecular Biology* 8, 209–16.

Swanson, E. B. & Erickson, L. R. (1989) Haploid tranformation in *Brassica napus* using an octopine-producing strain of *Agrobacterium tumefaciens*. *Theoretical and Applied Genetics* 78, 831–5.

Tautorus, T. E., Attree, S. M., Fowke, L. C. & Dunstan, D. I. (1990) Somatic embryogenesis from immature and mature zygotic embryos, and embryo regeneration from protoplasts in black spruce (*Picea mariana* Mill.) *Plant Science* 67, 115–24.

Ten Hoopen, H. J. G., Van Gulik, W. M. & Meijer, J. J. (1990) Possibilities, problems, and pitfalls of large-scale plant cell cultures. In: Nijkamp, H. J. J., Van Der Plas, L. W. H. & Van Aartrijk, J. (eds), *Progress in Plant Cellular and Molecular Biology*. Kluwer Academic Publishers, Dordrecht. pp. 673–81.

Thomas, J. C., Guiltinan, M. J., Bustos, S., Thomas, T. & Nessler, C. (1989) Carrot (*Daucas carota*) hypocotyl transformation using *Agrobacterium tumefaciens*. *Plant Cell Reports* 8, 354–7.

Timmers, A. C. J., de Vries, S. C. & Schel, J. H. N. (1989) Distribution of membrane-bound calcium and activated calmodulin during somatic embryogenesis of carrot (*Daucus carota* L.). *Protoplasma* 153, 24–9.

Tomes, D. T. & Smith, O. S. (1985) The effect of parental genotype on initiation of embryogenic callus from elite maize (*Zea mays* L.) germplasm. *Theoretical and Applied Genetics* 70, 505–9.

Toriyama, K., Hinata, K. & Sasaki, T. (1986) Haploid and diploid plant regeneration from protoplasts of anther callus in rice. *Theoretical and Applied Genetics* 73, 16–19.

Toriyama, K., Arimoto, Y., Uchimiya, H. & Hinata, K. (1988) Transgenic rice plants after direct gene transfer into protoplasts. *Bio/Technology* 6, 1072–4.

Trigiano, R. N., Beaty, R. M. & Graham, E. T. (1988) Somatic embryogenesis from immature embryos of redbud (*Cercis canadensis*). *Plant Cell Reports* 7, 148–50.

Trigiano, R. N., Gray, D. J., Conger, B. V. & McDaniel, J. K. (1989) Origin of direct somatic embryos from cultured leaf segments of *Dactylis glomerata*. *Botanical Gazette* 150, 72–7.

Trolinder, N. L. & Xhixian, C. (1989) Genotype specificity of the somatic embryogenesis response in cotton. *Plant Cell Reports* 8, 133–6.

Trulson, A. J., Simpson, R. B. & Shahin, E. A. (1986) Transformation of cucumber (*Cucumis sativus* L.) plants with *Agrobacterium rhizogenes*. *Theoretical and Applied Genetics* 73, 11–15.

Tulecke, W. (1987) Somatic embryogenesis in woody plants. In: Bonga, J. M. & Durzan, D. J. (eds), *Cell and Tissue Culture in Forestry,* vol. 2, *Specific Principles and Methods: Growth and Developments.* Martinus Nijhoff Publishers, Dordrecht, Boston, Lancaster, pp. 61–91.

Umbeck, P., Johnson, G., Barton, K. & Swain, W. (1987) Genetically transformed cotton (*Gossypium hirsutum* L.). *Bio/Technology* 5, 263–6.

Vardi, A. & Galun, E. (1988) Recent advances in protoplast culture of horticultural crops: *Citrus. Scientia Horticulturae* 37, 217–30.

Vardi, A., Spiegel-Roy, P. & Galun, E. (1982) Plant regeneration from *Citrus* protoplasts, variability in methodological requirements among cultivars and species. *Theoretical and Applied Genetics* 62, 171–176.

Vasil, I. K. (1985) Somatic embryogenesis and its consequences in the Graminae. In: Henke, R. R., Hughes, K. W., Constantin, M. J. & Hollaender, A. (eds), *Tissue Culture in Forestry and Agriculture* Plenum Press, New York, pp. 31–47.

Vasil, V. & Vasil, I. K. (1980) Isolation and culture of cereal protoplasts. II. Embryogenesis and plantlet formation from protoplasts of *Pennisetum americanum. Theoretical and Applied Genetics* 56, 97–9.

Vasil, V. & Vasil, I. K. (1987) Formation of callus and somatic embryos from protoplasts of a commercial hybrid of maize (*Zea mays* L.). *Theoretical and Applied Genetics* 73, 793–8.

Vasil, V., Wang, D. & Vasil, I. K. (1983) Plant regeneration from protoplasts of *Pennisetum purpureum* Schum. (Napier grass). *Zeitschrift für Pflanzenphysiologie* 111, 319–25.

Walker, C. C. (1989) Growth of embryogenic Norway spruce cultures in a bioreactor. *A-190 Independent Study* The Institute of Paper Science and Technology, Atlanta, GA. p. 46.

Walker, K. A. & Sato, S. J. (1981) Morphogenesis in callus tissue of *Medicago sativa:* the role of ammonium ion in somatic embryogenesis. *Plant Cell, Tissue and Organ Culture* 1, 109–21.

Walker, K. A., Wendeln, M. L. & Jaworski, E. G. (1979) Organogenesis in callus tissue of *Medicago sativa* the temporal separation of induction processes from differentiation processes. *Plant Science Letters* 16, 23–30.

Walton, P. D. & Brown, D. C. W. (1988) Screening of *Medicago* wild species for callus formation and the genetics of somatic embryogenesis. *Journal of Genetics* 67, 95–100.

Wan, Y., Sorensen, E. L. & Liang, G. H. (1988) Genetic control of *in vitro* regeneration in alfalfa *Medicago sativa* L. *Euphytica* 39, 3–10.

Wenck, A. R., Conger, B. V., Trigiano, R. N. & Sams, C. E. (1988) Inhibition of somatic embryogenesis in orchardgrass by endogenous cytokinins. *Plant Science* 88, 990–2.

Wetherell, D. F. (1984) Enhanced adventive embryogenesis resulting from plasmolysis of cultured wild carrot cells. *Plant Cell, Tissue and Organ Culture* 3, 221–7.

Wetzstein, H. Y., Ault, J. R. & Merkle, S. A. (1989) Further characterization of somatic embryogenesis and plantlet regeneration in pecan (*Carya illinoensis*). *Plant Science* 64, 193–201.

Williams, E. G. & Maheswaran, G. (1986) Somatic embryogenesis: factors influencing coordinated behaviour of cells as an embryogenic group. *Annals of Botany* 57, 443–62.

Williams, E. G., Collins, G. B. & Myers, J. R. (1990) Clovers (*Trifolium* spp.) In: Bajaj, Y. P. S. (ed.), *Biotechnology in Agriculture and Forestry*, vol. 10, *Legumes and Oilseed Crops, I.* Springer Verlag, Berlin, Heidelberg, pp. 242–87.

Willman, M. R., Schroll, S. M. & Hodges, T. K. (1989) Inheritance of somatic embryogenesis and plantlet regeneration from primary (Type 1) callus in maize. *In Vitro Cell and Developmental Biology* 25, 95–100.

Wilson, S. B., King, P. J. & Street, H. E. (1971) Studies of the growth in culture of plant cells. XII. A versatile system for the large scale batch or continuous culture of plant cell suspension. *Journal of Experimental Botany* 22, 177.

Yamada, M., Yang, Z. & Tang, D. (1986) Plant regeneration from protoplast-derived callus of rice (*Oryza sativa* L.). *Plant Cell Reports* 5, 85–8.

Young, R., Kaul, V. & Williams, E. G. (1987) Clonal propagation *in vitro* from immature embryos and flower buds of *Lycopersicon peruvianum* and *L. esculentum*. *Plant Science* 52, 237–42.

Chapter 8
Chloroplast and Mitochondrial Genomes: Manipulation Through Somatic Hybridization

G. Pelletier
*Laboratoire de Biologie Cellulaire, INRA, CNRA,
Route de St Cyr, 78026 Versailles Cedex, France*

Introduction

Early this century non-Mendelian inheritance was the subject of several reports questioning the classical rules of heredity established by Mendel and Morgan. It was postulated that particles other than the nuclear chromosomes constituted new categories of genetic information. The situation became clear with the discovery that plastids and mitochondria contain DNA molecules, and the demonstration that non-Mendelian mutations are correlated with changes in organelle DNAs (for a review see Grun, 1976).

In higher organisms, non-Mendelian inheritance of these genomes is characterized by an unequal transmission from paternal and maternal parents to the progeny. Plant breeders recognized the importance of such heredity when they discovered cytoplasmic male sterility (a case of genetic emasculation transmitted exclusively through the female parent), which was then used in several crop species to produce F_1 hybrid seeds on a commercial scale (Duvick, 1959).

The functions of these organelles in photosynthesis, respiration and production of metabolites became established. The structure of their genomes also became better understood, although conventional sexual methods remained inadequate for true genetic studies. About 15 years ago the techniques of protoplast fusion and plant regeneration were considered to be the long awaited tools for manipulating cytoplasmic genomes. They were applied to model species belonging to the genera *Nicotiana* or *Petunia*. General principles were established, and these were further confirmed and applied in several crop species to solve problems posed by plant breeders.

Inheritance of cytoplasmic genomes

Plastids and mitochondria contain relatively small genomes compared with the nucleus, less than one per thousand on a kilobase-pairs basis. These DNA molecules code for no more than approximately 10% of the polypeptides which are necessary to ensure organelle functioning, the remaining 90% or so being produced by nuclear genes and imported through the membranes of the organelle. Genetic information is duplicated several times in each organelle and a plant cell may contain hundreds or thousands of cytoplasmic genomes, for which it may therefore be considered as highly polyploid. At each cell generation, these genomes are transmitted to the progeny cells by a random distribution of organelles. However, male and female gametes, which participate equally to form the egg nucleus, have an asymmetric role in cytoplasmic genome transmission, not only because of differences in cell size and the number of organelles in the egg cell and in the sperm cell, but conceivably by some exclusion phenomenon which would occur following fertilization. Considerable variation has been described among genotypes for the frequency of paternal transmission of plastids in species exhibiting biparental inheritance of these organelles (Tilney-Basset & Abdel-Wahab, 1979) and there is evidence in favour of an active genetic control of exclusion.

Plastid inheritance in higher plants seems to have evolved from paternal or predominantly paternal transmission, as in gymnosperms, to predominantly or exclusively maternal inheritance in angiosperms. Among gymnosperms, paternal transmission of plastids has been described in *Sequoia* (Neale *et al.*, 1989), and predominantly paternal inheritance occurs in *Cryptomeria* (Ohba *et al.*, 1971), *Pinus* (Wagner *et al.*, 1987, 1989; Neale & Sederoff, 1989; White, 1990), *Pseudotsuga* (Neale *et al.*, 1986) and *Picea* (Szmidt *et al.*, 1988; Stine *et al.*, 1989). Biparental inheritance (equivalent participation of each gamete) has been described in *Larix* (Szmidt *et al.*, 1987).

In angiosperms, Smith (1989) reviewed cases of biparental inheritance of plastids and concluded that about one-third of genera, at least occasionally, inherit plastids biparentally. Among these genera *Medicago* (Smith *et al.*, 1986; Lee *et al.*, 1988; Schumann & Hancock, 1989; Masoud *et al.*, 1990), *Pelargonium, Borrago, Secale* (Kirk & Tilney-Bassett, 1978) and *Pennisetum* (Krishna-Rao & Koduru, 1978) include species of agronomic interest. Even in genera where plastids are universally known to be inherited in strictly uniparental-maternal fashion, occasional paternal transmission has been observed. In *Petunia* nuclear mutants have been selected which favour this type of inheritance (Cornu & Dulieu, 1988). In *Nicotiana*, the selection of streptomycin-resistant cells in calli derived from germinating seeds obtained in crosses in which the pollinator was a genotype bearing a plastid-encoded mutation for streptomycin resistance allowed Medgyesy *et al.*

(1986) and Horlow *et al.* (1990) to demonstrate the presence of paternally derived plastids in some seedlings.

The mitochondrial genome behaves more regularly and as a general rule is maternally inherited both in gymnosperms and angiosperms. The existence of stable cytoplasmic male sterility, a trait encoded by the mitochondrial genome (Lonsdale, 1987) in a wide range of crop species and its use in breeding hybrid varieties is evidence for strict maternal inheritance of this genome. In other genera with paternal or biparental plastid inheritance, for example *Pinus* (Neale & Sederoff, 1989) or *Oenothera* (Brennicke & Schwemmle, 1984), strictly maternal mitochondrial inheritance was observed. In experiments where occasional paternal transmission of plastids was observed, the mitochondrial genome is of maternal origin (Medgyesy *et al.*, 1986; Horlow *et al.*, 1990). Nevertheless, recent molecular studies in *Sequoia* (Neale *et al.*, 1989) have demonstrated a regular paternal inheritance in this genus, where plastids are also transmitted by the pollen. Recently, an isolated case of biparental inheritance of the mitochondrial genome in *Brassica* has been reported by Erickson & Kemble (1990).

To summarize these data on organelle inheritance, it appears that mitochondria and plastids behave independently during fertilization, as shown in particular by species with paternal or biparental plastid inheritance. But, as maternal inheritance is the most frequent case for both organelles in the majority of crop species, the cytoplasmic genetic information may be considered in practice, as a single unit in sexual reproduction.

During evolution and species divergence, cytoplasmic genomes and nuclear genomes must have evolved in a co-ordinate fashion so that normal functioning of these essential organelles was maintained. Uniparental (maternal) inheritance means that plastids and mitochondria have to evolve without the opportunity for genetic exchange with another cytoplasm. This situation has been confirmed by molecular analysis (Palmer, 1985), which has revealed that in flowering plants plastid genomes evolved very slowly, in general organization as well as in gene sequences. Plant mitochondrial DNA evolved differently: rapid changes in structure, organization and size are apparent, but sequence evolution has been even slower than in chloroplasts (Palmer & Herbon, 1988).

There has been evolution for a close adaptation of the broad nuclear information content with the narrow cytoplasmic content, which results in visible cytoplasmic effects when alloplasmic combinations are made (the nucleus of one species associated with the more or less remote cytoplasm of another one). In general these effects are different anomalies concerning the vegetative organs and more often the reproductive organs. Among them, cytoplasmic male sterility (Edwarson, 1970; Kaul, 1988) has been largely exploited by breeders, but, in many species, cytoplasms showing deleterious side-effects were never used because of the lack of suitable genetic techniques for modifying them, as in the case of wheat (Wilson & Driscoll, 1983), for example.

Protoplast fusion appeared as a tool for systematically obtaining a parasexual biparental inheritance of organelle genomes. From the fusion product in which nuclei, plastids and mitochondria from both parents are present in a single cell, new associations between the nucleus of one species and organelles from a remote one could be produced. Moreover, interactions at the DNA level between organelle genomes would be possible, allowing the creation of new genomes of interest in plant breeding.

Protoplast fusion and the exchange of cytoplasms

The product of cell fusion between two parents is a heterokaryocyte with two or more nuclei in a cytoplasm containing plastids and mitochondria of the two partners. A prerequisite to genetic exchange or addition between parental genomes is the breakage or the fusion of membranes surrounding nuclei, plastids and mitochondria. The disappearance of the nuclear membrane occurs at the beginning of mitosis and it is at this stage that hybrid nuclei are produced if the two nuclei divide at the same time (Constabel *et al.*, 1975). The frequency of nuclear hybrid formation from heterokaryocytes can be as high as 80–100% (Medgyesy *et al.*, 1980; Glimelius & Bonnett, 1981; Gleba *et al.*, 1984) or relatively or extremely low (Gleba *et al.*, 1984; Flick *et al.*, 1985; Kushnir *et al.*, 1987), mainly depending upon the combined genotypes.

If the aim of the experiment is to combine the nucleus of one parent with the entire cytoplasm of the other, and if the frequency of nuclear hybridization is high, one can irradiate the cytoplasm donor by x- or gamma rays before fusion (Zelcer *et al.*, 1978; Aviv & Galun, 1980). These treatments have no deleterious or mutagenic effects on organelle genomes recovered in the progeny of irradiated protoplasts, probably because these genomes are present at a high copy number in each plant cell (Chapter 9).

Another way to eliminate one nucleus is to have recourse to the production of cytoplasts (Bradley, 1983), which are subprotoplasts from which the nucleus has been physically removed. The technique remains difficult because preparations of cytoplasts are often contaminated by small nucleated protoplasts.

To retain only the cytoplasmic organelles of the irradiated parent, Medgyesy *et al.* (1980) proposed making use of the metabolic inhibition method first described in mammalian cell genetics (Wright, 1978). Chemicals like iodoacetate or iodoacetamide are used in order to eliminate the organelles of the other parent. The method combining irradiation of the cytoplasm donor parent and metabolic inactivation of the nuclear recipient parent (the donor–recipient method; Sidorov *et al.*, 1981; Galun & Aviv, 1986) is now routinely used as illustrated below.

Nevertheless, the products obtained are not always exactly those expected by the theory. Some nuclear information of the irradiated parent may persist as individual or translocated chromosomes (Menczel *et al.*, 1987; Sidorov *et al.*, 1987) which will eventually be detected by a careful analysis of the plant at meiosis (A. M. Chèvre, personal communication). Moreover, as we will see below, if the inhibition of mitochondria of the nuclear donor is incomplete, rearrangements of the mitochondrial genome may occur.

Sexual crossing remains the ideal way to combine the nucleus of one species with the entire cytoplasm of another. The conventional method consists of a first cross with the cytoplasm donor as the female parent, followed by a series of backcrosses by the same recurrent pollinator. This method is time-consuming. Spontaneous androgenesis, a form of apomixis in which the male gamete develops directly into an embryo in the embryo sac without nuclear fusion with the female nucleus, allows one to reach the same goal in one step (Goodsell, 1961; Pelletier *et al.*, 1987; Chen & Heneen, 1989), but this is still a rare occurrence and needs suitable genetic markers for the screening of large populations of plants.

Protoplast fusion and the recombination of plastid genomes

The most frequent behaviour of plastid populations mixed in a fusion product is a random segregation of parental types in successive cell generations in accordance with the Michaelis (1967) theory. The segregation pattern of chloroplasts in cell populations derived from fusion products was studied by Akada & Hirai (1986). They showed that in small colonies cells with exclusively one or the other parental plastids already exist. Several reports have confirmed these findings in somatic hybrids (with hybrid nuclei) as well as in cytoplasmic hybrids (or cybrids) (Chen *et al.*, 1977; Belliard *et al.*, 1978; Aviv *et al.*, 1980; Glimelius & Bonnett, 1981; Menczel *et al.*, 1981; Schiller *et al.*, 1982; Bonnett & Glimelius, 1983). A prolonged heteroplastidic state can also sometimes be observed and confirmed by studying the progeny of a chimeral somatic hybrid, which will also contain chimeral plants, detectable by plastid DNA hybridization with specific probes or by the use of plastid mutants (Fluhr *et al.*, 1983; Gleba *et al.*, 1985; Thomzik & Hain, 1988).

Mixed populations of plastid genomes have been known for a long time in species with biparental plastid inheritance. In *Oenothera*, Chiu & Sears (1985) did not observe evidence of plastid recombination in a population of sexual hybrids. On the contrary, two reports from the same research group (Medgyesy *et al.*, 1985; Thanh & Medgyesy, 1989) have recently demonstrated the occurrence of plastid genome recombination in higher plants after somatic hybridization between two species of *Nicotiana* and between representatives of *Nicotiana* and *Solanum*. In both cases plastid mutants characterized by antibiotic or toxin resistance and albinism were used (Fig.

8.1). The scarcity of plastid recombinants (only two cases) prevents any accurate estimation of the frequencies of these recombinations in hetero-plastidic cells derived from fusion. It appears that the key to these successes was the use of strong selection pressure, only possible when plastid mutants are available. Molecular analysis of the plastid genomes obtained in the *Nicotiana* recombinant revealed that it resulted from a high number of crossovers (Fejes *et al.*, 1990).

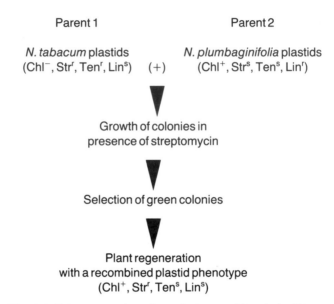

Parent 1   Parent 2

N. tabacum plastids   *N. plumbaginifolia* plastids
(Chl⁻, Str^r, Ten^r, Lin^s)   (+)   (Chl⁺, Str^s, Ten^s, Lin^r)

Growth of colonies in
presence of streptomycin

Selection of green colonies

Plant regeneration
with a recombined plastid phenotype
(Chl⁺, Str^r, Ten^s, Lin^s)

Fig. 8.1. Scheme employed to select recombinant plastid genomes in *Nicotiana* (Medgyesy *et al.*, 1985). The different plastid markers present in parental lines are: Chl⁺/Chl⁻ : normal/albino; Str^r/Str^s : streptomycin resistance/susceptibility; Ten^r/Ten^s : tentoxin resistance/susceptibility; Lin^r/Lin^s : lincomycin resistance/susceptibility.

Up to now, these examples are the only ones where plastid genes were exchanged between species. Direct gene transfer into plastids, described in *Chlamydomonas* and using high-velocity microprojectiles (Boynton *et al.*, 1988; Chapter 5), has not yet been achieved in higher plants, although transient expression of a chimeric gene was obtained in chloroplasts of *Nicotiana tabacum* with this technique (Daniell *et al.*, 1990).

Protoplast fusion and the recombination of mitochondrial genomes

Induced mutations affecting the mitochondrial genome of higher plants are difficult to select. Only two cases of cell lines resistant to oligomycin (a specific inhibitor of the mitochondrial ATPase complex) are known, without evidence so far of any modification of a mitochondrial structural gene sequence (Durand, 1987; Aviv & Galun, 1988).

Cytoplasmic male sterility, a trait which is found in a wide range of species (Kaul, 1988), is the only genetic marker known to be encoded by the mitochondrial genome (Lonsdale, 1987) and it is used by cell biologists in protoplast fusion experiments to follow the fate of mitochondrial genomes.

The lack of any selection pressure at the cell level is not an impediment to the recovery of mitochondrial recombinants. Interparental recombinations generally occur at a very high frequency, and were described very early in the history of plant somatic hybridization (Belliard *et al.*, 1978, 1979; Galun *et al.*, 1982; Boeshore *et al.*, 1983; Nagy *et al.*, 1983). The molecular study of novel mitochondrial restriction fragments in *Petunia* (Rothenberg *et al.*, 1985) and in *Brassica* (Vedel *et al.*, 1986) somatic hybrids or cybrids brought a definite confirmation of interparental DNA exchange. The result is a rearranged mitochondrial genome with a unique combination of parental sequences in each regenerated cybrid.

These fusion experiments involved cytoplasmic male-sterile (CMS) plants. Phenotypically, CMS plants may be characterized by the stage at which disorders are visible during floral morphogenesis. In Texas male-sterile genotypes of maize, as in the majority of CMS plants used for breeding, the abortion of microspores occurs between early meiosis and the first pollen mitosis. In *Nicotiana*, male sterilities are induced by the cytoplasm of different *Nicotiana* species (Gerstel, 1980), and in these alloplasmic combinations stamens are transformed into feminized or petal-like structures, or are totally absent. Each cytoplasm is recognizable by a specific floral morphology.

One interesting aspect of mitochondrial recombination in fusion between normal and male-sterile *Nicotiana* is that it creates new flower morphologies different from both parents (Belliard *et al.*, 1978, 1979; Pelletier, 1986). Moreover, by fusing *Nicotiana* protoplasts derived from plants with different cytoplasms, Aviv & Galun (1986) and Köfer *et al.* (1991) were able to regenerate male-fertile plants. These studies could lead to the isolation of different mitochondrial fragments involved in flower modifications found in these systems (absence of stamens, pistiloidy, petaloidy, incision of petals, anther filament length) since each of these different features seems to behave as a single independent trait (Köfer *et al.*, 1990).

Such recombinations between mitochondrial genomes have been found in a majority of species combinations, opening the possibility for exchange of mitochondrial genes between species.

Application of somatic fusion to organelle manipulation in crop species

As illustrated in Fig. 8.2, there are several possibilities resulting from protoplast fusion. These possibilities derive from the application of simple rules. Plastid genomes mixed together in a single cell exceptionally undergo recombination and segregate in further cell generations, sometimes very quickly when one parental type is less, or absolutely not, competitive compared with the other. In contrast, mitochondrial genomes very often undergo interparental recombination and, after segregation, one recombined genome, among many others, is randomly retained in a regenerated cybrid plant or in its progeny. The most frequent cybrid constitutions are therefore those where mitochondrial genomes are recombinant ones associated with one or other parental plastid genome (cases 7 and 8 in Fig. 8.2).

In order to prevent the transmission of one parental organelle genome, one could use differentiated cells for protoplast isolation. For example, Pental *et al.* (1989) observed that mitochondria, but not plastids, are transmitted from microspore protoplasts to the somatic hybrids formed on fusion with mesophyll protoplasts.

Whether it is possible for a plastid or a mitochondrial genome to coexist in the same cell with the nuclear genome of a distantly related species is still an open question. A case of apparent incompatibility is described by Bonnett & Glimelius (1990). They obtained cybrids containing *Nicotiana tabacum* nucleus and *Petunia hybrida* chloroplasts (Glimelius & Bonnett, 1986). These cybrids contained recombined mitochondria or *N. tabacum* mitochondria, but *P. hybrida* mitochondria were never found. Pental *et al.* (1986) obtained the combination of *P. hybrida* nucleus and *N. tabacum* chloroplasts. *Nicotiana* nuclei and *Atropa* (Kushnir *et al.*, 1987) or *Salpiglossis* (Thanh *et al.*, 1988) chloroplast combinations have also been produced. In contrast, *Nicotiana* nuclei and *Solanum* chloroplasts are incompatible, as demonstrated by Thanh & Medgyesy (1989), who regenerated green cybrid plants from this combination only after chloroplast recombination. It is likely that any new organellar combination is under the restrictive control of the nucleus with which it has to be associated. More information will be available in the future about the impact of other remote associations when progeny of symmetric or asymmetric interspecific nuclear somatic hybrids obtained in several families like Brassicaceae and Solanaceae are observed for their agronomic performance and characterized for their cytoplasmic composition (Glimelius, 1988).

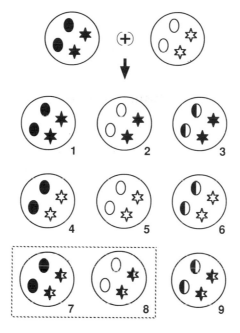

Fig. 8.2. The nine theoretical products of protoplast fusion between parents differing by plastid (● or ○) and mitochondrial (★ or ☆) genomes, considering three possibilities for each genome: exclusion of one or the other parental genome and interparental recombination (◑ or ★). Combinations 7 and 8 are the most frequently observed.

Transfer of male-sterile cytoplasms to different cultivars

Protoplast fusion is a tool to transfer in one step a male-sterile cytoplasm to different fertile varieties and so save several years of backcrosses needed by the conventional sexual method.

Rice

Yang *et al.* (1988, 1989), Akagi *et al.* (1989) and Kyozuka *et al.* (1989) succeeded in regenerating cybrid plants possessing the nucleus of a fertile variety and mitochondria derived by recombination from both parents but retaining the CMS trait from the male-sterile partner. These authors used the donor–recipient method described by Sidorov *et al.* (1981).

Carrot

Tanno-Suenaga *et al.* (1988) transferred the CMS trait (brown anther type) to a fertile cultivar by the donor–recipient method and Ichikawa & Imamura (1990) proposed an improved method using a recipient nuclear genotype in which a dominant selectable marker, kanamycin resistance, was introduced by *Agrobacterium*-mediated gene transfer (Chapter 3).

Brassica

Barsby *et al.* (1987b) transferred cytoplasmic male sterility of the Polima type to different oilseed rape cultivars and Primard *et al.* (1988) introduced into winter oilseed-type rape an improved CMS cybrid cytoplasm previously obtained by somatic hybridization.

Sugarbeet

Original procedures for protoplast culture in the presence of *n*-propylgallate, a lipoxygenase inhibitor, and the regeneration of entire plants from protoplasts (Krens *et al.*, 1990) open up possibilities for the transfer of cytoplasms from different CMS sources to different nuclear backgrounds. The method would be of particular interest for sugarbeet which is biennial, and where hybrid production world-wide is based on only the Owen (1945) CMS source, which represents a vulnerable situation.

Tobacco

Kumashiro *et al.* (1988) succeeded in creating a new type of cytoplasmic male sterility by fusing *N. tabacum* protoplasts with irradiated *N. africana* protoplasts, a combination which is sexually incompatible, particularly when the latter species is used as the maternal parent.

Citrus

Male sterility in cultivars with potential parthenocarpy would be of great economic interest. Vardi *et al.* (1989) succeeded in regenerating citrus cybrids with organelles (plastids and some mitochondrial sequences) from microcitrus. If male sterility is revealed among such plants, they would constitute a new tool for breeding seedless citrus varieties.

Improvement of cytoplasmic male sterility systems in Cruciferae

Brassica crops have great economic importance as vegetables (*Brassica oleracea, B. campestris*) or for oil production (*B. napus, B. juncea, B.*

campestris). They represent the third largest oil source after soybean and palm for human consumption and industrial uses. F_1 hybrid varieties have been produced in these species using self-incompatibility systems, but never on a large scale, and there was a real need for an efficient system of control of pollination through male sterility. Cytoplasmic male sterility discovered in radish (*Raphanus sativus*) by Ogura (1968) was introduced into *B. oleracea* and *B. napus* by Bannerot *et al.* (1974, 1977) through interspecific sexual crosses, but the resulting plants, although perfectly male-sterile, were practically unusable because of chlorophyll deficiency at low temperatures and flower malformations causing inefficient nectar production and reduced female fertility.

Somatic fusion between *Brassica* protoplasts bearing 'Ogura' radish organelles and protoplasts with *Brassica* organelles was performed to exchange radish chloroplasts for *Brassica* chloroplasts in order to recover normal chlorophyll synthesis (Pelletier *et al.*, 1983; Jarl, 1988). Male-sterile normal-greening plants were screened among regenerated plants. They were shown to contain recombined mitochondrial genomes (Chétrit *et al.*, 1985; Vedel *et al.*, 1986; G. Pelletier, D. Lancelin & M. Férault, unpublished results). Among these cybrids, some were selected by breeders because they were shown in field tests with their progeny to be corrected for the other defects related to floral morphology: they produced enough nectar to be as attractive to bees as fertile plants and possessed normal female fertility (Pelletier *et al.*, 1988; Fig. 8.3). As shown in Fig. 8.4 these cybrids are easily characterized in Southern hybridizations by the presence of a specific Sal I fragment of 4.3 kb and by the loss of an Ogura-specific Sal I fragment of 5.1 kb, which both hybridize to the same probe. The 4.3 Sal I fragment contains part of a specific Ogura region bearing the sequence responsible for male sterility (Bonhomme *et al.*, 1991).

That nuclear cytoplasm interactions between different species as in the case of *Brassica* nuclei and *Raphanus* cytoplasm lead to developmental abnormalities is well known by plant geneticists, who observed similar abnormalities in interspecific crosses between *Oenothera* species (Kirk & Tilney-Bassct, 1978) or in intergeneric crosses between *Triticum* and *Aegilops* (Panayotov, 1980). The results with *Brassica* cybrids confirm that, from a plant breeding perspective, only a small part of a foreign genome needs to be introgressed in the cultivated species when it bears a trait of interest like cytoplasmic male sterility, and that entire organelle genomes bring, together with the desirable trait, other genes whose interaction with the nucleus of the cultivated species has a high probability of leading to agronomical defects. These results demonstrate the practical importance of mitochondrial recombination, which allows the elimination of these unfavourable traits. These new cytoplasms are now used by breeders, who will be able to produce original F_1 hybrid varieties in the next few years in *Brassica napus*, *B. oleracea* and *B. campestris*, with an 'improved' system of

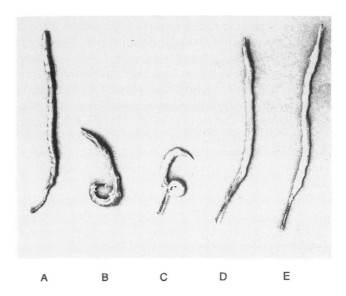

A   B   C   D   E

Fig. 8.3. Comparison of fruit morphology of *Brassica oleracea* plants differing by cytoplasmic organelles. A: *B. oleracea* organelles; B: *B. napus* plastids and Ogura/*B. napus* recombined mitochondria; C: *B. oleracea* plastids and Ogura (*R. sativus*) mitochondria; D and E: *B. oleracea* plastids and Ogura/*B. oleracea* recombined mitochondria. Only in these last two cases can a normal seed production be expected from CMS plants.

pollination control. Other combinations of cytoplasmic organelles in the same family were obtained by several groups and could have agronomic interest. Barsby *et al.* (1987a) combined Polima CMS and cytoplasmic triazine resistance in *Brassica napus* by fusing protoplasts from CMS rapeseed with rapeseed protoplasts bearing the cytoplasm of *B. campestris*. Jourdan *et al.* (1989b) synthesized male-sterile triazine-resistant *Brassica napus* by somatic hybridization between Ogura cytoplasmic male-sterile *B. oleracea* and atrazine-resistant *B. campestris*. The same group (Jourdan *et al.*, 1989a) also transferred triazine-resistant chloroplasts from *B. napus* to cauliflower, where this herbicide resistance trait could be of economic interest.

Conclusions

If we consider the prospects for manipulation of cytoplasmic genomes, two questions may be asked:

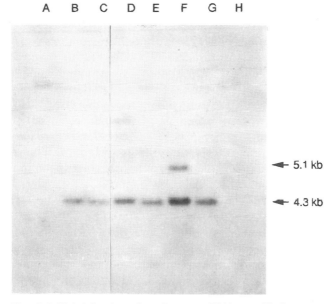

Fig. 8.4. Hybridization of an Ogura mtDNA-specific fragment to total DNA restricted by Sal I from: A: *B. oleracea*; B, C, D, E: male-sterile cybrids between *B. oleracea* and Ogura; F: Ogura radish; G: male-sterile cybrid between Ogura and *B. napus*; H: *B. napus*. Male-sterile cybrids shown here are those recovering a normal flower and fruit morphology. They retained the 4.3 kb Ogura Sal I fragment and lost the 5.1 kb Ogura Sal I fragment. The probe used in this hybridization is a part of the 4.3 kb Sal I fragment which is correlated with male sterility. (From Bonhomme *et al.*, 1991, and D. Lancelin, M. Férault, L. Boulidard & G. Pelletier, submitted.)

1. What sort of technical progress is still possible?
2. What kinds of improvement of cultivated species have we not yet considered?

On the technical side, considerable progress is required in order to increase the number of species where it is possible to regenerate plants from proto-plasts at a frequency sufficient to allow recovery of cybrids. We have reason to be optimistic about this problem if we consider the significant progress accomplished with the cereals, a group of species previously considered to be recalcitrant. During the past ten years, it has been shown that somatic embryos and plants can be obtained from protoplasts isolated from embryo-genic cell suspension cultures in pearl millet (Vasil & Vasil, 1980), rice

(Fujimura *et al.*, 1985), maize (Rhodes *et al.*, 1988) and wheat (Vasil *et al.*, 1990). Progress appears relatively slow, but is significant (see also Chapter 7).

Genetic modifications obtained by organelle recombination occur as uncontrolled phenomena and researchers would like to have more precise tools to modify organelle gene sequences, as is now possible for nuclear genes by DNA transfer. DNA uptake and integration are being attempted for organelle genomes by the particle-gun technique (Chapter 5). Another approach could be to take advantage of the natural import processes of macromolecules into these organelles. Vestweber & Schatz (1989) demonstrated the *in vitro* uptake of DNA by yeast mitochondria when this DNA is covalently linked to a mitochondrial precursor protein including a targeting sequence. Other instances are given in Chapter 9.

More technical progress must be achieved in order to have a greater understanding of the functioning of plastid and mitochondrial genes. The use of novel techniques from a plant breeding perspective also requires increased knowledge of the genetic resources available from such genomes. To some extent, breeders have neglected the importance of cytoplasmic genes. Most associations between nuclei and entire cytoplasms of distantly related species failed to give better agronomic performances than homospecific combinations. However, results presented in this chapter illustrate the interest in partial genome transfer compared with entire genome transfer, and cytoplasmic information must now be considered in crop improvement as an association of separable elements. This is extremely important considering that cytoplasms of cultivated varieties have, in general, a very narrow genetic basis.

References

Akada, S. & Hirai, A. (1986) Chloroplast genomes in hybrid calli derived from cell fusion: a novel system to study chloroplast segregation in hybrid calli. In: Mantell, S. H., Chapman, G. P. & Street, P. F. S. (eds), *The Chondriome Chloroplast and Mitochondrial Genomes*. Longmans Press, Essex, England, pp. 290–8.

Akagi, H., Sakamoto, M., Negishi, T. & Fujimura, T. (1989) Construction of rice cybrid plants. *Molecular and General Genetics* 215, 501–6.

Aviv, D. & Galun, E. (1980) Restoration of fertility in cytoplasmic male sterile (CMS) *Nicotiana sylvestris* by fusion with X-irradiated *N. tabacum* protoplasts. *Theoretical and Applied Genetics* 58, 121–7.

Aviv, D. & Galun, E. (1986) Restoration of male fertile *Nicotiana* by fusion of protoplasts derived from two different cytoplasmic male sterile cybrids. *Plant Molecular Biology* 7, 411–17.

Aviv, D. & Galun, E. (1988) Transfer of cytoplasmic organelles from an oligomycin-resistant *Nicotiana* cell suspension into tobacco protoplasts yielding oligomycin-resistant cybrid plants. *Molecular and General Genetics* 215, 128–33.

Aviv, D., Fluhr, F., Edelman, M. & Galun, E. (1980) Progeny analysis of the interspecific somatic hybrids *Nicotiana tabacum* (CMS) + *Nicotiana sylvestris* with respect to nuclear and chloroplast markers. *Theoretical and Applied Genetics* 56, 145–50.

Bannerot, H., Boulidard, L., Cauderon, Y. & Tempé, J. (1974) Transfer of cytoplasmic male sterility from *Raphanus sativus* to *Brassica oleracea*. *Proceedings of Eucarpia Meeting Cruciferae* 25, 52–4.

Bannerot, H., Boulidard, L. & Chupeau, Y. (1977) Unexpected difficulties met with the radish cytoplasm. *Eucarpia Cruciferae Newsletter* 2, 16.

Barsby, T. L., Chuong, P. V., Yarrow, S. A., Wu, S. C., Coulmans, M., Kemble, R. J., Powell, A. D., Beversdorf, W. P. & Pauls, K. P. (1987a) The combination of Polima (CMS) and cytoplasmic triazine resistance in *Brassica napus*. *Theoretical and Applied Genetics* 73, 809–14.

Barsby, T. L., Yarrow, S. A., Kemble, R. J. & Grant, I. (1987b) The transfer of cytoplasmic male sterility to winter type oilseed rape (*Brassica napus* L.) by protoplast fusion. *Plant Science* 53, 243–8.

Belliard, G., Pelletier, G., Vedel, F. & Quetier, F. (1978) Morphological characteristics and chloroplast DNA distribution in different cytoplasmic parasexual hybrids of *Nicotiana tabacum*. *Molecular and General Genetics* 165, 231–7.

Belliard, G., Vedel, F. & Pelletier, G. (1979) Mitochondrial recombination in cytoplasmic hybrids of *Nicotiana tabacum* by protoplast fusion. *Nature* 281, 401–3.

Boeshore, M. L., Lifshitz, I., Manson, M. & Izhar, S. (1983) Novel composition of mitochondrial genomes in *Petunia* somatic hybrids derived from cytoplasmic male sterile and fertile plants. *Molecular and General Genetics* 190, 459–67.

Bonhomme, S., Budar, F., Férault, M. & Pelletier, G. (1991) A 2.5 kb NcoI fragment of Ogura radish mitochondrial DNA is correlated with cytoplasmic male sterility in Brassica hybrids. *Current Genetics* 19, 121–7.

Bonnett, H. T. & Glimelius, K. (1983) Somatic hybridization in *Nicotiana*: behaviour or organelles after fusion of protoplasts from male fertile and male sterile cultivars. *Theoretical and Applied Genetics* 65, 213–17.

Bonnett, H. T. & Glimelius, K. (1990) Cybrids of *Nicotiana tabacum* and *Petunia hybrida* have an intergeneric mixture of chloroplasts from *P. hybrida* and mitochondria identical or similar to *N. tabacum*. *Theoretical and Applied Genetics* 79, 550–5.

Boynton, J. E., Gillham, N. W., Harris, E. H., Hosle, J. P., Johnson, A. M., Jones, A. R., Randolph-Anderson, B. L., Robertson, D., Klein, T. M., Shark, K. B. & Sanford, J. C. (1988) Chloroplast transformation in *Chlamydomonas* with high velocity microprojectiles. *Science* 240, 1534–8.

Bradley, P. M. (1983) The production of higher plant subprotoplasts. *Plant Molecular Biology Reporter* 1, 117–23.

Brennicke, A. & Schwemmle, B. (1984) Inheritance of mitochondrial DNA in *Oenothera berteriana* and *Oenothera odorata* hybrids. *Zeitschrift für Naturforschung, section C, Biosciences* 39, 191–2.

Chen, B. Y. & Heneen, W. K. (1989) Evidence for spontaneous diploid androgenesis in *Brassica napus* L. *Plant Sexual Reproduction* 2, 15–17.

Chen, K., Wildman, S. G. & Smith, M. H. (1977) Chloroplast DNA distribution in parasexual hybrids as shown by polypeptide composition of fraction 1 protein. *Proceedings of the National Academy of Sciences USA* 74, 5109–12.

Chétrit, P., Mathieu, C., Vedel, F., Pelletier, G. & Primard, C. (1985) Mito-chondrial DNA polymorphism induced by protoplast fusion in Cruciferae. *Theoretical and Applied Genetics* 69, 361–6.

Chiu, W. L. & Sears, B. B. (1985) Recombination between chloroplast DNAs does not occur in sexual crosses of *Oenothera*. *Molecular and General Genetics* 198, 525–8.

Constabel, F., Dudits, D., Gamborg, D. L. & Kao, K. N. (1975) Nuclear fusion in intergeneric heterocaryons. *Canadian Journal of Botany* 53, 2092–5.

Cornu, A. & Dulieu, H. (1988) Pollen transmission of plastid-DNA under genotypic control in *Petunia hybrida* HORT. *Journal of Heredity* 79, 40–4.

Daniell, H., Vivekananda, J., Nielsen, B. L., Ye, G. N., Tewari, K. K. & Sanford, J. C. (1990) Transient foreign gene expression in chloroplasts of cultured tobacco cells after biolistic delivery of chloroplast vectors. *Proceedings of the National Academy of Sciences USA* 87, 88–92.

Durand, J. (1987) Isolation of antibiotic resistant variants in a higher plant, *Nicotiana sylvestris*. *Plant Science* 51, 113–18.

Duvick, D. N. (1959) The use of cytoplasmic male sterility in hybrid seed production. *Economic Botany* 13, 167–95.

Edwarson, J. R. (1970) Cytoplasmic male sterility. *Botanical Review* 36, 341–420.

Erickson, L. & Kemble, R. (1990) Paternal inheritance of mitochondria in rapeseed (*Brassica napus*). *Molecular and General Genetics* 222, 135–9.

Fejes, E., Engler, D. & Maliga, P. (1990) Extensive homologous chloroplast DNA recombination in the pt 14 *Nicotiana* somatic hybrid. *Theoretical and Applied Genetics* 79, 28–32.

Flick, C. E., Kut, S. A., Bravo, J. E., Gleba, Y. Y. & Evans, D. A. (1985) Segrega-tion of organelle trait following protoplast fusion in *Nicotiana*. *Bio/Technology* 3, 555–60.

Fluhr, R., Aviv, D., Edelman, M. & Galun, E. (1983) Cybrids containing mixed and sorted-out chloroplasts following interspecific somatic fusions in *Nicotiana*. *Theoretical and Applied Genetics* 65, 289–94.

Fujimura, T., Sakurai, M., Akagi, H., Negishi, T. & Hirose, A. (1985) Regenera-tion of rice plants from protoplasts. *Plant Tissue Culture Letter* 2, 74–5.

Galun, E. & Aviv, D. (1986) Organelle transfer. *Methods in Enzymology* 118, 565–611.

Galun, E., Arzee-Gonen, P., Fluhr, R., Edelman, M. & Aviv, D. (1982) Cytoplas-mic hybridization in *Nicotiana*: mitochondrial DNA analysis of progenies result-ing from fusion between protoplasts having different organelle constitutions. *Molecular and General Genetics* 186, 50–6.

Gerstel, D. U. (1980) Cytoplasmic male sterility in *Nicotiana* (a review). *North Carolina Technical Bulletin USA* 263, 1–31.

Gleba, Y. Y., Kolesnik, N. N., Meshkene, I. V., Cherep, N. N. & Parokonny, A. S. (1984) Transmission genetics of the somatic hybridization process in *Nicotiana*. I. Hybrids and cybrids among the regenerates from cloned protoplast fusion products. *Theoretical and Applied Genetics* 69, 121–8.

Gleba, Y. Y., Komarnitsky, I. K., Kolesnik, N. N., Meshkene, I. & Martyn, G. I. (1985) Transmission genetics of the somatic hybridization process in *Nicotiana*. II. Plastome heterozygotes. *Molecular and General Genetics* 198, 476–81.

Glimelius, K. (1988) Potentials of protoplast fusion in plant breeding programmes. *Plant Cell, Tissue and Organ Culture* 12, 159–68.

Glimelius, K. & Bonnett, H. T. (1981) Somatic hybridization in *Nicotiana:* restoration of photoautotrophy to an albino mutant with defective plastids. *Planta* 153, 497–503.

Glimelius, K. & Bonnett, H. T. (1986) *Nicotiana* cybrids with *Petunia* chloroplasts. *Theoretical and Applied Genetics* 72, 794–8.

Goodsell, S. F. (1961) Male sterility in corn by androgenesis. *Crop Science* 1, 227–8.

Grun, P. (1976) *Cytoplasmic Genetics and Evolution.* Columbia University Press, New York.

Horlow, C., Goujaud, J., Lepingle, A., Missonier, C. & Bourgin, J. P. (1990) Transmission of paternal chloroplasts in tobacco (*Nicotiana tabacum*). *Plant Cell Reports* 2, 249–52.

Ichikawa, H. & Imamura, J. (1990) A highly efficient selection method for somatic hybrids which uses an introduced dominant selectable marker combined with iodoacetamide treatment. *Plant Science* 67, 227–35.

Jarl, C. I. (1988) 'Photosynthesis defective male sterile rapeseed: characterization and correction.' Thesis, Department of Genetics and Breeding, Swedish University of Agricultural Sciences, Svalöv, Sweden.

Jourdan, P. S., Earle, E. D. & Mutschler, M. A. (1989a) Atrazine resistant cauliflower obtained by somatic hybridization between *Brassica oleracea* and ATR–*B. napus. Theoretical and Applied Genetics* 78, 271–9.

Jourdan, P. S., Earle, E. D. & Mutschler, M. A. (1989b) Synthesis of male sterile triazine-resistant *Brassica napus* by somatic hybridization between cytoplasmic male sterile *B. oleracea* and atrazine-resistant *B. campestris. Theoretical and Applied Genetics* 78, 445–55.

Kaul, M. L. M. (1988) *Male Sterility in Higher Plants.* Monographs on Theoretical and Applied Genetics 10, Springer-Verlag, Berlin, Heidelberg, New York, London, Paris, Tokyo.

Kirk, J. T. O. & Tilney-Bassett, R. E. A. (1978) *The Plastids: Their Chemistry, Structure, Growth and Inheritance.* W. H. Freeman, London.

Köfer, W., Glimelius, K. & Bonnett, H. T. (1990a) Modifications of floral development in tobacco induced by fusion of protoplasts of different male sterile cultivars. *Theoretical and Applied Genetics* 79, 97–102.

Köfer, W., Glimelius, K. & Bonnett, H. T. (1990b) Restoration of normal stamen development and pollen formation by fusion of different cytoplasmic male-sterile cultivars of *Nicotiana tabacum. Theoretical and Applied Genetics* 81, 390–6.

Krens, F. A., Jamar, D., Rouwendal, G. J. A. & Hall, R. D. (1990) Transfer of cytoplasm from new Beta CMS sources to sugar beet by asymmetric fusion. 1. Shoot regeneration from mesophyll protoplasts and characterization of regenerated plants. *Theoretical and Applied Genetics* 79, 390–6.

Krishna-Rao, M. & Koduru, P. R. K. (1978) Biparental plastid inheritance in *Pennisetum americanum. Journal of Heredity* 69, 327–30.

Kumashiro, T., Asahi, T. & Komari, T. (1988) A new type of cytoplasmic male sterile tobacco obtained by fusion between *Nicotiana tabacum* and X-irradiated *N. africana* protoplasts. *Plant Science* 55, 247–54.

Kushnir, S. G., Shlumukov, L. R., Pogrebnyak, N. J., Berger, S. & Gleba, Y. Y. (1987) Functional cybrid plants possessing a *Nicotiana* genome and an *Atropa* plastome. *Molecular and General Genetics* 209, 159–63.

Kyozuka, J., Kaneda, T. & Shimamoto, K. (1989) Production of cytoplasmic male sterile rice (*Oryza sativa* L.) by cell fusion. *Bio technology* 7, 1171–4.

Lee, D. J., Blake, T. K. & Smith, S. E. (1988) Biparental inheritance of chloroplast DNA and the existence of heteroplasmic cells in alfalfa. *Theoretical and Applied Genetics* 76, 545–9.

Lonsdale, D. (1987) Cytoplasmic male sterility: a molecular perspective. *Plant Physiology and Biochemistry* 25, 265–72.

Masoud, S. A., Johnson, L. B. & Sorensen, E. L. (1990) High transmission of paternal plastid DNA in alfalfa plants demonstrated by restriction fragment polymorphism analysis. *Theoretical and Applied Genetics* 79, 49–55.

Medgyesy, P., Menczel, L. & Maliga, P. (1980) The use of cytoplasmic streptomycin resistance: chloroplast transfer from *Nicotiana tabacum* into *Nicotiana sylvestris* and isolation of their somatic hybrids. *Molecular and General Genetics* 179, 693–8.

Medgyesy, P., Fejes, E. & Maliga, P. (1985) Interspecific chloroplast recombination in a *Nicotiana* somatic hybrid. *Proceedings of the National Academy of Sciences USA* 82, 6960–4.

Medgyesy, P., Pay, A. & Marton, L. (1986) Transmission of paternal chloroplasts in *Nicotiana*. *Molecular and General Genetics* 204, 195–8.

Menczel, L., Nagy, F., Kiss, Z. R. & Maliga, P. (1981) Streptomycin resistant and light sensitive somatic hybrids of *Nicotiana tabacum* + *Nicotiana knightiana*: correlation of resistance to *N. tabacum* plastids. *Theoretical and Applied Genetics* 59, 191–5.

Menczel, L., Morgan, A., Brown, S. & Maliga, P. (1987) Fusion mediated combination of Ogura-type cytoplasmic male sterility with *Brassica napus* plastids using X irradiated CMS protoplasts. *Plant Cell Reports* 6, 98–101.

Michaelis, P. (1967) The investigation of plasmone segregation by the pattern analysis. *Nucleus* 10, 1–14.

Nagy, F., Lazar, G., Menczel, L. & Maliga, P. (1983) A heteroplasmic state induced by protoplast fusion is a necessary condition for detecting rearrangements in *Nicotiana* mitochondrial DNA. *Theoretical and Applied Genetics* 66, 203–7.

Neale, D. B. & Sederoff, R. R. (1989) Paternal inheritance of chloroplast DNA and maternal inheritance of mitochondrial DNA in loblolly pine. *Theoretical and Applied Genetics* 77, 212–16.

Neale, D. B., Wheeler, N. C. & Allard, R. W. (1986) Paternal inheritance of chloroplast DNA in Douglar fir. *Canadian Journal of Forest Research* 16, 1152–4.

Neale, D. B., Marshall, K. A. & Sederoff, R. R. (1989) Chloroplast and mitochondrial DNA are paternally inherited in *Sequoia sempervirens* D. Don Endl. *Proceeding of the National Academy of Sciences USA* 86, 9347–9.

Ogura, H. (1968) Studies of the new male sterility in Japanese radish with special reference to the utilization of this sterility towards the practical raising of hybrid seeds. *Memoirs of the Faculty of Agriculture, Kayoshima University*, 6, 39–78.

Ohba, K., Iwakawa, M., Okada, Y. & Murai, M. (1971) Paternal transmission of a plastid anomaly in some reciprocal crosses of Sugi, *Cryptomeria Japonica* D. Don. *Silvae Genetica* 20, 101–7.

Owen, F. V. (1945) Cytoplasmically inherited male sterility in sugar beet. *Journal of Agricultural Research* 71, 423–40.

Palmer, J. (1985) Comparative organization of chloroplast genomes. *Annual Review of Genetics* 19, 325–54.

Palmer, J. D. & Herbon, L. A. (1988) Plant mitochondrial DNA evolves rapidly in structure, but slowly in sequence. *Journal of Molecular Evolution* 28, 87–97.

Panayotov, I. (1980) New cytoplasmic male sterility sources in common wheat: their genetical and breeding considerations. *Theoretical and Applied Genetics* 56, 153–60.

Pelletier, G. R. (1986) Plant organelle genetics through somatic hybridization. In: Miflin, B. (ed.), *Oxford Surveys of Plant Molecular and Cell Biology.* Oxford University Press, Oxford, pp. 97–121.

Pelletier, G., Primard, C., Vedel, F., Chétrit, P., Remy, R., Rousselle, P. & Renard, M. (1983) Intergeneric cytoplasmic hybridization in Cruciferae by protoplast fusion. *Molecular and General Genetics* 191, 244–50.

Pelletier, G., Ferault, M., Goujaud, J., Vedel, F. & Caboche, M. (1987) The use of rootless mutants for the screening of spontaneous androgenetic and gynogenetic haploids in *Nicotiana tabacum*: evidence for the direct transfer of cytoplasm. *Theoretical and Applied Genetics* 75, 13–15.

Pelletier, G., Primard, C., Ferault, M., Vedel, F., Chétrit, P., Renard, M. & Delourme, R. (1988) Use of protoplasts in plant breeding: cytoplasmic aspects. *Plant Cell, Tissue and Organ Culture* 12, 173–80.

Pental, D., Hamill, J. D., Pirrie, A. & Cocking, E. C. (1986) Somatic hybridization of *Nicotiana tabacum* and *Petunia hybrida*: recovery of plants with *Petunia hybrida* nuclear genome and *Nicotiana tabacum* chloroplast genome. *Molecular and General Genetics* 202, 342–7.

Pental, D., Pradhan, A. K. & Mukhopadhyay, A. (1989) Transmission of organelles in triploid hybrids produced by gametosomatic fusions of two *Nicotiana* species. *Theoretical and Applied Genetics* 78, 547–52.

Primard, C., Lepingle, A., Masson, J., Lancelin, D., Chèvre, A. M. & Pelletier, G. (1988) Transfer of male sterile cybrid cytoplasms into winter rape cultivar of *Brassica napus* by protoplast fusion. *Cruciferae Newsletter* 13, 78–9.

Rhodes, C. A., Pierce, D. A., Mettler, I. J., Mascarenhas, D. & Detmar, J. J. (1988) Genetically transformed maize plants from protoplasts. *Science* 240, 204–7.

Rothenberg, M., Boeshore, M. L., Hanson, M. R. & Izhar, S. (1985) Intergenomic recombination of mitochondrial genomes in a somatic hybrid plant. *Current Genetics* 9, 615–18.

Schiller, B., Herrmann, R. G. & Melchers, G. (1982) Restriction endonuclease analysis of plastid DNA from tomato, potato and some of their somatic hybrids. *Molecular and General Genetics* 186, 453–9.

Schumann, C. M. & Hancock, J. F. (1989) Paternal inheritance of plastids in *Medicago sativa. Theoretical and Applied Genetics* 78, 863–6.

Sidorov, V. A., Menczel, L., Nagy, F. & Maliga, P. (1981) Chloroplast transfer in *Nicotiana* based on metabolic complementation between irradiated and iodoacetate treated protoplasts. *Planta* 152, 341–5.

Sidorov, V. A., Zubko, M. K., Kuchko, A. A., Komarnitsk, I. K. & Gleba, Y. Y. (1987) Somatic hybridization in potato: use of γ-irradiated protoplasts of *Solanum pinnatisectum* in genetic reconstruction. *Theoretical and Applied Genetics* 74, 364–8.

Smith, S. E. (1989) Biparental inheritance of organelles. *Plant Breeding Review* 6, 361–93.

Smith, S. E., Bingham, E. T. & Fulton, R. W. (1986) Transmission of chlorophyll deficiencies in *Medicago sativa*: evidence for biparental inheritance of plastids. *Journal of Heredity* 77, 35–8.

Stine, M., Sears, B. B. & Keathley, D. E. (1989) Inheritance of plastids in interspecific hybrids of blue spruce and white spruce. *Theoretical and Applied Genetics* 78, 768–74.

Szmidt, A. E., Alden, T. & Hallgren, J. E. (1987) Paternal inheritance of chloroplast DNA in *Larix*. *Plant Molecular Biology* 9, 59–64.

Szmidt, A. E., El-Kassaby, Y. A., Sigurgeirsson, A. & Alden, T. (1988) Classifying seedlots of *Picea sitchensis* and *P. glauca* in zones of introgression, using restriction analysis of chloroplast DNA. *Theoretical and Applied Genetics* 76, 841–5.

Tanno-Suenaga, L., Ichikawa, H. & Imamura, J. (1988) Transfer of the CMS trait in *Daucus Carota* L. by donor–recipient protoplast fusion. *Theoretical and Applied Genetics* 76, 855–60.

Thanh, N. D. & Medgyesy, P. (1989) Limited chloroplast gene transfer via recombination overcomes plastome genome incompatibility between *Nicotiana tabacum* and *Solanum tuberosum*. *Plant Molecular Biology* 12, 87–93.

Thanh, N. D., Pay, A., Smith, M. A., Medgyesy, P. & Marton, L. (1988) Intertribal chloroplast transfer by protoplast fusion between *Nicotiana tabacum* and *Salpiglossis sinuata*. *Molecular and General Genetics* 213, 186–90.

Thomzik, J. E. & Hain, R. (1988) Transfer and segregation of triazine tolerant chloroplasts in *Brassica napus* L. *Theoretical and Applied Genetics* 76, 165–71.

Tilney-Bassett, R. A. E. & Abdel-Wahab, O. A. L. (1979) Maternal effects and plastid inheritance. In: Newth, D. R. & Balls, M. (eds), *Maternal Effects in Development*. British Society of Developmental Biology Symposium 4, Cambridge University Press, London, pp. 29–45.

Vardi, A., Arzee-Gonen, P., Frydman-Shani, A., Bleichman, S. & Galun, E. (1989) Protoplast fusion mediated transfer of organelles from *Microcitrus* into *Citrus* and regeneration of novel alloplasmic trees. *Theoretical and Applied Genetics* 78, 741–7.

Vasil, I. K., Vasil, V. & Redway, F. (1990) Plant regeneration from embryogenic calli, cell suspension cultures and protoplasts of *Triticum aestivum* L. (wheat). In: Nijkamp, M. J. J., Van der Plas, L. H. W. & Van Aartrijk, J. (eds), *Current Plant Science and Biotechnology in Agriculture*, vol. 9, *Progress in Plant Cellular and Molecular Biology*. Kluwer Academie, Dordrecht, Boston, London, pp. 33–7.

Vasil, V. & Vasil, I. K. (1980) Isolation and culture of cereal protoplasts. II. Embryogenesis and plantlet formation from protoplasts of *Pennisetum americanum*. *Theoretical and Applied Genetics* 56, 97–9.

Vedel, F., Chétrit, P., Mathieu, C., Pelletier, G. & Primard, C. (1986) Several different mitochondrial DNA regions are involved in intergenomic recombination in *Brassica napus* cybrid plants. *Current Genetics* 11, 17–24.

Vestweber, D. & Schatz, G. (1989) DNA protein conjugates can enter mitochondria via the protein import pathway. *Nature* 338, 170–2.

Wagner, D. B., Furnier, G. R., Saghai-Maroof, M. A., Williams, S. M., Dancik, B. P. & Allard, R. W. (1987) Chloroplast DNA polymorphisms in lodgepole and jack pines and their hybrids. *Proceedings of the National Academy of Sciences USA* 84, 2097–100.

Wagner, D. B., Gouindaraju, D. R., Yeatman, C. W. & Pitel, J. A. (1989) Paternal chloroplast DNA inheritance in a diallel cross of Jack Pine (*Pinus banksiana* Lamb). *Journal of Heredity* 80, 483–5.

White, E. E. (1990) Chloroplast DNA in *Pinus monticola* survey of within species variability and detection of heteroplasmic individuals. *Theoretical and Applied Genetics* 79, 251–5.

Wilson, P. & Driscoll, C. J. (1983) Hybrid wheat. In: Frankel, R. (ed.), *Heterosis: Reappraisal of Theory and Practice*. Monographs on Theoretical and Applied Genetics 6, Springer Verlag, Berlin, Heidelberg, New York, Tokyo, pp. 94–123.

Wright, W. E. (1978) The isolation of heterocaryons and hybrids by a selective system using irreversible biochemical inhibitors. *Experimental Cell Research* 112, 395–407.

Yang, Z. Q., Shikanai, T. & Yamada, Y. (1988) Asymmetric hybridization between cytoplasmic male-sterile (CMS) and fertile rice (*Oryza sativa* L.) protoplasts. *Theoretical and Applied Genetics* 76, 801–8.

Yang, Z. Q., Shikanai, T., Mori, K. & Yamada, Y. (1989) Plant regeneration from cytoplasmic hybrids of rice (*Oryza sativa* L.). *Theoretical and Applied Genetics* 77, 305–10.

Zelcer, A., Aviv, D. & Galun, E. (1978) Interspecific transfer of cytoplasmic male sterility by fusion between protoplasts of normal *Nicotiana sylvestris* and x ray irradiated protoplasts of male sterile *N. tabacum*. *Zeitschrift für Pflanzenphysiologie* 90, 397–407.

Chapter 9
Modification of the Chloroplast Genome with Particular Reference to Herbicide Resistance

Ray J. Rose
Department of Biological Sciences, University of Newcastle, New South Wales 2308, Australia

Introduction

The chloroplast genome is the most thoroughly studied of the three plant genomes. Chloroplast DNA (cpDNA) has been completely sequenced in two angiosperms, tobacco (Shinozaki *et al.*, 1986) and rice (Hiratsuka *et al.*, 1989), and in the liverwort *Marchantia polymorpha* (Ohyama *et al.*, 1986). Furthermore, most of the cpDNA-encoded proteins that are associated with photosynthesis have been mapped to the chloroplast chromosome (Shinozaki *et al.*, 1988; Sugiura, 1989). Although the chloroplast genome is the best known of the plant genomes it is the least amenable to manipulation. Two major reasons for this are the absence of a suitable vector (Haring & De Block, 1990) and the rarity of intermolecular recombination (Sears, 1980; Rose *et al.*, 1990). It is perhaps not surprising, then, that the chloroplast genome is more conserved than the nuclear or mitochondrial genomes (Palmer, 1985a).

As key photosynthetic genes are highly conserved (Zurawski & Clegg, 1987), it might be asked why means of modifying cpDNA are being sought. Two examples suffice to indicate that there are good agricultural reasons why a capacity for manipulating cpDNA is important. Resistances to photosystem II-inhibiting herbicides can involve the 32 kDa plastoquinone-binding protein of photosystem II (PSII) (Hirschberg & McIntosh, 1983), which is a psbA gene product (Whitfeld & Bottomley, 1983; Table 9.1). The rbcL gene (Table 9.1), which encodes the large (catalytic) subunit of ribulose bisphosphate carboxylase/oxygenase, is also a prime target for manipulation (Andrews, 1988; McFadden & Daniell, 1988).

In this chapter currently available and potential strategies for the

Table 9.1. Coding by the chloroplast genome.

Component	Gene (CP genes)	Number CP encoded	Number NUC encoded
Protein synthesizing apparatus			
rRNA	rDNA	4	0
tRNA	trn	30	0
ribosomal proteins (SS)	rps	12	24–26
ribosomal proteins (LS)	rpl	8	34–36
initiation factor 1	infA	1	0
Thylakoid membranes			
PSI complex	psa (A, B, C)	3	7*
PSII complex	psb (A, B, C, D, E, F, G, H, I, K, L)	11	5*
CYT b–f complex	pet (A, B, D)	3	3
ATP synthase CF_0	atp (A, B, E)	3	2
ATP synthase CF_1	atp (F, H, J)	3	1
RuBP carboxylase	rbc (L)	1	1
Others			
NADH dehydrogenase	ndh (A, B, C, D, E, F)	6	?
RNA polymerase	rpo (A, B, C_1, C_2)	4	1*
Open reading frames		38*	
Possible total genes		**127***	

Based on the tabulated data of Dyer (1984), Shinozaki *et al.* (1988), Sugiura (1989), Ghanotakis & Yocum (1990) and the studies of Capel & Bourque (1982), Ohyama *et al.* (1986), Shinozaki *et al.* (1986), Hiratsuka *et al.* (1989), Steppuhn *et al.* (1989) and Hu & Bogorad (1000). *indicates best estimates, SS = small subunit, LS = large subunit, CP = chloroplast, NUC = nucleus.

manipulation of the chloroplast genome are explored. Herbicide resistance in relation to the psbA gene is used to illustrate the application of cpDNA modification to an agricultural character important to plant breeders.

Inheritance of plastids

Plastids are most commonly maternally inherited. Based on reciprocal crosses with a relatively small number of species, about two-thirds of higher plant genera show maternal plastid inheritance, with one-third of genera showing at least occasional biparental inheritance (Smith, 1989). The recent studies by Corriveau & Coleman (1988) using the fluorochrome 4',6-diamidino-2-phenylindole (DAPI) indicate 14% of higher plant genera have cpDNA present in their pollen. It would seem then that somewhere between 14 and 33% of plants commonly show biparental inheritance. Some of this variation is accounted for by some genera showing only occasional biparental plastid inheritance, and plastid inheritance patterns are genotype-dependent (Smith, 1989). When inheritance is biparental the amount of paternal transmission is variable, with some species showing almost complete paternal transmission (Wagner *et al.*, 1987; Lee *et al.*, 1988; Strauss *et al.*, 1989; Masoud *et al.*, 1990; Chapter 8). Tilney-Basset (1988) has shown that plastid inheritance patterns in *Pelargonium* are under nuclear genetic control, with transmission ranging from completely maternal to paternal depending on the nuclear genotype.

There are two important points that can be made from the studies of plastid inheritance. First, a significant number of plants show biparental plastid inheritance, which allows different crossing strategies compared with maternal inheritance. Second, there is the potential for selection for biparental inheritance (Smith, 1989).

The chloroplast genome – an overview

The chloroplast genome has been the subject of extensive reviews (Herrmann & Possingham, 1980; Whitfeld & Bottomley, 1983; Dyer, 1984; Palmer, 1985b; Shinozaki *et al.*, 1988; Sugiura, 1989) and its main features are as follows.

Structure

Most chloroplast genomes are in the size range of 120–160 kb and are closed circular molecules (Palmer, 1985b) with an inverted repeat as part of their structure (Whitfeld & Bottomley, 1983; Palmer, 1985b), as shown in Fig. 9.1. The inverted repeat is, however, absent in some legumes (Palmer & Thompson, 1982; Palmer *et al.*, 1987) and conifers (Lidholm *et al.*, 1988; Strauss *et al.*, 1988). It has been argued that the inverted repeat acts to reduce variation through its effect on intramolecular recombination (Palmer & Thompson, 1982). However, it is apparent from the extreme cpDNA variation in plants regenerated from cultured wheat pollen (Day & Ellis, 1984) that the inverted repeat does not necessarily preclude extensive variation.

Chloroplast DNA molecules are usually present in groups of several molecules in nucleoids. The organization of cpDNA into nucleoids has been reviewed recently (Rose, 1988). The supercoiling and folding of the cpDNA to form nucleoids is dependent on basic histone-like proteins. The nucleoids

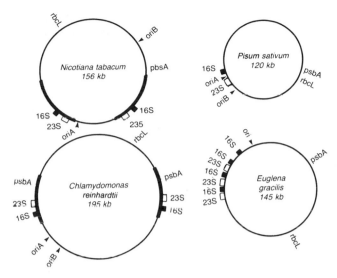

Fig. 9.1. Physical and gene maps of cpDNA from *Nicotiana tabacum, Pisum sativum, Chlamydomonas reinhardtii* and *Euglena gracilis* showing map location of 16S (closed boxes) and 23S (open boxes) rRNA genes, inverted repeats (thick lines), psbA and rbcL genes, and putative origins of replication (*ori*). The notation *ori*A and *ori*B denotes that on the cpDNA diagrammed there are two putative origins, with the designation A or B being arbitrary. The map diagrams of *C. reinhardtii* and *E. gracilis* cpDNA are modified from Palmer (1985b). The map diagram of *P. sativum* is also modified from Palmer (1985b), with the *ori* regions from Meeker *et al.* (1988). The map diagram of *N. tabacum* is modified from Shinozaki *et al.* (1986).

are bound to chloroplast membranes at precise locations depending on species and stage of plastid development. Such an arrangement helps ensure segregation of cpDNA at plastid division (Fig. 9.2).

Coding

The chloroplast genome encodes many of the components of its own protein synthesizing apparatus. These components are the ribosomal and transfer RNAs, a number of ribosomal proteins and a protein synthesis initiation factor (Table 9.1). It also appears that cpDNA encodes chloroplast RNA polymerase subunits (Sugiura, 1989; Hu & Bogorad, 1990). The chloroplast protein synthesizing system translates the cpDNA-encoded photosynthetic proteins. These cpDNA genes are listed in Table 9.1. Four of the five major

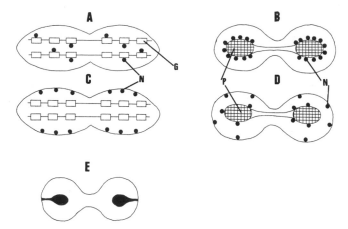

Fig. 9.2. Idealized diagram showing nucleoid segregation in dividing chloroplasts, etioplasts and proplastids. A, SN-type chloroplast; B, etioplast from plant with SN-type chloroplasts; C, PS-type chloroplast; D, etioplast from plant with PS-type chloroplasts; E, proplastid from plant with SN-type or PS-type chloroplasts. The plastid envelope is shown as a single line. SN refers to the scattered nucleoid arrangement and PS to the peripheral scattered nucleoid arrangement (Kuroiwa *et al.,* 1981). N, nucleoid; P, prolamellar body, G, granum (from Rose, 1988).

thylakoid protein complexes contain a number of cpDNA-encoded proteins, including the psbA protein of PSII. The large subunit of ribulose bisphosphate carboxylase is also encoded by cpDNA. The centrality of cpDNA to the photosynthetic process and hence plant productivity is clear.

Copy number

In most cases every plastid of a plant contains multiple copies of the same type of cpDNA. In spinach, chloroplasts in mesophyll cells contain 57 to 353 cpDNA copies (depending on leaf age) whereas root proplastids may contain as few as ten (Scott & Possingham, 1980). Even in this latter case root cells with about ten plastids (Possingham, 1980) will contain 100 cpDNA molecules. Spinach mesophyll cells each contain up to 200 chloroplasts (Possingham & Rose, 1976), giving a huge cpDNA copy number per cell. The cpDNA copy number per plastid changes in plant development in a predictable way (Scott & Possingham, 1983). In manipulating the chloroplast genome a key requirement is that the modified cpDNA must amplify and become dominant.

Heteroplasmy

Heteroplasmy of cpDNA in higher plants is not common; however, it does occur in the important forage legume alfalfa and in conifers (Govindaraju *et al.*, 1988; Johnson & Palmer, 1989). These latter species have predominantly paternal transmission (Wagner *et al.*, 1987; Lee *et al.*, 1988; Strauss *et al.*, 1989; Masoud *et al.*, 1990) and, as indicated previously, lack an inverted repeat. If the basis of heteroplasmy in higher plants were understood, then complementation between genomes might be feasible. Stable heteroplasmy has not been produced in somatic hybridizations carried out to date (Rose *et al.*, 1990).

Recombination

Recombination of cpDNAs has been observed following somatic hybridization (Medgyesy *et al.*, 1985; Thanh & Medgyesy, 1989). However, it must be considered a rare event, requiring good selection markers for detection (Rose *et al.*, 1990). There seems little doubt that the situation is quite different in mitochondria, where recombination following somatic hybridization has been commonly observed (Rose *et al.*, 1990).

Methods for the manipulation of the chloroplast genome

There are a number of different approaches to the manipulation of the chloroplast genome. These range from the established methods of sexual hybridization through to current research into the direct transformation of the chloroplast genome.

Sexual hybridization

Species with maternal plastid inheritance

In plants having maternal inheritance it is possible to transfer a new chloroplast type by hybridization and extensive backcrossing. The principle of the method has been clearly illustrated by Kaul (1988) in his extensive consideration of cytoplasmic male sterility. The female parent with the chloroplasts to be transferred is hybridized to the male parent with the required nuclear genotype. The hybrid plant produced is repeatedly backcrossed as a female to the male parent with the required nuclear genes. This plant breeding strategy is valuable where a plastid character has been identified which warrants complete replacement of the existing plastid. However, this approach is time-consuming and results in at least some change in the nuclear genotype.

Species with biparental plastid inheritance

In those species having biparental inheritance of plastid genomes cytoplasmic hybrids can be produced, which is not possible in those species with uniparental maternal plastid inheritance. An example of an agricultural species showing a high paternal plastid transmission is alfalfa (Lee *et al.*, 1988; Masoud *et al.*, 1990). Where mixed cytoplasms are present recombination is possible but, based on somatic hybridization studies (Medgyesy *et al.*, 1985; Thanh & Medgyesy, 1989) and observations on sexual hybrids of *Oenothera* (Chiu & Sears, 1985), it is a rare event.

In the case of biparental inheritance where a mixed population of plastids is present, sorting out would normally be expected to occur, but in some species heteroplasmy is possible (Rose *et al.*, 1986), allowing the potential for organelle complementation (Smith, 1989). Biparental inheritance gives much more flexibility to a breeding programme for improved cytoplasmic characters. However, unidirectional plastid transfer without backcrossing steps requires organelle transfer without nuclear transfer by one of the parents, and these systems are not available (Smith, 1989).

Production of chloroplast DNA mutations

A great many cpDNA mutations have arisen spontaneously or have been induced by mutagens or nuclear 'mutator' genes (Borner & Sears, 1986). In the first instance they were identified as cpDNA mutations by their non-Mendelian inheritance. These mutations represent valuable genetic resources for the dissection of the chloroplast genome as well as for research on the manipulation and transformation of the chloroplast genome, where they have particular value as markers. Borner & Sears (1986) have collated information on the non-Mendelian mutants that are available. These mutations have been shown to affect the chloroplast translational apparatus, photosystem I, photosystem II, the cytochrome f/b6 complex, carbon fixation or the ATP synthase.

Somatic hybridization

Somatic hybridization is discussed in detail in Chapter 8. Here I wish to emphasize the value of somatic hybridization compared with sexual hybridization in transferring chloroplast genomes by cybrid formation (there is no nuclear hybridization in this case). Such advantages are discussed more fully by Rose *et al.* (1990). Cybrid plants contain the nucleus of one species and part or all of the cytoplasm derived from another. Some advantages of cybrid production to plant breeding are:

1. Chloroplasts can be transferred in a single step. Alloplasmic plants have

traditionally been produced by recurrent backcrossing after hybridization (Feldman & Sears, 1981). This has particular relevance to the majority of flowering plants which have uniparental maternal inheritance, as indicated above.
2. Chloroplasts may also be transferred beyond the normal boundaries defined by sexual hybridization. Cybrid production can occur across wider genetic distances than nuclear hybridization. This is clearly illustrated in the studies in *Nicotiana*, where even intertribal cybrids have been produced (Table 9.2).

Table 9.2. Cybrids obtained by interspecific and intergeneric protoplast fusion of specified donor cytoplasm with *N. tabacum*.

Donor of chloroplast	Sexual hybrid	Fertility of cybrid	Somatic hybrid regeneration	Fertility of somatic hybrid	Source
N. debneyi	Yes female fertile	female fertile	not attempted	—	a, b
N. suaveolens	yes (rare) female fertile	female fertile	no	—	c, d
Solanum tuberosum	no	yes[A]	yes	infertile	e
Atropa belladonna	no	yes	yes	infertile	f
Petunia hybrida	no	yes	yes	infertile	g
Salpiglossis sinuata	no	yes	yes	infertile	h

From Rose *et al.,* 1990.
Cybrid formation involving *N. tabacum* containing different cytoplasms has not been included in this table. The data have been obtained from:

a: Asahi *et al.,* 1988;   b: Kumashiro & Kubo, 1986;
c: Thomas & Rose, 1988;   d: Thomas *et al.,* 1988;
e: Thanh & Medgyesy, 1989;   f: Kushnir *et al.,* 1987;
g: Glimelius & Bonnett, 1986; h: Thanh *et al.,* 1988.

[A]Recombinant cpDNA, no cybrids with *S. tuberosum* cpDNA obtained.

3. In sexual reproduction the cytoplasmic contribution of the male parent is usually very small. Somatic hybridization offers the singular advantage of producing a heterokaryon with approximately equal contributions from

both parents. Cybrid production allows a wide array of combinations among the two chloroplast types and two mitochondrial types present in the initial heterokaryon. Sorting out can produce a greater number of cytoplasmic organelle combinations than is easily achieved by conventional crossing (Kumar & Cocking, 1987; Chuong *et al.*, 1988).

In both somatic hybrids and cybrids there is a rapid sorting out of the parental chloroplast types during plant regeneration. The mechanism for rapid segregation is not certain, but a reduction in the number of chloroplasts per cell in the early post-fusion divisions may account for this phenomenon (Rose *et al.*, 1990). Biased segregation can occur and can frequently be related to an unequal input of plastids, but in some cases plastid–mito-chondrial–nuclear incompatibilities are likely to be involved (Rose *et al.*, 1990). Heteroplasmic regenerates can be produced but sorting out will eventually occur in subsequent generations (Flick *et al.*, 1985). As noted previously, cpDNA recombination does occur in somatic hybrids or cybrids, but it is a rare event (Medgyesy *et al.*, 1985; Thanh & Medgyesy, 1989).

Tissue culture

Tissue culture has been used to modify the chloroplast genome in two ways.

Selection

Maliga and co-workers have treated tissue with mutagens and utilized antibiotics in the culture medium to select for cpDNA mutations (Maliga, 1984). Resistant mutants have been developed against drugs such as streptomycin, or lincomycin, which blocks protein synthesis on prokaryotic ribosomes. These results emphasize the importance of maintaining selection pressure to obtain a uniform population of mutant plastids. This then indicates that given the appropriate chloroplast markers, there can be selection for specific cpDNAs. Furthermore, this procedure has the potential to provide mutant selectable genes suitable for cpDNA transformation, as illustrated in the recent studies of Moll *et al.* (1990).

Culture-induced variation

Mitochondria undergo culture-induced or somaclonal variation. Plants regenerated from culture show changes in restriction patterns in a number of species, e.g. maize (Gengenbach *et al.*, 1981) and potato (Kemble & Shepard, 1984), but not in others, such as alfalfa (Rose *et al.*, 1986) and *Brassica napus* (Kemble *et al.*, 1988). Such changes in cpDNA have not been reported (Kemble & Shepard, 1984; Rose *et al.*, 1986). However, in the case of cpDNA, another phenomenon has been described. Regeneration from tissue culture has shown apparent change in restriction patterns but such changes have been due to the presence of two types of cpDNA in the parent plant and the detection of only one type in the regenerate (Rose *et al.*, 1986; Johnson & Palmer, 1989). This is useful to know in devising breeding strategies for species such as alfalfa where this selective loss occurs (Rose *et al.*, 1986).

Chloroplast DNA transformation

Although there has been some progress in the transformation of the chloroplast genome, a number of fundamental problems remain. Even if DNA could enter the plastid and be integrated into the chloroplast genome, it is necessary for the recombinant plastid to become dominant in the regenerated plant (Cornelissen *et al.*, 1987). Progress made to date is reviewed in this section. Prospects for the transformation of chloroplasts have also been considered by Loffelhardt (1987) and Haring & De Block (1990).

The Ti plasmid of Agrobacterium tumefaciens

In 1985 De Block *et al.* reported the transformation of plastids for a chimeric gene with the promoter region of the nopaline synthase gene and the coding sequence of the chloramphenicol acetyltransferase (CAT) gene. There appear to be two reasons for this. First, the promoter can drive transcription in eukaryotes and prokaryotes (Herrera-Estrella *et al.*, 1983) and chloramphenicol blocks protein synthesis on prokaryotic ribosomes including those of chloroplasts (Ellis, 1969), thereby acting as a suitable selection agent. However, the transformed plastids were eventually lost, as determined by culture on chloramphenicol of tissue from a regenerated plant, presumably because the transformed plastids were at a selective disadvantage in the regenerated plant (Cornelissen *et al.*, 1987). More prolonged culture at the callus phase, where there are low plastid numbers (Thomas & Rose, 1983), may have overcome this problem. However, this plastid transformation system has not been reproducible (Cornelissen *et al.*, 1987).

Schell's laboratory has followed up this approach to chloroplast transformation by designing much more specific constructs (Cornelissen *et al.*, 1987). A specific plastid promoter was used (psbA) fused to the neomycin phosphotransferase (NPT II) gene. Furthermore, the NPT II gene was flanked at its 3' end by the 3' end part of the rbcL gene to facilitate homologous recombination without disruption of an essential function. Unfortunately, although kanamycin-resistant calli were produced, the introduced sequences were present in the nucleus rather than the cpDNA.

It appears that the NPT II gene is not a suitable marker for chloroplast transformation as plants having NPT II activity in the chloroplast (targeted by an rbcS transit peptide) have lower kanamycin resistance compared with plants with NPT II activity localized in the cytoplasm (Haring & De Block, 1990). Plants with the CAT gene localized to the chloroplast are resistant to higher concentrations of chloramphenicol than are plants with the CAT gene located in the nucleus (Haring & De Block, 1990), making CAT a better gene for the selection of chloroplast transformants. However, using this improved selection procedure and several different constructs to optimize insertion and expression in the cpDNA, no chloramphenicol-resistant calli or shoots were obtained using either *Agrobacterium*-mediated or direct gene transfer (Haring & De Block, 1990).

At this time, then, there is some evidence that cpDNA transformation is feasible using *Agrobacterium*, but it occurs at too low a frequency to be reliable using the techniques discussed in this section. The situation is further complicated by the large cpDNA copy numbers and the factors that influence chloroplast segregation mechanisms, as discussed elsewhere (Rose, 1988; Rose *et al.*, 1990). An understanding of how the copy number of the transformed cpDNA can be regulated should be a goal of future research.

Microprojectile techniques

Microprojectile techniques have been described in Chapter 5 of this volume. In principle these techniques offer the capacity to transform cpDNA in higher plants by the direct delivery of modified cpDNA into the chloroplast. Moreover, this technique has been successfully utilized in *Chlamydomonas* (Boynton *et al.*, 1988), where transformation of chloroplasts depended upon the introduction of cloned wild-type genes into organelle mutants with stringent defects in photosynthesis. The mutant studied in most detail was a mutant (*ac-u-c*-2-21) of the chloroplast *atpB* gene. This mutant has a 2.5 kb deletion in the Bam 10 restriction fragment spanning the *atpB* gene. Transformation occurred by bombarding layers of cells on Petri plates with tungsten microprojectiles coated with the wild-type Bam 10 fragment cloned in pBR313. Green transformed colonies occurred on a lawn of dead, bleached cells. In most transformants the 2.5 kb deletion was returned to normal size by a replacement event with the normal 7.6 kb fragment. Twelve transformants had a restored restriction fragment 50 to 100 base pairs smaller or larger than that of the wild-type. About 25% of the transformants contained unintegrated donor plasmid, which was eventually lost during subculture on selective plates.

The suggested mechanism for the replacement of the mutant Bam fragment with the wild-type Bam fragment is a double crossover or gene conversion event between regions of homology beyond the ends of the mutant fusion fragment and the wild-type Bam 10 fragment carried by pBR313 (Boynton *et al.*, 1988). The restored wild-type *atpB* gene remains as part of the chloroplast gene in all transformants and is expressed and inherited in the usual uniparental fashion. Few mutations that eliminate chloroplast gene function in higher plants are known but resistances to antibiotics and herbicides may provide alternatives for selection purposes.

The transformation results obtained by Boynton *et al.* (1988) have been confirmed by Blowers *et al.* (1989). Furthermore, the bacterial NPT II gene fused to the rbcL promoter was integrated into the inverted repeat region of the Bam 10 restriction fragment. The NPT II gene was found to be expressed and to be stably maintained in the chloroplast chromosome.

Electroporation techniques

Electroporation techniques have been able to successfully transform nuclear genes in higher plants (see Chapter 4) but have not been utilized in chloroplast transformation approaches. Once again the question is whether there is an intrinsic difficulty in DNAs moving into the chloroplast from the cytoplasm compared with the cytoplasm to nucleus, or whether it is a question of a lack of suitable stringent selection procedures for chloroplast transformation. Experiments carried out with isolated plastids have shown that DNA can be taken up (see below), suggesting that selection could be the major problem. An interesting recent experiment with protoplasts has involved the use of an electroporation transformation procedure for introducing proteins with the appropriate transit peptides into chloroplasts (Teeri *et al.*, 1989), providing a fast method for studying protein targeting in intact plant cells. An extension of this idea is presented below.

The use of transit peptides

Although transit peptides have not been used as a means of delivering DNA into chloroplasts, such an approach has been tried in animal mitochondria. Transit peptides covalently linked to DNA have enabled short pieces of DNA to be introduced (Vestweber & Schatz, 1989). A double-stranded 24-base pair piece of DNA had its 5' end covalently linked to the C terminus of a mitochondrial precursor protein. It would be necessary to obtain recombination of the mtDNA with this introduced piece of DNA for transformation, unless the introduced DNA were self-replicating. How much DNA can be successfully introduced using this technique remains to be ascertained. If the limit is a relatively few nucleotides, then the technique may have limited use. In plant cells the constructs could conceivably be introduced using electroporation, and they would then be able to enter all the chloroplasts of a cell. The transit peptide methodology would seem to offer an elegant approach to both chloroplast and mitochondrion transformation, and additional experimentation is warranted.

Viral vectors

There have been some interesting reports of the entry of viral nucleic acids into chloroplasts, prompting their candidature for chloroplast transformation (Groning *et al.*, 1987). However, a good deal of further work is required to fully evaluate these possibilities. The *Abutilon* mosaic virus is a gemini virus with a single-stranded DNA genome. Groning *et al.* (1987) have found that this viral DNA is present specifically within intact plastids. The mechanism of transport of this viral genome across the chloroplast envelope is of great interest. Schoelz & Zaitlin (1989) have provided evidence that tobacco mosaic virus RNA enters the chloroplast *in vivo*. However, it seems unlikely that it replicates within the chloroplast, as no subgenomic RNA was identified (Schoelz & Zaitlin, 1989). Nevertheless, this represents an example of nucleic acid transport across the chloroplast envelope.

Uptake of DNA into isolated chloroplasts

A general strategy for the manipulation of cpDNA in higher plants is to transform isolated chloroplasts and to return these chloroplasts to isolated protoplasts for plant regeneration (Daniell & McFadden, 1987; Weber *et al.*, 1989; Haring & De Block, 1990). Daniell & McFadden (1987) have reported the uptake and expression of pUC derivatives into isolated cucumber etioplasts. The pUC derivatives carried genes for the large and small subunits of ribulose bisphosphate carboxylase from *Anacystis nidulans* or CAT from *Escherichia coli*. DNA binding and uptake by the cucumber etioplasts were greatly increased by EDTA treatment. Weber *et al.* (1989) have employed a UV-laser microbeam to manipulate isolated chloroplasts of *Brassica napus* (L.). Holes of a very small diameter are cut into the chloroplast membrane with the laser microbeam. DNA is introduced into the chloroplast through the laser holes. The membranes reseal and the DNA taken up by the chloroplasts is resistant to externally applied DNase.

Early studies of chloroplast uptake into protoplasts were based on the use of polyethylene glycol (Carlson, 1973; Bonnett & Erikson, 1974) and damage or loss of the chloroplast envelope would have precluded subsequent chloroplast division. Injection of isolated chloroplasts into protoplasts (Haring & De Block, 1990) should result in chloroplasts that replicate. However, there is no guarantee that such introduced plastids will become the dominant type. This latter problem is in principle the same problem that occurs following protoplast fusion and one we have discussed in detail (Rose *et al.*, 1990). In some cases it may be enough to maximize the ratio of introduced foreign chloroplasts to endogenous chloroplasts. However, 'incompatibility' phenomena that are imperfectly defined preclude some chloroplasts becoming dominant. Chloroplasts can be transferred over relatively wide genetic distances (Rose *et al.*, 1990) and it may be that a 'universal donor' chloroplast that has minimal incompatibility problems is feasible.

Some outstanding problems

SELECTION PROTOCOLS. Suitable selection constructs are available to enable selection for antibiotic resistance (e.g. streptomycin) and herbicide resistance due to mutations in genes encoding chloroplast proteins. Suitable promoters active in chloroplasts can be used. Furthermore, a range of cpDNA mutations have been identified (Borner & Sears, 1986). The experience with *Agrobacterium* suggests that it is important not to rely on promoter type alone (Cornelissen *et al.*, 1987) as nuclear insertion might still allow the fortuitous utilization of nuclear promoters. However, more experiments are clearly required in this area. Attention to the type of cells being cultured may also be of significance in facilitating appropriate selection (i.e. autotrophic as opposed to heterotrophic).

CHLOROPLAST DNA COPY NUMBER. A lack of knowledge is a limitation to devising strategies designed to ensure the dominance of a transformed cpDNA or strategies involving an independent replicon in the chloroplast. This is an area which should improve as knowledge about *ori* regions in cpDNA increases (Fig. 9.1). In *Euglena gracilis* (Koller & Delius, 1982; Ravel-Chapuis *et al.*, 1982) and *Chlamydomonas reinhardtii* (Waddell *et al.*, 1984) the replication origins have been located upstream of the 16S rRNA gene. In the case of *Chlamydomonas* a second origin was spaced 6.5 kb from the first origin. In higher plants putative *ori* regions are close to the rRNA genes but have a different relative position from those of *Euglena* and *Chlamydomonas*. In pea (which lacks an inverted repeat), one D-loop region maps in the spacer region between the 16S and 23S rRNA gene and one downstream of the 23S rRNA gene (Meeker *et al.*, 1988). In *Petunia hybrida* there is evidence based on homology with the *Euglena* origin that an origin region is located in the small single-copy region close to one of the inverted repeats (de Haas *et al.*, 1986). A similar situation occurs in *Nicotiana tabacum* (Fig. 9.1). In the case of *Chlamydomonas* two back-to-back promoters have been located in a 79 bp restriction fragment within the D-loop region (Wu *et al.*, 1986). In an idealized situation promoter enhancement of the replication of transformed cpDNAs would ensure dominance. Independent replicons with membrane attachment as for cpDNA (Rose, 1988) are also feasible. Membrane attachment is implicated in both segregation (Rose, 1988) and replication (Lindbeck & Rose, 1990). The use of replicons would require very sophisticated transformation procedures.

THE SITE OF DNA INSERTION. Transformation of cpDNA requires that DNA insertions are made in such a way that existing gene functions are not impaired. Delineating an area of cpDNA where suitable marker genes can be inserted will be a future focus for research (Haring & De Block, 1990). However, given the detailed knowledge now available on the sequence and genetic map of cpDNA (Sugiura, 1989), this should not be a limiting problem for transformation. The NPT II gene has been stably introduced into *Chlamydomonas* cpDNA after insertion into the inverted repeat region (Blowers *et al.*, 1989).

Targeting proteins to the chloroplast

Chloroplast proteins that are encoded by the nucleus are synthesized as larger precursors on cytoplasmic ribosomes, and then imported by the chloroplast. These polypeptides include N-terminal leader sequences or transit peptides that facilitate entry through the chloroplast envelope (Mishkind & Scioli, 1988) and, in certain instances, into the thylakoid (Wallace *et al.*, 1989). The two types of transit peptide are quite distinct; each is removed by a specific peptidase as the target membrane is penetrated.

It is possible to construct chimeric proteins including a transit peptide (or peptides) linked to a protein that is not normally imported by the chloroplast. For example, the bacterial NPT II gene has been spliced behind the promoter and transit peptide encoding sequences of the gene for the small subunit of ribulose bisphosphate carboxylase (Van den Broeck *et al.*, 1985). Cheung *et al.* (1988) have used this approach to insert a mutant Q_B protein into chloroplasts and so confer atrazine resistance (discussed below). The transcriptional regulatory sequences and the transit peptide-encoding sequence of a nuclear rbcS gene were spliced to the coding region of the psbA gene isolated from the cpDNA of the atrazine-resistant biotype of *Amaranthus*. Tobacco plants were transformed for this construct, using the Ti plasmid vector pGV3850. Some of the transformed plants showed greater tolerance of atrazine than the control plants. The modified Q_B protein was shown to be inserted into the thylakoid membrane by immunochemical analysis of photosynthetic membrane proteins.

Whether these atrazine-tolerant transgenic plants would survive under field applications of atrazine has not yet been tested. In the transformed plants, some PSII reaction centres would contain mutant Q_B proteins and others would not (Cheung *et al.*, 1988). Details on these latter points are required before the efficacy of this procedure can be fully assessed.

Herbicide resistance encoded by chloroplast DNA

The *s*-triazines (e.g. atrazine and simazine) and the urea group of herbicides (e.g. dichlorophenyl dimethylurea or diuron) are herbicidal because they inhibit photosynthesis (Mazur & Falco, 1989). These herbicides block electron transport at PSII by binding to the Q_B or herbicide-binding protein (also known as the DI protein) encoded by the psbA gene of cpDNA (Steinback *et al.*, 1981; Zurawski *et al.*, 1982).

The Q_B protein

The Q_B protein has a relative molecular weight of 32 kDa and is part of the PSII complex which is embedded in the thylakoid membrane (Mattoo & Edelman, 1987). The Q_B protein binds a plastoquinone which is the secondary electron-accepting plastoquinone (Vermaas *et al.*, 1984) and passes electrons to the free plastoquinone pool (van Rensen, 1982). The PSII herbicides bind to the Q_B protein and so displace plastoquinone (Vermaas *et al.*, 1984), thereby blocking photosynthetic electron flow. The degradation and synthesis of the Q_B protein is light-regulated (Ohad *et al.*, 1985). The 32 kDa Q_B protein is synthesized as a 34 kDa precursor, which is post-translationally processed and assembled into PSII complexes in stacked thylakoids (Roy, 1981; Mattoo & Edelman, 1987). It is a chloroplast gene product (Steinback *et al.*, 1981; Zurawski *et al.*, 1982).

The psbA gene

The psbA gene which encodes the Q_B protein was first sequenced by Zurawski *et al.* (1982) in spinach and *Nicotiana debneyi*. The psbA gene has since been cloned and sequenced in a variety of photosynthetic organisms. It encodes a hydrophobic protein with 353 amino acids in spinach and *Nicotiana debneyi* (Zurawski *et al.*, 1982), rice and tobacco (Sugiura, 1989). The psbA gene is highly conserved (Schuster *et al.*, 1986). It lacks introns in land plants (Sugiura, 1989) but *Euglena* (Karabin *et al.*, 1984) and *Chlamydomonas* (Erickson *et al.*, 1984) have four introns. The common relative position of the psbA gene on the chloroplast chromosome of flowering plants is usually found in the large single-copy region near to the end of one of the inverted repeats (Livore *et al.*, 1989), as shown in Fig. 9.1.

Mutant psbA genes

Mutant psbA genes are present in herbicide-resistant weed biotypes and in several micro-organisms selected for herbicide resistance (Ohad & Hirschberg, 1990). Resistance to the *s*-triazines of the type which is inherited in plastid DNA is now known for over 40 weed species (Jacobs *et al.*, 1988) after having been first reported in 1970 in *Senecio vulgaris* (Ryan, 1970). Triazine resistance, for example, has also been found in *Brassica* (Darr *et al.*, 1981), *Chenopodium* (Solymosi & Lehoczxi, 1989), *Poa* (Gasquez & Darmency, 1983), *Solanum* (Goloubinoff *et al.*, 1984) and *Amaranthus* (Hirschberg & McIntosh, 1983). The mutations result in the substitution of serine 264 with glycine or alanine (Hirschberg & McIntosh, 1983; Goloubinoff *et al.*, 1984; Erickson *et al.*, 1984). In higher plants only transversion from serine to glycine in amino acid position 264 of the 32 kDa protein has been shown. This has been reported in the dicotyledons

Amaranthus hybridus (Hirschberg & McIntosh, 1983), *Solanum nigrum* (Goloubinoff *et al.*, 1984), *Chenopodium album* (Bettini *et al.*, 1987) and *Brassica napus* (Reith & Strauss, 1987), and in the monocotyledon *Poa annua* (Barros & Dyer, 1988). Studies using cyanobacteria and site-specific mutagenesis indicate that the change from serine to glycine is associated with *s*-triazine resistance, whereas the change from serine to alanine is responsible for resistance to phenylurea herbicides (Ohad & Hirschberg, 1990). Clearly these are sources of resistance which can be exploited for breeding purposes.

It should be emphasized that naturally occurring atrazine resistance in a number of crop species, e.g. rice, maize and wheat, is due to detoxification by glutathione S-transferases (Oxtoby & Hughes, 1989). However, molecular manipulation may be complicated because of the need to control other factors such as the concentration of glutathione (Oxtoby & Hughes, 1989).

Breeding strategies utilized for the manipulation of the psbA gene

A number of the strategies that can be used for the manipulation of the chloroplast genome have been used to obtain herbicide resistance through psbA mutant genes. In this section examples are given of how the various strategies to alter the chloroplast genome have been used to introduce herbicide resistance into crop plants. Hirschberg & McIntosh (1983) have pointed out that herbicide resistance has arisen in weed species showing a high degree of phenotypic variation and it is unlikely that such variants will be found in crop plants.

Sexual hybridization

In some cases resistant weeds are sexually compatible with crop plants and this has enabled the introduction of atrazine-resistant chloroplasts in crops such as *Brassica* and *Setaria* (Mazur & Falco, 1989). In the case of *Brassica* an atrazine-resistant cultivar was released in 1986 (Mazur & Falco, 1989). This atrazine-resistant Canola is the first example of a herbicide-resistant crop to be produced using chloroplasts encoding a mutant psbA gene. As is commonly the case, the triazine resistance was maternally inherited and the *Brassica campestris* chloroplasts were transferred to *Brassica napus* by backcrossing the resistant plant as the female parent to *B. napus* (Beversdorf *et al.*, 1980). After eight generations the nuclear genome is almost isogenic (Oxtoby & Hughes, 1989). Such approaches are continuing in the genus *Brassica* (Ayotte *et al.*, 1989).

Somatic hybridization

In some crops such as *Brassica* and *Solanum* the utilization of atrazine-resistant germplasm has been made possible by somatic hybridization (Pelletier *et al.*, 1988; Chapter 8). Chloroplast-encoded atrazine resistance is available in the wild species *Brassica campestris* (Darr *et al.*, 1981) and *Solanum nigrum* (Gressel *et al.*, 1984). Both cybrids and hybrids have been produced (Gressel *et al.*, 1984; Barsby *et al.*, 1987; Pelletier *et al.*, 1988; Jourdan *et al.*, 1989). As discussed above, a single-step transfer is possible.

Selection from cell culture

Atrazine resistance has been selected for, using *Nicotiana* photomixotrophic cell cultures (Cseplo *et al.*, 1985). The resistance has been shown to be maternally inherited and is therefore most probably due to a mutation in the psbA gene. As the number of plants that can be regenerated from cells and tissue culture increases, this strategy will be open to a wide range of plants, although there is no guarantee that resistance will be directed against cellular characteristics such as uptake which may not be present in the regenerated plant.

Nuclear transformation

As discussed in detail earlier, transgenic plants showing increased tolerance to atrazine have been produced (Cheung *et al.*, 1988) by inserting a mutant psbA gene into the nucleus. Entry of the gene product into the chloroplast was made possible by transit peptide sequences ligated to the psbA gene. More detailed analysis of this system is now required in order to assess its field suitability.

Chloroplast DNA transformation

There is a great deal of understanding of the molecular detail of the psbA gene and its product the Q_B protein which would provide a refined approach to the genetic engineering for this type of herbicide resistance, if cpDNA could be transformed. Although this is now looking increasingly feasible, it is a complex area which will take some time to reach fruition. With the transformation of the cpDNA of *Chlamydomonas* a beginning has been made. Furthermore, it has been demonstrated in principle in the cyanobacterium *Anacystis nidulans* R2 that herbicide resistance can be genetically engineered. The transformability of *A. nidulans* has been exploited and a mutated gene conferring diuron resistance introduced into the genome of wild-type cells (Golden & Haselkorn, 1985). The mutation consists of a single nucleotide change that replaces serine with alanine in amino acid position 264 of the wild-type 32 kDa protein. The cloned fragment introduced appears to transform the wild-type cells by homologous recombination.

Solymosi & Lehoczki (1989) have pointed out that all PSII herbicide-resistant weed biotypes sequenced to date have the same amino acid subsitution at the same site in the psbA gene product. Substitution at other places in the sequence would be expected to lead to different resistances and co-resistances and this could be readily exploited given the knowledge of the psbA gene. An atrazine-resistant biotype of *Amaranthus bouchonii* with diuron resistance has been identified and is the first vascular plant to have diuron resistance (Solymosi & Lehoczki, 1989). This illustrates the potential for manipulation of the psbA gene.

Yield penalties associated with resistance to atrazine

Weed biotypes that are resistant to triazine herbicides unfortunately have lower photosynthetic efficiencies compared with their susceptible counterparts, which results in a yield penalty of 10–30% (Gressel & Ben-Sinai, 1985; Havaux, 1989; Mazur & Falco, 1989; Oxtoby & Hughes, 1989). The yield penalty in *Senecio vulgaris* can be directly related to the slow-down of electron transfer from Q_A^- to Q_B, resulting in an increase in the steady-state concentration of Q_A^- (Havaux, 1989). As discussed by Havaux (1989), the change in electron transfer characteristics underlies the 20% decline in CO_2 fixation reported for *Senecio vulgaris* (Holt *et al.*, 1981). These data suggest that photosynthesis at low light intensities is an important contributing factor to the reduced growth and competitive fitness of atrazine-resistant plants (Ort & Baker, 1988).

A yield penalty due to the inhibition of photosynthesis may, however, be compensated for by the decreased competition from weeds. Furthermore, there are sometimes other advantages. In the case of oilseed rape (*Brassica napus*) where atrazine resistance has been introduced (Beversdorf *et al.*, 1980), spraying with atrazine has the advantage of killing off the wild mustard with which it may outcross, resulting in an unacceptable content of erucic acid in its oil (Barros & Dyer, 1988).

Studies in *Chlamydomonas reinhardtii* have shown that there may be more subtle ways of obtaining resistance to herbicides which act on the 32 kDa protein by having mutations affecting amino acids at sites other than the serine 264. Amino acid residue 264 (serine–glycine), as discussed above, is the basis of atrazine resistance in those higher plants analysed to date. The *Chlamydomonas* studies carried out by Erickson *et al.* (1985) were directed at the cross-resistance which often accompanies resistance to the commonly used *s*-triazine herbicides. The studies showed that three different amino acid residues in the 32 kDa protein can be independently altered to produce three different patterns of resistance to *s*-triazine and urea-type herbicides. Erickson *et al.* (1985) found that the resistant strain DCMU4 with a 100× atrazine resistance and a 10× DCMU resistance (compared with wild strains) had a glycine rather than a serine at amino acid position 264. This strain had an altered electron transport as in higher plants. Interestingly, mutant Dr2 had a 2× atrazine resistance and a 15× DCMU resistance but showed normal electron transport. Mutant Dr2 had a valine rather than an isoleucine at amino acid position 219 (Erickson *et al.*, 1985).

The *Chlamydomonas* data suggest that there are amino acid changes in the 32 kDa protein which could confer herbicide resistance in higher plants without an effect on electron flow. In this regard it is relevant that *Amaranthus bouchonii* has recently been reported to show co-resistance to atrazine and diuron (Solymosi & Lehoczki, 1989), and the sequence of the 32 kDa protein would be of interest. Clearly, however, the genetic manipulation of the 32 kDa protein in conjunction with an appropriate transformation system offers wide possibilities for testing the manipulation of herbicide resistance with minimized effects on photosynthesis.

Future prospects

The genetic manipulation of the psbA gene has already been shown by traditional (Beversdorf *et al.*, 1980) and more recent technologies (Gressel & Ben-Sinai, 1985) to produce herbicide-resistant plants that can provide agriculturally useful cultivars. Although the engineering of the gene has been demonstrated in principle in cyanobacteria (Golden & Haselkorn, 1985), it will be some time before such refined techniques are available in higher plants. However, there are grounds for optimism given the array of approaches now being initiated, as discussed in this chapter, and the progress in *Chlamydomonas*. The successful manipulation of the psbA gene will open up this technology to other chloroplast genes. Refined techniques for cpDNA manipulation in higher plants will require more understanding of the control of cpDNA copy number so that engineered cpDNA molecules can readily dominate cpDNA populations. This latter objective, if achieved, could signal the onset of technology which may enable the utilization of independent replicons in the chloroplast. As pointed out by Jourdan *et al.* (1989) and in Chapter 1, it is important that herbicide resistance should be used in an environmentally sound way.

Acknowledgements

My own research on chloroplast DNA at the University of Newcastle has been supported by the Australian Research Council.

References

Andrews, T. J. (1988) Catalysis by cyanobacterial ribulose-bisphosphate carboxylase large subunits in the complete absence of small subunits. *Journal of Biological Chemistry* 263, 12213–19.

Asahi, K., Kumashiro, T. & Kubo, T. (1988) Constitution of mitochondrial and chloroplast genomes in male sterile tobacco obtained by protoplast fusion of *Nicotiana tabacum* and *N. debneyi*. *Plant and Cell Physiology* 29, 43–9.

Ayotte, R., Harney, P. M. & Souza Machado, V. (1989) The transfer of triazine resistance from *Brassica napus* L. to *B. oleracea* and recovery of an 18-chromosome, triazine resistant BC$_3$. *Euphytica* 40, 15–19.

Barros, M. D. C. & Dyer, T. A. (1988) Atrazine resistance in the grass *Poa annua* is due to a single base change in the chloroplast gene for the D1 protein of photosystem II. *Theoretical and Applied Genetics* 75, 610–16.

Barsby, T. L., Chuong, P. V., Yarrow, S. A., Wu, S.-C., Coumans, M., Kemble, R. J., Powell, A. D., Beversdorf, W. D. & Pauls, K. P. (1987) The combination of Polima cms and cytoplasmic triazine resistance in *Brassica napus*. *Theoretical and Applied Genetics* 73, 809–14.

Bettini, P., McNally, S., Sevignac, M., Darmency, H., Gasquez, J. & Dron, M. (1987) Atrazine resistance in *Chenopodium album*: low and high levels of resistance to the herbicide are related to the same chloroplast psbA gene mutation. *Plant Physiology* 84, 1442–6.

Beversdorf, W. D., Weiss-Lerman, J., Erickson, L. R. & Souza Machado, V. (1980) Transfer of cytoplasmically inherited triazine resistance from bird's rape to cultivated oil seed rape (*Brassica campestris* and *Brassica napus*). *Canadian Journal of Genetics and Cytology* 22, 167–72.

Blowers, A. D., Bogorad, L., Shark, K. B. & Sanford, J. C. (1989) Studies on *Chlamydomonas* chloroplast transformation: foreign DNA can be stably maintained in the chromosome. *Plant Cell* 1, 123–32.

Bonnett, H. T. & Eriksson, T. (1974) Transfer of algal chloroplasts into protoplasts of higher plants. *Planta* 120, 71–9.

Borner, T. & Sears, B. B. (1986) Plastome mutants. *Plant Molecular Biology Reporter* 4, 69–92.

Boynton, J. E., Gillham, N. W., Harris, E. H., Hosler, J. P., Johnson, A. M., Jones, A. R., Randolph-Anderson, B. L., Robertson, D., Klein, T. M., Shark, K. B. & Sanford, J. C. (1988) Chloroplast transformation in *Chlamydomonas* with high velocity microprojectiles. *Science* 240, 1534–8.

Capel, M. S. & Bourque, D. P. (1982) Characterization of *Nicotiana tabacum* chloroplast and cytoplasmic ribosomal proteins. *Journal of Biological Chemistry* 257, 7746–55.

Carlson, P. S. (1973) The use of protoplasts for genetic research. *Proceedings of the National Academy of Sciences USA* 70, 598–602.

Cheung, A. Y., Bogorad, L., Van Montagu, M. & Schell, J. (1988) Relocating a gene for herbicide tolerance: a chloroplast gene is converted into a nuclear gene. *Proceedings of the National Academy of Sciences USA* 85, 391–5.

Chiu, W.-L. & Sears, B. B. (1985) Recombination between chloroplast DNAs does not occur in sexual crosses of *Oenothera*. *Molecular and General Genetics* 198, 525–8.

Chuong, P. V., Beversdorf, W. D., Powell, A. D. & Pauls, K. P. (1988) Somatic transfer of cytoplasmic traits in *Brassica napus* L. by haploid protoplast fusion. *Molecular and General Genetics* 211, 197–201.

Cornelissen, M. J., De Block, M., Van Montagu, M., Leemans, J., Schreier, P. H. & Schell, J. (1987) Plastid transformation: a progress report. In: Hohn, T. &

Schell, J. (eds), *Plant DNA Infectious Agents*. Springer-Verlag, Vienna and New York, pp. 311–21.

Corriveau, J. L. & Coleman, A. W. (1988) Rapid screening method to detect potential biparental inheritance of plastid DNA and results for over 200 angiosperm species. *American Journal of Botany* 75, 1443–58.

Cseplo, A., Medgyesy, P., Hideg, E., Demeter, S., Marton, L. & Maliga, P. (1985) Triazine resistant *Nicotiana* mutants from photomyxotrophic cell cultures. *Molecular and General Genetics* 200, 508–10.

Daniell, H. & McFadden, B. A. (1987) Uptake and expression of bacterial and cyanobacterial genes by isolated cucumber etioplasts. *Proceedings of the National Academy of Sciences USA* 84, 6349–53.

Darr, S., Souza Machado, V. & Arntzen, C. J. (1981) Uniparental inheritance of a chloroplast photosystem II polypeptide controlling herbicide binding. *Biochimica et Biophysica Acta* 643, 219–28.

Day, A. & Ellis, T. H. N. (1984) Chloroplast DNA deletions associated with wheat plants regenerated from pollen: possible basis for maternal inheritance of chloroplasts. *Cell* 39, 359–68.

De Block, M., Schell, J. & Van Montagu, M. (1985) Chloroplast transformation by *Agrobacterium tumefaciens*. *EMBO Journal* 4, 1367–72.

De Haas, J. M., Boot, K. J. M., Haring, M. A., Kool, A. J. & Nijkamp, H. J. H. (1986) A *Petunia hybrida* chloroplast DNA region, close to one of the inverted repeats, shows sequence homology with the *Euglena gracilis* chloroplast DNA region that carries the putative replication origin. *Molecular and General Genetics* 202, 48–54.

Dyer, T. A. (1984) The chloroplast genome: its nature and role in development. In: Baker, N. R. & Barber, J. (eds), *Topics in Photosynthesis*, vol. 5, *Chloroplast Biogenesis*. Elsevier, Amsterdam, pp. 24–69.

Ellis, R. J. (1969) Chloroplast ribosomes: stereospecificity of inhibition by chloramphenicol. *Science* 163, 477–8.

Erickson, J. M., Rahire, M., Bennoun, P., Delepelaire, P., Diner, B. & Rochaix, J.-D. (1984) Herbicide resistance in *Chlamydomonas reinhardtii* results from a mutation in the chloroplast gene for the 32-kilodalton protein of photosystem II. *Proceedings of the National Academy of Sciences USA* 81, 3617–21.

Erickson, J. M., Rahire, M., Rochaix, J.-D. & Mets, L. (1985) Herbicide resistance and cross resistance: changes at three distinct sites in the herbicide-binding protein. *Science* 25, 204–7.

Feldman, M. & Sears, E. R. (1981) The wild gene resources of wheat. *Scientific American* 244(1), 98–109.

Flick, C. E., Kut, S. A., Bravo, J. E., Gleba, Y. Y. & Evans, D. A. (1985) Segregation of organelle traits following protoplast fusion in *Nicotiana*. *Bio/Technology* 3, 555–60.

Gasquez, J. & Darmency, H. (1983) Variation of chloroplast properties between two triazine resistant biotypes of *Poa annua* L. *Plant Science Letters* 30, 99–106.

Gengenbach, B. G., Connelly, J. A., Pring, D. R. & Conde, M. F. (1981) Mitochondrial DNA variation in maize plants regenerated during tissue culture selection. *Theoretical and Applied Genetics* 59, 161–7.

Ghanotakis, D. F. & Yocum, C. F. (1990) Photosystem II and the oxygen-evolving complex. *Annual Review of Plant Physiology* 41, 255–76.

Glimelius, K. & Bonnett, H. T. (1986) *Nicotiana* cybrids with *Petunia* chloroplasts. *Theoretical and Applied Genetics* 72, 794–8.

Golden, S. S. & Haselkorn, R. (1985) Mutation to herbicide resistance maps within the psbA gene of *Anacystis nidulans* R2. *Science* 229, 1104–7.

Goloubinoff, P., Edelman, M. & Hallick, R. B. (1984) Chloroplast-coded atrazine resistance in *Solanum nigrum*: psbA loci from susceptible and resistant biotypes are isogenic except for a single codon change. *Nucleic Acids Research* 12, 9489–95.

Govindaraju, D. R., Wagner, D. B., Smith, G. P. & Dancik, B. P. (1988) Chloroplast DNA variation within individual trees of a *Pinus banksiana–Pinus contorta* sympatric region. *Canadian Journal of Forest Research* 18, 1347–50.

Gressel, J. & Ben-Sinai, G. (1985) Low intraspecific competitive fitness in a triazine-resistant, nearly nuclear-isogenic line of *Brassica napus*. *Plant Science* 38, 229–32.

Gressel, J., Cohen, N. & Binding, H. (1984) Somatic hybridisation of an atrazine resistant biotype of *Solanum nigrum* with *Solanum tuberosum*. *Theoretical and Applied Genetics* 67, 131–84.

Groning, B. R., Abouzid, A. & Jeske, H. (1987) Single-stranded DNA from a butilon mosaic virus is present in the plastids of infected *Abutilon sellovianum*. *Proceedings of the National Academy of Sciences USA* 84, 8996–9000.

Haring, M. A. & De Block, M. (1990) New roads towards chloroplast transformation in higher plants. *Physiologia Plantarum* 79, 218–20.

Havaux, M. (1989) Comparison of atrazine-resistant and -susceptible biotypes of *Senecio vulgaris* L: effects of high and low temperatures on the *in vivo* photosynthetic electron transfer in intact leaves. *Journal of Experimental Botany* 40, 849–54.

Herrera-Estrella, L., De Block, M., Messens, E., Hernalsteens, J.-P., Van Montagu, M. & Schell, J. (1983) Chimeric genes as dominant selectable markers in plant cells. *EMBO Journal* 2, 987–95.

Herrmann, R. G. & Possingham, J. V. (1980) Plastid DNA – the plastome. In: Reinhert, J. (ed.), *Chloroplasts*. Springer-Verlag, Berlin, pp. 45–96.

Hiratsuka, J., Shimada, H., Whittier, R., Ishibashi, T., Sakamoto, M., Mori, M., Kondo, C., Honji, Y., Sun, C.-R., Meng, B.-Y., Li, Y.-Q., Kanno, A., Nishizawa, Y., Hirai, A., Shinozaki, K. & Sugiura, M. (1989) The complete sequence of the rice (*Oryza sativa*) chloroplast genome: intermolecular recombination between distinct tRNA genes accounts for a major plastid DNA inversion during the evolution of the cereals. *Molecular and General Genetics* 217, 185–94.

Hirschberg, J. & McIntosh, L. (1983) Molecular basis of herbicide resistance in *Amaranthus hybridus*. *Science* 222, 1346–9.

Holt, J. S., Stemler, A. J. & Radosevich, S. R. (1981) Differential light responses of photosynthesis by triazine-resistant and triazine-susceptible *Senecio vulgaris* biotypes. *Plant Physiology* 67, 744–8.

Hu, J. & Bogorad, L. (1990) Maize chloroplast RNA polymerase: the 180-, 120- and 38-kilodalton polypeptides are encoded in chloroplast genes. *Proceedings of the National Academy of Sciences USA* 87, 1531–5.

Jacobs, B. F., Duesing, J. H., Antonovics, J. & Patterson, D. T. (1988) Growth performance of triazine-resistant and -susceptible biotypes of *Solanum nigrum* over a range of temperatures. *Canadian Journal of Botany* 66, 847–50.

Johnson, L. B. & Palmer, J. D. (1989) Heteroplasmy of chloroplast DNA in

Medicago. Plant Molecular Biology 12, 3–11.

Jourdan, P. S., Earle, E. D. & Mutschler, M. A. (1989) Atrazine-resistant cauliflower obtained by somatic hybridisation between *Brassica oleracea* and ATR–*B. napus. Theoretical and Applied Genetics* 78, 271–9.

Karabin, G. D., Farley, M. & Hallick, R. B. (1984) Chloroplast gene for M_r 32,000 polypeptide of photosystem II in *Euglena gracilis* is interrupted by four introns with conserved boundary sequences. *Nucleic Acids Research* 12, 5801–12.

Kaul, M. L. H. (1988) *Male Sterility in Higher Plants.* Springer-Verlag, Berlin.

Kemble, R. J. & Shepard, J. F. (1984) Cytoplasmic DNA variation in a potato protoclonal population. *Theoretical and Applied Genetics* 69, 211–16.

Kemble, R. J., Yarrow, S. A., Wu, S.-C. & Barsby, T. L. (1988) Absence of mitochondrial and chloroplast DNA recombinations in *Brassica napus* plants regenerated from protoplasts, protoplast fusions and anther culture. *Theoretical and Applied Genetics* 75, 875–81.

Koller, B. & Delius, H. (1982) Origin of replication in chloroplast DNA of *Euglena gracilis* located close to the region of variable size. *EMBO Journal* 1, 995–8.

Kumar, A. & Cocking, E. C. (1987) Protoplast fusion: a novel approach to organelle genetics in higher plants. *American Journal of Botany* 74, 1289–303.

Kumashiro, T. & Kubo, T. (1986) Cytoplasmic transfer of *Nicotiana debneyi* to *N. tabacum* by protoplast fusion. *Japanese Journal of Breeding* 36, 39–48.

Kuroiwa, T., Suzuki, T., Ogawa, K. & Kawano, S. (1981) The chloroplast nucleus: distribution, number, size and shape and a model for the multiplication of the chloroplast genome during chloroplast development. *Plant and Cell Physiology* 22, 381–96.

Kushnir, S. G., Shlumukov, L. R., Pogrebnyak, N. J., Berger, S. & Gleba, Y. (1987) Functional cybrid plants possessing a *Nicotiana* genome and an *Atropa* plastome. *Molecular and General Genetics* 209, 159–63.

Lee, D. J., Blake, T. K. & Smith, S. E. (1988) Biparental inheritance of chloroplast DNA and the existence of heteroplasmic cells in alfalfa. *Theoretical and Applied Genetics* 76, 545–9.

Lidholm, J., Szmidt, A. E., Hallgren, J.-E. & Gustafsson, P. (1988) The chloroplast genomes of conifers lack one of the rRNA-encoding inverted repeats. *Molecular and General Genetics* 212, 6–10.

Lindbeck, A. G. C. & Rose, R. J. (1990) Thylakoid-bound chloroplast DNA from spinach is enriched for replication forks. *Biochemical and Biophysical Research Communications* 172, 204–10.

Livore, A., Scheuring, C. & Magill, C. (1989) The rice psb-A chloroplast gene has a standard location. *Current Genetics* 16, 447–51.

Loffelhardt, W. (1987) Engineering chloroplasts: prospects and limitations. *Physiologia Plantarum* 69, 735–41.

McFadden, B. A. & Daniell, H. (1988) Binding, uptake and expression of foreign DNA by cyanobacteria and isolated etioplasts. *Photosynthesis Research* 19, 23–37.

Maliga, P. (1984) Isolation and characterisation of mutants in plant cell culture. *Annual Review of Plant Physiology* 35, 519–42.

Masoud, S. A., Johnson, L. B. & Sorensen, E. L. (1990) High transmission of paternal plastid DNA in alfalfa plants demonstrated by restriction fragment polymorphic analysis. *Theoretical and Applied Genetics* 79, 49–55.

Mattoo, A. K. & Edelman, M. (1987) Intramembrane translocation and posttranslational palmitoylation of the chloroplast 32-kDa herbicide-binding protein. *Proceedings of the National Academy of Sciences USA* 84, 1497–501.

Mazur, B. J. & Falco, S. C. (1989) The development of herbicide resistant crops. *Annual Review of Plant Physiology and Plant Molecular Biology* 40, 441–70.

Medgyesy, P., Fejes, E. & Maliga, P. (1985) Interspecific chloroplast recombination in a *Nicotiana* somatic hybrid. *Proceedings of the National Academy of Sciences USA* 82, 6960–4.

Meeker, R., Nielsen, B. & Tewari, K. K. (1988) Localization of replication origins in pea chloroplast DNA. *Molecular and Cellular Biology* 8, 1216–23.

Mishkind, M. L. & Scioli, S. E. (1988) Recent developments in chloroplast protein transport. *Photosynthesis Research* 19, 153–84.

Moll, B., Polsby, L. & Maliga, P. (1990) Streptomycin and lincomycin resistances are selective plastid markers in cultured *Nicotiana* cells. *Molecular and General Genetics* 221, 245–50.

Ohad, I., Kyle, D. J. & Hirschberg, J. (1985) Light dependent degradation of Q_B-protein in isolated pea thylakoids. *EMBO Journal* 4, 1655–9.

Ohad, N. & Hirschberg, J. (1990) A similar structure of the herbicide binding site in photosystem II of plants and cyanobacteria is demonstrated by site specific mutagenesis of the psbA gene. *Photosynthesis Research* 23, 73–9.

Ohyama, K., Fukuzawa, H., Kohchi, T., Shirai, H., Sano, T., Sano, S., Umesono, K., Shiki, Y., Takeuchi, M., Chang, Z., Aota, S.-I., Inokuchi, H. & Ozeki, H. (1986) Chloroplast gene organization deduced from complete sequence of liverwort *Marchantia polymorpha* chloroplast DNA. *Nature* 322, 572–4.

Ort, D. R. & Baker, N. R. (1988) Consideration of photosynthetic efficiency at low light as a major determinant of crop photosynthetic performance. *Plant Physiology and Biochemistry* 26, 555–65.

Oxtoby, E. & Hughes, M. A. (1989) Breeding for herbicide resistance using molecular and cellular techniques. *Euphytica* 40, 173–80.

Palmer, J. D. (1985a) Chloroplast DNA and molecular phylogeny. *BioEssays* 2, 263–7.

Palmer, J. D. (1985b) Comparative organisation of chloroplast genomes. *Annual Review of Genetics* 19, 325–54.

Palmer, J. D. & Thompson, W. F. (1982) Chloroplast DNA rearrangements are more frequent when a large inverted repeat sequence is lost. *Cell* 29, 537–50.

Palmer, J. D., Osorio, B., Aldrich, J. & Thompson, W. F. (1987) Chloroplast DNA evolution among legumes: loss of a large inverted repeat occurred prior to other sequence rearrangements. *Current Genetics* 11, 275–86.

Pelletier, G., Primard, C., Ferault, M., Vedel, F., Chetrit, P., Renard, M. & Delourme, R. (1988) Use of protoplasts in plant breeding: cytoplasmic aspects. *Plant Cell Tissue and Organ Culture* 12, 173–80.

Possingham, J. V. (1980) Plastid replication and development in the life cycle of higher plants. *Annual Review of Plant Physiology* 31, 113–29.

Possingham, J. V. & Rose, R. J. (1976) Chloroplast replication and chloroplast DNA synthesis in spinach leaves. *Proceedings of the Royal Society of London Series B* 193, 295–305.

Ravel-Chapuis, P., Heizmann, P. & Nigon, V. (1982) Electron microscopic localization of the replication origin of *Euglena gracilis* chloroplast DNA. *Nature* 300, 78–81.

Reith, M. & Strauss, N. A. (1987) Nucleotide sequence of the chloroplast gene responsible for triazine resistance in canola. *Theoretical and Applied Genetics* 73, 357–63.

Rose, R. J. (1988) The role of membranes in the segregation of plastid DNA. In Boffey, S. A. & Lloyd, D. (eds), *Society for Experimental Biology Seminar Series 35: Division and Segregation of Organelles.* Cambridge University Press, Cambridge, UK, pp. 171–95.

Rose, R. J., Johnson, L. B. & Kemble, R. J. (1986) Restriction endonuclease studies on the chloroplast and mitochondrial DNAs of alfalfa (*Medicago sativa* L.) protoclones. *Plant Molecular Biology* 6, 331–8.

Rose, R. J., Thomas, M. R. & Fitter, J. F. (1990) The transfer of cytoplasmic and nuclear genomes by somatic hybridisation. *Australian Journal of Plant Physiology* 17, 303–21.

Roy, H. (1981) Post-translational assembly of chloroplast proteins. *What's New in Plant Physiology* 12, 41–4.

Ryan, G. F. (1970) Resistance of common groundsel to simazine and atrazine. *Weed Science* 18, 614–16.

Schoelz, J. E. & Zaitlin, M. (1989) Tobacco mosaic virus enters chloroplasts *in vivo*. *Proceedings of the National Academy of Sciences USA* 86, 4496–500.

Schuster, G., Pecker, I., Hirschberg, J., Kloppstech, K. & Ohad, I. (1986) Transcription control of the 32 kDa-Q_B protein of photosystem II in differentiated bundle sheath and mesophyll chloroplasts of maize. *FEBS Letters* 198, 56–60.

Scott, N. S. and Possingham, J. V. (1980) Chloroplast DNA in expanding spinach leaves. *Journal of Experimental Botany* 31, 1081–92.

Scott, N. S. & Possingham, J. V. (1983) Changes in chloroplast DNA levels during growth of spinach leaves. *Journal of Experimental Botany* 34, 1756–67.

Sears, B. B. (1980) Elimination of plastids during spermatogenesis and fertilization in the plant kingdom. *Plasmid* 4, 233–55.

Shinozaki, K., Ohme, M., Tanaka, M., Wakasugi, T., Hayashida, N., Matsubayashi, T., Zaita, N., Chunwongse, J., Obokata, J., Yamaguchi-Shinozaki, K., Ohto, C., Torazawa, K., Meng, B. Y., Sugita, M., Deno, H., Kamogashira, T., Yamada, K., Kusuda, J., Takaiwa, F., Kato, A., Tohdoh, N., Shimada, H. & Sugiura, M. (1986) The complete nucleotide sequence of the tobacco chloroplast genome: its gene organisation and expression. *EMBO Journal* 5, 2043–9.

Shinozaki, K., Hayashida, N. & Sugiura, M. (1988) *Nicotiana* chloroplast genes for components of the photosynthetic apparatus. *Photosynthesis Research* 18, 7–31.

Smith, S. E. (1989) Biparental inheritance of organelles and its implications in crop improvement. *Plant Breeding Reviews* 6, 361–93.

Solymosi, P. & Lehoczki, E. (1989) Co-resistance of atrazine-resistant *Chenopodium* and *Amaranthus* biotypes to other photosystem II inhibiting herbicides. *Zeitschrift für Naturforschung* 44c, 119–27.

Solymosi, P. & Lehoczxi, E. (1989) Characterization of a triple (atrazine–pyrazon–pyridate) resistant biotype of common lambsquarters. *Journal of Plant Physiology* 134, 685–90.

Steinback, K. E., McIntosh, L., Bogorad, L. & Arntzen, C. J. (1981) Identification of the triazine receptor protein as a chloroplast gene product. *Proceedings of the National Academy of Sciences USA* 78, 7463–7.

Steppuhn, J., Hermans, J., Nechushtai, R., Hermann, G. S. & Herrmann, R. G. (1989) Nucleotide sequences of cDNA clones encoding the entire precursor polypeptide for subunit VI and of the plastome-encoded gene for subunit VII of the photosystem I reaction center from spinach. *Current Genetics* 16, 99–108.

Strauss, S. H., Palmer, J. D., Howe, G. & Doerksen, A. H. (1988) Chloroplast genomes of two conifers lack a large inverted repeat and are extensively rearranged. *Proceedings of the National Academy of Sciences USA* 85, 3898–902.

Strauss, S. H., Neale, D. B. & Wagner, D. B. (1989) Genetics of the chloroplasts in conifers: biotechnology research reveals some surprises. *Journal of Forestry* 87, 11–17.

Sugiura, M. (1989) The chloroplast chromosomes in land plants. *Annual Review of Cell Biology* 5, 51–70.

Teeri, T. H., Patel, G. K., Aspegren, K. & Kauppinen, V. (1989) Chloroplast targeting of neomycin phosphotransferase II with a pea transit peptide in electroporated barley mesophyll protoplasts. *Plant Cell Reports* 8, 187–90.

Thanh, N. D. & Medgyesy, P. (1989) Limited chloroplast gene transfer via recombination overcomes plastome–genome incompatibility between *Nicotiana tabacum* and *Solanum tuberosum*. *Plant Molecular Biology* 12, 87–93.

Thanh, N. D., Pay, A., Smith, M. A., Medgyesy, P. & Marton, L. (1988) Intertribal chloroplast transfer by protoplast fusion between *Nicotiana tabacum* and *Salpiglossis sinuata*. *Molecular and General Genetics* 213, 87–93.

Thomas, M. R. & Rose, R. J. (1983) Plastid number and plastid structural changes associated with tobacco mesophyll protoplast culture and plant regeneration. *Planta* 158, 329–38.

Thomas, M. R. & Rose, R. J. (1988) Enrichment for *Nicotiana* heterokaryons after protoplast fusion and subsequent growth in agarose microdrops. *Planta* 175, 396–402.

Thomas, M. R., Fitter, J. T. & Rose, R. J. (1988) Transfer of cytoplasmic genomes using protoplast fusion and a physical selection procedure in *Nicotiana*. *Proceedings of the Australian Biochemical Society* 20, 85.

Tilney-Bassett, R. A. E. (1988) Inheritance of plastids in *Pelargonium*. In Boffey, S. A. & Lloyd, D. (eds), *Society for Segregation of Organelles*. Cambridge University Press, Cambridge, UK, pp. 171–95.

Van den Broeck, G., Timko, M. P., Kausch, A. P., Cashmore, A. R., van Montagu, M. & Herrera-Estrella, L. (1985) Targeting of a foreign protein to chloroplasts by fusion to the transit peptide from the small subunit of ribulose-1,5-bisphosphate carboxylase. *Nature* 313, 358–63.

van Rensen, J. J. S. (1982) Molecular mechanisms of herbicide action near photosystem II. *Physiologia Plantarum* 54, 515–21.

Vermaas, W. F. J., Renger, G. & Arntzen, C. J. (1984) Herbicide/quinone binding interactions in photosystem II. *Zeitschrift für Naturforschung* 39c, 368–73.

Vestweber, D. & Schatz, G. (1989) DNA–protein conjugates can enter mitochondria via the protein import pathway. *Nature* 338, 170–2.

Waddell, J., Wang, X.-M. & Wu, M. (1984) Electron microscopic localization of the chloroplast DNA replicative origins in *Chlamydomonas reinhardtii*. *Nucleic Acids Research* 12, 3843–56.

Wagner, D. B., Furnier, G. R., Saghai-Maroof, M. A., Williams, S. M. & Dancik, B. P. (1987) Chloroplast DNA polymorphisms in lodgepole and jack pines and their hybrids. *Proceedings of the National Academy of Sciences USA* 44, 2097–100.

Wallace, T. P., Stewart, A. C., Pappin, D. & Howe, C. J. (1989) Gene sequence for the 9 kDa component of Photosystem II from the cyanobacterium *Phormidium laminosum* indicates similarities between cyanobacterial and other leader sequences. *Molecular and General Genetics* 216, 334–9.

Weber, G., Monajembashi, S., Greulich, K.-O. & Wolfrum, J. (1989) Uptake of DNA in chloroplasts of *Brassica napus* (L.) facilitated by a UV-laser microbeam. *European Journal of Cell Biology* 43, 73–9.

Whitfeld, P. R. & Bottomley, W. (1983) Organisation and structure of chloroplast genes. *Annual Review of Plant Physiology* 34, 279–310.

Wu, M., Kong, X. F. & Kung, S. D. (1986) Prokaryotic promoters in the chloroplast DNA replication origin of *Chlamydomonas reinhardtii*. *Current Genetics* 10, 819–22.

Zurawski, G. & Clegg, M. T. (1987) Evolution of higher-plant chloroplast DNA-encoded genes: implications for structure–function and phylogenetic studies. *Annual Review of Plant Physiology* 38, 391–418.

Zurawski, G., Bohnert, H. J., Whitfeld, P. R. & Bottomley, W. (1982) Nucleotide sequence of the gene for the M_r 32,000 thylakoid membrane protein from *Spinacia oleracea* and *Nicotiana debneyi* predicts a totally conserved primary translation product of M_r 38,950. *Proceedings of the National Academy of Sciences USA* 79, 7699–703.

Chapter 10
Breeding for Resistance to Insects

John A. Gatehouse
*Department of Biological Sciences, University of
Durham, South Road, Durham DH1 3LE, UK*

Introduction

Crop damage caused by insects is a major economic factor in agriculture in tropical and temperate regions of the world. Large stands of a single crop species have long been recognized as favouring dramatic increases in pest species, with the numbers of herbivorous insects increasing according to the area occupied by the crop (Marchal, 1908; Strong, 1979). Failure to prevent insect attack can have devastating consequences, both for plants in the field (Fig. 10.1) and for stored plant products. For example, when stored cowpea seeds are not treated to kill the seed weevil *Callosobruchus maculatus*, as much as 100% of the seed can be destroyed in 5 months (Singh, 1978). The expenditure involved in preventing insect attack is very considerable (Fig. 10.2).

Hill (1987) lists ten orders of insects plus arachnid mites as important pests of major crops, causing damage to every part of the growing plant, and to plant products following harvest. Of the insects, the most important are Lepidoptera, Coleoptera, Homoptera, Diptera and Orthoptera. In the Lepidoptera (moths and butterflies) it is normally the larval stages (caterpillars) that are responsible for crop damage, usually to the green aerial parts of the plant. Although many species attack one plant species only, some of the more serious agricultural pests are polyphagous.

Coleoptera (beetles) also have a larval stage that is usually the major cause of damage to plants; in this case, almost any part of the plant can be attacked, with many species specializing on a particular organ, such as seeds or roots. A given species is often specific for a particular crop. Homoptera (plant bugs, including aphids, scales and hoppers) do not pass through a larval stage, and both immature forms and adults damage plants by sucking sap from leaves, stems and shoots. In Diptera (flies) the larval stage is the familiar maggot; fruits are a favoured food source, although bulbs, roots and

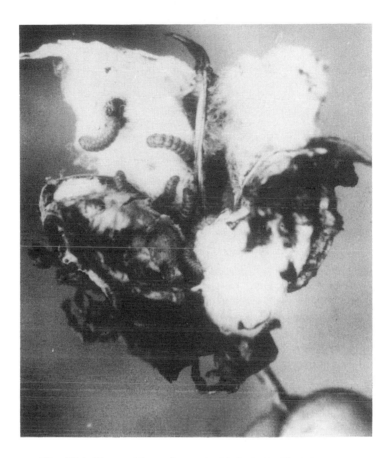

Fig. 10.1. The problem. Cotton boll infested with pink bollworms (*Pectinophora gossypiella*); note damage to squares.

leaves are also attacked. Orthoptera (grasshoppers, crickets, locusts, etc.) are polyphagous in their habits, and both adults and immature nymphs cause damage.

Control of insect pests has been an integral part of the development of agricultural practices. A build-up of pest species can generally be reduced by making the plant community more diverse, but care must be taken that alternative food sources for the same pests are not provided (see, for example, Pimental *et al.*, 1977). Traditionally, physical containment of the crop,

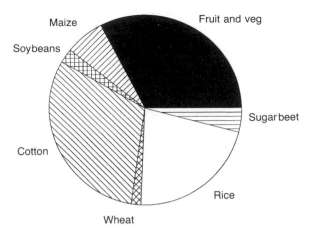

Fig. 10.2. Total world insecticide usage, divided according to crops (1988 figures; data supplied by Agricultural Genetics Company). Total expenditure $6075 million.

removal of insects, and cultural practices such as crop rotation, inter-cropping and time of sowing have provided a degree of control. These measures continue to be used in peasant communities and low-input agricultural systems. However, modern high-intensity agriculture, which has been responsible for the large increases in yield of food production achieved this century, is dependent on the use of chemical pesticides. In view of the undoubted effectiveness of chemicals in killing insects, this is not surprising. The drawbacks to this strategy have taken some time to become apparent.

First, the use of chemical pesticides is highly inefficient. Most of the applied chemical is wasted either in runoff or in subsequent washing from the plant surfaces, and it has been estimated that as much as 98% of sprayed pesticides end up in the soil. It may also be difficult to deliver pesticides to the most vulnerable part of the plant, e.g. the roots or the insides of stems or fruits.

Secondly, pesticides and their residues are often highly toxic to beneficial organisms. A plant species in its natural environment is in balance with the various organisms, including insects, that depend upon it for food. These insects in turn provide food for predatory insects, arachnids, birds, fish and mammals. This 'natural order' is often overlooked when considering crop plants and the use of pesticides that may be toxic to non-target organisms. Soil condition may decline, reducing yields. Human toxicity is also a major concern.

Thirdly, the very strong selection pressure on insect populations imposed by insecticides causes resistance to such compounds to be rapidly acquired. For example, the Colorado potato beetle has been subjected to intensive treatment with pesticides, as a result of which many populations are now resistant to nearly all permitted insecticides (Hare, 1990). Resistance to the synthetic pyrethroids was observed within 2–4 years of their introduction and widespread use. As insects become more resistant, applications of insecticides are repeated more frequently in attempts to eliminate them.

Fourthly, over-use of pesticides and herbicides can decrease the vigour of the crop, and actually make it more susceptible to insect attack. For example, maize plants treated with herbicide were found to be infested by about twice as many aphids and corn borers (Oka & Pimental, 1976).

For all of these reasons, the over-use of broad-spectrum insecticides is an agricultural practice that is not sustainable, as has been recognized in the banning of organochlorine insecticides such as DDT in many countries. Great efforts have been made to find safe and selective pesticides, although little has been done to wean the farmers of the industrialized countries away from their reliance on exogenously applied chemicals as the answer to every need.

Alternatives to chemical pesticides for insect control are available, and have been successfully employed in the control of major pests. Two of these methods, biological control using predators and parasites of the pest species, and the prevention of mating between fertile adult insects using pheromone traps or the release of sterile insects, are reviewed elsewhere (see Hill, 1983, 1987). A third method, improving the inherent resistance of the crop plant to insect attack, forms the major focus of this chapter.

Inherent resistance of plants to insect attack

Plants and phytophagous insects have been co-existing and evolving under selection pressures provided by each other for approximately 130 million years in the case of flowering plants, and for at least twice as long for more primitive genera (Ehrlich & Raven, 1964; Boulter *et al.*, 1990b). It is therefore not surprising that plants have evolved a variety of physical and chemical defences against insects. Plants are normally protected by combinations of physical and chemical mechanisms. For example, trichomes on the leaves of *Solanum berthaultii* provide a physical barrier to aphid infestation, and also release two chemical defences: a volatile terpenoid, β-farnesene, which acts as an alarm pheromone and causes aphids to stop feeding, and a phenolic exudate linked to a phenolase/peroxidase enzyme system, which sets to a sticky mass and immobilizes an aphid when disturbed by contact (Gregory *et al.*, 1986).

Similarly, many seeds are defended by a physical barrier, the testa, which also contains condensed tannins as a chemical defence, while the embryo is further protected by toxic or antimetabolic constituents.

Physical mechanisms

Physical mechanisms include the production of barriers such as lignified tissue, cuticular waxes, spines, hairs or sticky exudates, all of which can prevent an insect reaching vulnerable tissue that it can feed on. Besides outer barriers, such as toughened seed-coats, barriers can also occur inside plant tissue.

An example of an internal physical defence mechanism is the development of silica deposits in leaves and stems of grasses and cereals; this is a common adaptation to xerophytic conditions, but can also be present in some varieties of species normally grown under semi-arid conditions. The presence of these deposits deters attack by insects whose larvae bore into the tissues, exemplified in the case of protection of Italian ryegrass from shoot flies (Moore, 1984). The silica deposit-forming phenotype, and thus resistance, can readily be bred into commercial lines of a particular species from the resistant varieties.

Hairy surfaces are equally effective in deterring some insects; for example, cotton jassids (a leafhopper) have been effectively eliminated as a pest of cotton in Africa and India by the post-war introduction of pubescent cotton varieties (Hill, 1987), and the hooked trichomes of some varieties of bean plants effectively prevent the attack of aphids (which become impaled on them). Active defences are also possible, as when latexes or resins are exuded from wound sites, although in this case the physical defence is usually supplemented by the presence of insecticidal chemicals in the latex or resin (Harborne, 1988).

Since insect eggs often have to be laid on or in the future food source for larvae to survive, prevention of oviposition by physical (or chemical) means is an effective defence against certain species.

Physical defence mechanisms have been exploited by plant breeders in the past, since the phenotype is often easily determined by visual inspection or by a simple test that can be carried out in the field. However, many physical mechanisms are polygenic in character, and are not easily transferred by conventional breeding methods.

Chemical mechanisms

Insecticidal compounds are found across the complete spectrum of secondary metabolites produced in different plant species. Some act as antifeedants, by preventing the insect from recognizing the plant tissue as a suitable food source. Others kill the insect either directly or by interfering with normal developmental processes (toxins and antimetabolites). Examples are given in Table 10.1.

Many insects use taste and smell cues to recognize their preferred food source. Interfering with the insect's biochemical perception of the plant is

Table 10.1. Plant secondary metabolites exhibiting insecticidal activity.

Compound	Plant source	Reference
Non-protein antimetabolites		
Alkaloids		
2,5-dihydroxymethyl 3,4-dihydroxypyrrolidine (DMDP)	*Lonchocarpus*	Evans *et al.*, 1985
Castanospermine	*Castanospermum australe*	Nash *et al.*, 1986
Non-protein amino acids		
p-aminophenylalanine	*Vigna*	Birch *et al.*, 1986
Terpenoids		
pyrethroids	Compositae	Mann, 1987
juvabione	*Abies balsamea*	Mann, 1987
gossypol	*Gossypium hirsutum*	Berardi & Goldblatt, 1980
Rotenoids (isoflavanoids)	*Lonchocarpus salvadorensis*	Birch *et al.*, 1985
Tannins	*Vicia faba*	Boughdad *et al.*, 1986
Polysaccharides		
pectosans	*Phaseolus vulgaris*	Ishii, 1952
heteropolysaccharides	*Phaseolus vulgaris*	Applebaum *et al.*, 1970 Gatehouse *et al.*, 1987
Glucosinolates	Cruciferae	Erickson & Feeny, 1974
Cyanogenic glycosides	*Lotus corniculatus*	Mann, 1987
Protein antimetabolites		
Lectins	*Phaseolus vulgaris*	Janzen *et al.*, 1976
Arcelin	*Phaseolus vulgaris*	Gatehouse *et al.*, 1984 Osborn *et al.*, 1986 Minney *et al.*, 1990
Protease inhibitors	*Vigna unguiculata*	Gatehouse *et al.*, 1979 Hilder *et al.*, 1987
	Solanaceae	Johnson *et al.*, 1990
α-Amylase inhibitor	*Phaseolus vulgaris*	Gatehouse *et al.*, 1987
	Phaseolus vulgaris	Ishimoto & Kitamura, 1988

therefore an effective primary means of defence. Insect taste receptors react favourably to amino acids and sugars, but generally unfavourably towards many secondary metabolites. However, most pest species are specialists, and will have evolved to use the secondary metabolites of their food source as a positive signal for consumption or oviposition.

Toxins and antimetabolites inside plant tissues can form a second line of defence, and be effective against non-specialists as well as specialists. The plant can sometimes prevent or retard a population build-up of a pest species by interfering with its moulting, metamorphosis or reproductive capacity, or by providing a poor nutrient balance that slows its development.

The modes of action of some of the major types of plant products that affect insects are outlined below. The reader is referred to articles by Fraenkel (1969), Dethier (1972), Schoonhoven (1972), Feeny (1975), Swain (1977) and Mann (1987) for more detailed information.

Terpenoids and steroids

These compounds share a common origin in the biosynthetic pathway to mevalonate, and are generally hydrophobic molecules containing only carbon, hydrogen and oxygen. The lower terpenoids (C_{10} and C_{15}) are often volatile, and are major components of the scents and odours which attract or repel insects. Flower oils such as geraniol attract pollinating insects, whereas volatile terpenes produced in other parts of the plant, such as the mixture of myrcene and α- and β-pinenes produced in pine bark resin, can repel. Such compounds often provide signals which allow specialists to locate their preferred food plants, as when the bark beetle *Dendroctonus brevicornis* is attracted to Ponderosa pine by the very terpenoid mixture that repels other insects (Wood, 1982).

The cotton boll weevil (*Anthonomus grandis*) is similarly attracted to cotton by the mixture of α-pinene, limonene and caryophyllene released by the plant. Other C_{10} terpenoids are potent insecticides, e.g. the pyrethroids, originally isolated from members of the family Compositae. Certain C_{15} terpenoids mediate a more subtle interaction between plant and insect; juvabione is such a compound, synthesized by Balsam fir (*Abies balsamea*). This acts as an analogue of an insect juvenile hormone, and can prevent the metamorphosis from larval stage to adult of bugs such as *Pyrrhocoris apterus* when they are fed on, or in contact with, material containing the compound. Although not preventing initial attack, this resistance mechanism would prevent an insect population from building up rapidly.

Steroids are derived from C_{30} terpenoids, and are a vital component of cellular membranes, as well as having many hormonal roles; almost all insects depend on plants for their sterols, since they lack the capacity to synthesize the steroid nucleus. Certain plant species contain insect moulting hormones (although the role of these compounds in insect resistance is uncertain), and others contain highly toxic steroidal alkaloids (see below) and saponins.

Phenolics, flavanoids and tannins

These compounds, whose structure is based on the aromatic alcohol phenol, are predominantly hydrophilic, and often contain *O*-linked sugar residues. They are derived from the aromatic amino acid phenylalanine (in some grasses also from tyrosine) via the enzymatic deamination catalysed by phenylalanine-ammonia lyase (PAL); flavanoids contain a second aromatic nucleus derived (ultimately) from acetate and added by the enzyme chalcone synthase. Both PAL and chalcone synthase are tissue-specific, inducible enzymes and have been shown to be involved in the plant's response to environmental stress and pathogens (Smith *et al.*, 1977; Swain, 1977; Hahlbrook & Schiel, 1989). Tannins are of two types: hydrolysable tannins contain simple phenolic acids combined with the sugar glucose, whereas condensed tannins are polymers of flavanoid monomers (strictly, hydroxyflavanols). Plentiful evidence exists for the importance of flavanoids in plant selection by insects, mediated by their taste; for example the flavanoids isoquercetin and morin are vital in the recognition of mulberry (*Morus* spp.) by the silkworm *Bombyx mori* (Hamamura *et al.*, 1962). However, flavanoids, phenolics and tannins all generally have a bitter and repellent taste, and act as feeding deterrents for non-adapted insects (Harborne, 1988). Many of these compounds are toxic when fed in bio-assay in artificial diet. Condensed tannins are particularly effective in giving protection against phytophagous insects, as in the case of tannins from the seed-coat of field bean, *Vicia faba*. These tannins are strongly antimetabolic to larvae of the seed weevil *Callosobruchus maculatus* (Boughdad *et al.*, 1986), possibly by complexing with digestive enzymes and preventing nutrient utilization, or by acting as a feeding deterrent (Klocke & Chan, 1982).

Alkaloids and non-protein amino acids

These compounds are generally nitrogenous bases, produced from amino acids; a very large diversity of structural types exists. Many alkaloids are highly toxic, such as strychnine, and are thus very effective at deterring opportunistic attack. However, the strong selection pressure caused by such an effective defence has, in many cases, resulted in specialists having evolved to overcome the defence mechanism, as in the case of the Danaid butterflies, which feed on plants containing pyrrolizidine alkaloids. The reasons for toxicity of some alkaloids have been elucidated. Alkaloidal sugar analogues, such as castanospermine and DMDP (2,5-dihydroxymethyl-3,4-dihydroxypyrrolidine) have been shown to be potent inhibitors of insect glycosidases (Evans *et al.*, 1985; Nash *et al.*, 1986), whereas non-protein amino acids such as canavanine (found in several legume species, including *Dioclea* spp.) are incorporated into proteins in place of their homologues (arginine, for canavanine) and produce non-functional protein molecules (Rosenthal *et al.*, 1976; Rosenthal, 1982).

Miscellaneous secondary metabolites

Many other types of secondary metabolites have been shown to play a role in the interaction between plants and insects. Glucosinolates (thioglucosides) and the volatile allyl isothiocyanates released from them by hydrolysis have been shown to act both as a feeding deterrent (and toxin) to non-pest species (Erickson & Feeny, 1974) and as a feeding stimulus to pest species (Dethier, 1972) in crucifers. Cyanogenic glycosides are a well-known toxic component in many strains of clover. Heteropolysaccharides, as seed components, have also been reported as having toxic effects on insects (Applebaum *et al.*, 1970; Applebaum & Guez, 1972; Gatehouse *et al.*, 1987).

Proteins

Although proteins are not in the strict sense classed as secondary metabolites, it has become clear that certain types of protein have similar functions to classical secondary metabolites, including protection of the plant against insects.

Protease inhibitors are low-molecular-weight proteins found in many different plant species and tissues, especially in leaves and seeds (Richardson, 1977; Belitz & Weder, 1990). These proteins are stable to heat and extremes of pH, and are resistant to proteolysis. Most of the well-characterized inhibitors are of the Bowman–Birk or 'double-headed' type. These are small proteins (about 80 amino acid residues) containing a high proportion of cysteine, which provides multiple intramolecular disulphide bonds. The molecule has two domains, each of which contains an enzyme inhibitory site. Hence these inhibitors can be active against one or more proteolytic enzymes, including bovine trypsin, chymotrypsin, elastase, subtilisin and others.

The role of protease inhibitors in the plant has been the subject of much debate, but there is now no doubt that a major function of these proteins is to confer insect resistance. They are toxic when fed in artificial diet to both coleopteran (Gatehouse *et al.*, 1979) and lepidopteran (Steffens *et al.*, 1978) species, and have been shown *in vitro* to inhibit insect gut proteolysis (Broadway & Duffy, 1986). However, not all inhibitors are equally effective against all insects, and it is evident that adaptation has occurred (Gatehouse *et al.*, 1985).

Protease inhibitors are actively synthesized in plants attacked by herbivorous insects, giving additional evidence for their protective role (Ryan, 1983, 1985). The exact mechanism of protease inhibitor toxicity is not yet fully clear; besides the obvious interference with utilization of dietary protein, effects on amino acid balance through over-production of digestive enzymes in the affected insect may occur.

Inhibitors of other digestive enzymes are also present in plant tissues, particularly seeds. Probably the most significant of these are amylase inhibitors (Garcia-Olmedo *et al.*, 1987; Belitz & Weder, 1990), some of which are structurally related to protease inhibitors. Amylase inhibitors from wheat and *Phaseolus vulgaris* have been shown to be toxic towards insects by bio-assay in artificial diet (Gatehouse *et al.*, 1986; Ishimoto & Kitamura, 1988), although once again, specificity of toxicity is observed.

Lectins, or carbohydrate-binding proteins, are another type of protein whose toxicity towards insects is fairly well established. Toxic effects appear to be mediated through binding of the lectin to glycoproteins in the gut wall, leading to its disruption (Pusztai *et al.*, 1979). Various lectins with different sugar-binding specificities may show selective toxicities. Nevertheless, lectins from several legume species have been shown to be toxic towards both Coleoptera and Lepidoptera (Shukle & Murdoch, 1983; Gatehouse *et al.*, 1984).

A class of proteins related to lectins is the ribosome-inactivating proteins (RIPs). Class II RIPs contain a subunit with lectin activity and a ribosome-inactivating subunit linked together and are highly toxic, whereas Class I RIPs contain only the ribosome-inactivating polypeptide and are much less toxic to insects (Gatehouse *et al.*, 1990a).

Plant breeding for insect resistance

In setting out to improve the insect resistance of crop plants by breeding, a first step is the realization that inherent resistance may have been lost during the course of breeding programmes aimed at producing 'improved' varieties. In some cases this has been done accidentally, as a result of a plant breeder aiming to maximize the yield of a crop; the final products of such breeding programmes have frequently lost much of their natural resistance to pests.

In a few cases the resistance has been deliberately bred out, in an attempt to get rid of compounds that limit the processing or utilization of the crop. An example is the presence of a phenolic compound, gossypol, in cotton. This was held to be deleterious, as gossypol is toxic to mammals, and prevents the use of cotton seed meal as an animal feed. The development of glandless cotton, which does not produce this compound, was once a long-term goal in cotton breeding. Unfortunately, when glandless lines were produced, they were found to be susceptible to attack by various insects, including the cotton budworm (*Heliothis virescens*), since gossypol forms an essential part of the plant's defences against these insects (Berardi & Goldblatt, 1980).

A need for improved resistance may also arise when crops are exposed to insect pests they have not previously had the opportunity of developing defences to. This can come about either through the introduction of crops to parts of the world where they are not native, or by the introduction of pests to new areas by the avoidance of quarantine measures. For example, the introduction of the potato into semi-desert areas of the western USA from Bolivia early in the nineteenth century exposed it to the Colorado potato beetle, to which it had no defence. This beetle is now the most serious pest of potatoes world-wide, having spread to Europe in the late nineteenth and early twentieth centuries by being shipped in cargoes of potatoes (Hill,

1987). Phytosanitary regulations have been found inadequate, in many cases, to prevent the spread of pests.

Sources of resistance genes

A successful breeding programme will depend on two basic requirements: a suitable source of 'resistance genes' and a method for assaying the resistance to insects developed by successive progeny. The following sources are available as a basis for a conventional breeding programme: lines or varieties of the crop in question, wild biotypes of the same species, compatible wild progenitors, or close relatives. The availability of germplasm collections for most major crops makes screening of a large collection of lines a fairly straightforward, if extremely time-consuming, procedure. If possible, it is often better to use indigenous knowledge and start by examining material that has been found (or claimed) to be insect-resistant in the field.

Many crops still contain enough diversity in their gene-pool to allow this approach to succeed; for example, screening over 5000 lines of cowpea (*Vigna unguiculata*) for resistance to the seed weevil *Callosobruchus maculatus* resulted in one line (Tvu 2027) that showed significant resistance and could be made the basis of a breeding programme (Gatehouse *et al.*, 1979). The increasing genetic homogeneity of many modern varieties of major crop plants makes it important to preserve older varieties which retain some of the genetic diversity of wild biotypes. Most plant breeding has concentrated on crosses between 'advanced' cultivated lines, which contain only a small fraction ($< 5\%$ for most crops) of the germplasm variability. Access to the full genetic diversity of a crop species may be made via the major international plant breeding institutes, which hold extensive germplasm collections, with different institutes specializing in particular crops. Collection of wild biotypes, or related and compatible species, requires considerable field-work and botanical expertise; wherever possible local collaborators should be sought. For most crops, international programmes to identify, collect and bulk up insect-resistant varieties are in progress, and form a vitally important genetic resource for plant breeders.

For such an approach to be successful, it is necessary that the resistance mechanism chosen does not make the crop toxic or otherwise unsuitable for agricultural purposes; there would be little point, for example, in breeding the production of a toxic alkaloid into crop plants. The previously quoted example of gossypol in cotton is also relevant in this context.

Insect bio-assay

A further problem in breeding programmes for insect resistance is that the bio-assay is time-consuming, space-demanding and labour-intensive if the only method available is to actually infest plants (either as individuals or as stands) with the insect pests. In addition, due to variability in both the insects and their food plants, obtaining statistically significant results for parameters such as insect survival, biomass and damage to the plant is frequently very difficult. If the segregation of polygenic characteristics is being followed, an insect bio-assay may give results so poor as to be unusable when individual plants are assayed. In many cases, however, this is the only reliable method available, e.g. in the case of sap-suckers such as aphids, and it may be necessary to bulk up genetically homogeneous populations of plants to allow bio-assays to be carried out. The development of screening techniques which avoid the problems and inherent variability of insect bio-assay is therefore desirable. If a clear biochemical or physical basis for the resistance can be established, then this phenotype can be screened for in crossing experiments. Alternatively, the development of RFLP marker technology may allow polymorphisms that are shown to be linked to resistance to be used as markers (Chapter 2). However, the identification of these polymorphisms and demonstration of their linkage will involve an extensive research programme of screening and bio-assay, so that the problem is moved to a different stage of the breeding programme rather than avoided.

Arcelin-mediated resistance towards the storage pest Zabrotes subfasciatus *in common bean*

The bruchid beetle *Zabrotes subfasciatus* is responsible for serious losses of stored bean seeds in Central and South America. Screening by bio-assay of a wide range of collected lines and wild biotypes by workers at CIAT (Schoonhoven *et al.*, 1983) showed that certain accessions showed virtually complete resistance to attack by this pest. Fortunately, the accessions, although wild biotypes, were fully compatible with agricultural varieties of *Phaseolus vulgaris*, and a conventional breeding strategy to transfer resistance into new agricultural varieties could be contemplated, and was commenced.

Concomitantly, seeds of the resistant accessions were analysed for various compounds which might play a role in conferring resistance. When the seed proteins were examined by SDS-polyacrylamide gel electrophoresis (SDS-PAGE), an immediate and obvious difference in band patterns was observed between resistant and susceptible accessions. All the resistant accessions contained polypeptides of M_r 32,000–36,000 as major components of their seed proteins, whereas none of the susceptible accessions contained these polypeptides (Fig. 10.3). The novel protein was named 'arcelin' (Romero Andreas *et al.*, 1986). Further screening of accessions showed that four different patterns of arcelin polypeptides were found, which were associated with different geographical origins of the accessions (Osborn *et al.*, 1986). The four forms were labelled arcelin[1-4], and whereas arcelin[1,2,4] were found in accessions resistant to *Z. subfasciatus*, arcelin[3] was not. In addition, examination of material from the early stages of a crossing

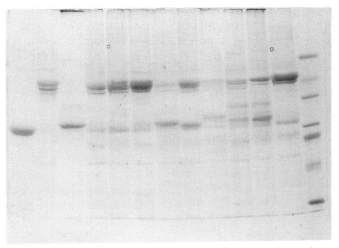

Fig. 10.3. SDS-polyacrylamide gel analysis of seed proteins extracted from single seeds of *Phaseolus vulgaris* lines. Lines are arbitrarily designated 1–9: track PHA, seed lectin (E_2L_2); G_{ii}, phaseolin (major storage protein); arc, arcelin (putative resistance factor). The presence of arcelin in lines 4, 6, 7 and 8 is readily scored, and was confirmed by Western blotting vs. anti-arcelin antibodies. These lines showed resistance towards the bruchid storage pest *Zabrotes subfasciatus*. Track M contains molecular weight markers, M_r 66,000; 45,000; 36,000; 29,000; 24,500; 20,100; 14,200 from top of gel. (Data provided by B. H. P. Minney.)

programme between an arcelin[1]-containing accession and a commercial line showed that the presence of arcelin[1] could be correlated with seed resistance. This allowed subsequent screening of material from the crossing programme to be carried out much more easily; as a preliminary screen, the presence of arcelin[1] polypeptides could be rapidly determined by SDS-PAGE analysis of a small piece of cotyledon, thus enabling potentially resistant seeds to be separated from susceptible seeds without the need for insect bio-assay.

Although the breeding programme could be carried out without direct evidence of arcelin toxicity or knowledge of the mechanism of toxicity, these data are necessary to establish screening for the presence of arcelin as a valid measure of resistance. Both arcelin[1] and arcelin[4] have been purified, and shown to be toxic to developing larvae of *Z. subfasciatus* when incorporated into artificial diets (Osborn *et al.*, 1988; Minney *et al.*, 1990), albeit only at high concentrations (> 7% of dry weight of the diet). In the case of arcelin[4], toxicity seems to depend on the protein's indigestibility by the insect, as shown by *in vitro* proteolysis assays, thus preventing amino acids vital for development being available. Supplementation of arcelin[4]-containing diets with digestible protein considerably decreased its toxic effects (Minney *et al.*, 1990). Although the original wild-type accessions show evidence for multiple mechanisms of resistance, it is clear that the presence of a high content of arcelin is in itself sufficient to confer resistance, and thus the screening method is valid.

Limitations to conventional breeding strategies

The major limitation to conventional strategies for breeding for insect resistance is the availability of suitable sources of insect resistance genes. It is often the case that an insect pest of the wild progenitor of a crop species has become a major pest of the cultivated crop, and that no resistance can be identified either in the crop or in its progenitor. It may be possible to find a related species with resistance which can be crossed successfully, provided the fertility of such a wide cross can be manipulated to enable fertile progeny to be produced, e.g. by using embryo rescue techniques. As previously stated, if resistance is a polygenic character, the breeding programme to transfer resistance into commercial crop lines may be difficult and lengthy; inheritance of resistance to seed weevils in cowpea was found to be complex (Redden *et al.*, 1983).

New methods for plant breeding

The following methods are described in more detail in other chapters and texts (e.g. Lindsey & Jones, 1989); their application to breeding for insect resistance is considered below.

Mutation breeding

Mutagenesis by chemical mutagens, or by radiation, has been used extensively to generate new varieties of crop plants. However, it has not (to the author's knowledge) been used to produce a successful insect-resistant crop plant variety. The reasons for this are several: the difficulty of screening for suitable mutations, the failure of such mutagenesis to generate positive changes to the genome (since inactivation of genes by mutagenesis is much more common than activation), and the large number of progeny that must be handled may be cited as examples.

Somaclonal variation

This technique is subject to the same drawbacks as mutation breeding, with the added disadvantage that not all crop plants can be taken through a protoplast or callus stage from which somatic embryos or adventitious shoots are regenerated. The problems of screening large numbers of progeny from a random mutagenesis programme for insect resistance are in general too great for such approaches to be feasible.

Protoplast fusion

This method is again dependent on a suitable tissue culture system for the crop plant being available but if this is the case, then it offers the possibility of transferring characteristics (albeit at random) across a wider species difference than would be possible using conventional techniques. A resistance mechanism possessed by one species may thus be transferred to another, provided the two species are closely enough related for the resulting fusion to be viable and capable of being regenerated to yield fertile plants. In practice, the technique has been limited to the production of interspecific hybrids between commercial crops and related wild species which cannot normally be crossed; however, this still means that the problem of lack of resistance genes within a species may be overcome. The ease with which potato (*Solanum tuberosum*) and related *Solanum* spp. may be regenerated from tissue culture, and their subsequent propagation via vegetative rather than sexual reproduction, make this a good system to work with; interspecific hybrids produced by protoplast fusion between *S. brevidens* (resistant to viral pathogens of commercial potato) and *S. tuberosum* have been made which allowed the viral resistance genes to be crossed into commercial potato. Interspecific hybrids between *S. berthaultii* and *S. tuberosum* have also been produced in an attempt to mobilize the resistance shown by the wild species towards aphids and Colorado potato beetle into commercial potato (Wright *et al.*, 1985).

Although this technique has considerable potential in a limited number of cases, the random nature of the transfer of characteristics, the necessity to grow on and screen large numbers of progeny, the necessity for a subsequent breeding programme to establish resistance in a commercial line, and, most significantly, the difficulty of regeneration from protoplasts for many crop species are significant drawbacks to this approach.

Genetic engineering of plants for insect resistance

The genetic engineering of plants allows single genes to be transferred to the target species from a wide variety of sources, including higher and lower plants, bacteria and animals. Plants have the necessary basic biosynthetic capacity to deal with the synthesis of foreign proteins, as has been shown by the successful production of functional bacterial enzymes (Jefferson, 1987) and mammalian immunoglobulins (Hiatt *et al.*, 1989) in transgenic plants. Genetic engineering thus overcomes the species barrier, and gives complete control over the character of the transgenic plant; not only can single genes be transferred, but their temporal and spatial expression can be predetermined by the use of suitable gene promoters in the constructs.

Two major types of technology have been developed for plant genetic engineering. Direct transfer techniques are those in which DNA is directly introduced into plant tissues, either by uptake of DNA into chemically or electrically treated protoplasts or microinjection of DNA into cell nuclei, or by such methods as soaking seeds in DNA solutions, injection of DNA into floral vascular bundles, or firing microparticles coated in DNA into plant tissue (Chapters 4, 5). Indirect transfer techniques use a vector – either a bacterial plasmid or a virus – to introduce the foreign DNA (Lindsey & Jones, 1989). In both cases, it is necessary to use a marker gene, usually a selectable marker such as antibiotic or herbicide resistance, but occasionally a screenable marker such as pigment production, to allow transformed tissue to be isolated, and thus the gene to be transferred must be attached to the marker gene in a gene construct. Where a vector is used, both genes must be incorporated into the vector in such a way that they do not impair its normal function.

The technique most often used to date, in terms of the transfer of agronomically useful genes to viable plants, has been introduction of foreign DNA via the Ti plasmid of the soil bacterium *Agrobacterium tumefaciens* (Chapter 3). The mechanism of the DNA transfer to plants caused by this bacterium and the methods necessary to exploit it have been extensively reviewed (e.g. Herrera-Estrella & Simpson, 1988). The commonly employed methods rely on infection of tissue pieces (usually leaf discs) with *Agrobacterium* carrying the gene to be transferred, and subsequent regeneration of plants from the microcalli formed at the infection sites. Selection is used to prevent regeneration of untransformed tissue. The ease of tissue culture of Solanaceae has made them favoured subjects for genetic engineering experiments based on infection by *Agrobacterium*. On the other hand, a drawback to this method is that the range of plants that can be readily transformed and regenerated is limited, and excludes major crop plants among the cereals, which cannot be directly infected by *Agrobacterium*. Present developments suggest that the technical problems involved in DNA transfer to a wide range of crops will be solved in the short term, although at present it is not clear in all cases which transformation technique will emerge as the method of choice.

Limitations of genetic engineering techniques

Besides the limited number of plant species which are amenable to genetic engineering at the present state of the technology, a further important consideration is the availability of the gene(s) to be transferred. If the insect-icidal compound is a primary gene product, i.e. a protein, there is (at least in theory) no technical barrier to obtaining and transferring its encoding gene. However, attempting to engineer a transgenic plant that produces an insecticidal compound which is the product of a multi-step specialized biochemical pathway is not at present feasible. Such pathways are often not fully characterized, and the enzymes carrying out individual steps are un-known or unpurified. Obtaining the multiple genes necessary is therefore difficult; transferring multiple genes and arranging their co-ordinate regula-tion to give a functional biochemical pathway have not yet been achieved. Thus, although compounds such as alkaloids would offer a high degree of protection against insect attack, they are not exploitable by this method. Other potential constraints and commercial considerations have been discussed elsewhere (Gatehouse *et al.*, 1990b).

Examples of plant genetic engineering for insect resistance

Use of a plant proteinase inhibitor gene

Evidence for the role of proteinase inhibitors in conferring endogenous resistance to insect attack has been outlined above. The cowpea (*Vigna unguiculata*) trypsin inhibitor (CpTI) is accumulated in cowpea seeds, and was identified as an insecticidal component by bio-assay. Increased amounts of the inhibitor in an artificial diet prevented development of the larvae of *Callosobruchus maculatus*, a coleopteran pest of stored cowpea seeds (Gatehouse *et al.*, 1979). Further bio-assays showed that the compound was also effective against other insects from the orders Coleoptera, Lepidoptera and Orthoptera (Table 10.2), including a number of economically important field and storage pests. An important consideration with an introduced gene is the potential toxicity of its product towards humans and other mammals, and in this respect CpTI has the advantage of proved insignificant toxicity, since it is already present in a commonly consumed foodstuff, cowpea seeds. The role of CpTI as an insecticidal protein had initially been based on the observation that a resistant line of cowpeas (provided by IITA, Nigeria) had an elevated content of CpTI; however, subsequent work has shown that resistance of cowpea seeds to their major storage pest is multi-mechanistic (Xavier-Filho *et al.*, 1989). The difference between resistant and susceptible lines of cowpea with respect to CpTI was found to be quantitative, not qualitative, which meant that all biochemical and molecular biological work could be, and was, done with proteins and DNA purified from ordinary com-mercial cowpea material, and not the resistant line. This was advantageous, in not having to grow up and cross with genetically heterogeneous material collected from the field, and in not compromising any rights pertaining to such collected material.

Table 10.2. Insect pests against which cowpea trypsin inhibitors (CpTI) are effective.

Order	Insect pest	Primary crops attacked
Field pests		
Lepidoptera	*Heliothis virescens*	Tobacco, cotton
	Heliothis zea	Maize, cotton, beans, tobacco
	Helicoverpa armigera	Cotton, beans, maize, sorghum
	Spodoptera littoralis	Maize, rice, cotton, tobacco
	Chilo partellus	Maize, sorghum, sugarcane, rice
	Autographa gamma	Sugarbeet, lettuce, cabbage, beans, potato
	Manduca sexta	Tomato, tobacco, potato
Orthoptera	*Locusta migratoria*	Polyphagous but preference for wild and cultivated grasses
Coleoptera	*Diabrotica undecimpunctata*	Maize
	Costelytra zealandica	Grasses, clover
	Anthonomus grandis	Cotton
Storage pests		
Coleoptera	*Callosobruchus maculatus*	Cowpea, soybean
	Tribolium confusum	Most flours

CpTI was purified from cowpea seeds and characterized (Gatehouse *et al.*, 1980; Hilder *et al.*, 1989) The protein is a mixture of isoinhibitors of the Bowman–Birk type described earlier. The major inhibitors in cowpea contain either two trypsin inhibitory sites, or one trypsin and one chymotrypsin site. A trypsin/chymotrypsin isoinhibitor was subjected to protein sequencing. A library of cDNA clones was produced from mRNA isolated from developing cowpea seeds, and a cDNA species containing the mature CpTI protein coding sequence was then identified by screening with an oligonucleotide probe based on the determined protein sequence (Hilder *et al.*, 1987). In order for this cDNA to be transcribed and translated in a transgenic plant, it had to be made into a functional gene by adding a 5' flanking region from the cauliflower mosaic virus (CaMV) genome ('35S' promoter), which provided an active, constitutively expressed promoter and a start of transcription, and a 3' flanking region from the *Agrobacterium tumefaciens* nopaline synthase gene (*nos* terminator), which provided a polyadenylation signal sequence and a transcription terminator. The synthetic gene was incorporated into an *Agrobacterium* 'binary' vector (a plasmid, capable of replication in *E. coli* and *A. tumefaciens*, containing a section of DNA transferred to the plant and the appropriate marker genes) and was transformed into tobacco plants, as a model system. A control transformation was carried out where the CpTI coding sequence was placed in the incorrect orientation relative to the CaMV promoter. The construct is summarized in Fig. 10.4.

The transgenic plants were tested for expression of the foreign protein by immunoassay; as expected, the 'correct' construct produced CpTI protein at

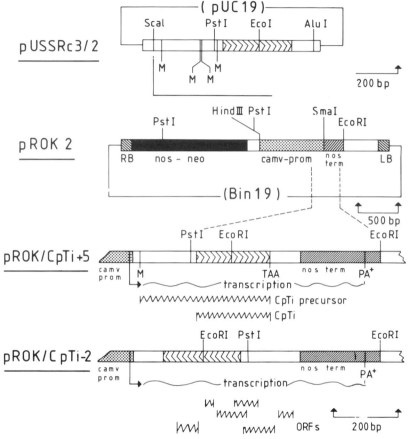

Fig. 10.4. Construction of the recombinant plasmid used to produce transgenic plants expressing CpTI (cowpea trypsin inhibitor). The plasmid pUSSRc3/2, containing a mature complete CpTI coding sequence (hatched area) and the preprotein 'leader' sequence (unfilled area, with met start codons indicated by M) was restricted with endonucleases Sca I and Alu I. The resulting fragment was ligated into the Sma I site of the binary vector pROK 2, a broad host-range plasmid containing, between the T-DNA border sequences from the *Agrobacterium tumefaciens* Ti plasmid, a selectable marker gene (*nos-neo*) conferring kanamycin resistance on transformed tissue, a strong constitutive promoter (camv-prom) and a terminator sequence from the *Agrobacterium* nopaline synthase gene. Ligation occurring in the correct orientation relative to the promoter gave the construct pROK/CpTI+5, used to produce CpTI-expressing plants; ligation in the incorrect orientation gave the construct pROK/CpTI−2, used to produce transformed control plants incapable of expressing CpTI. (From Hilder *et al.,* 1987, with permission.)

amounts up to 1% of total soluble leaf protein, whereas the control construct gave no expression. A transformant expressing CpTI at the highest content tested was cloned by vegetative propagation, and the resulting plants were used for insect bio-assays against first-instar larvae of the lepidopteran pest, *Heliothis virescens* (tobacco budworm). Analysis of the bio-assay showed a highly significant reduction in leaf damage, insect survival and insect biomass (approx. twofold, twofold and fourfold respectively over a 7-day feeding trial) in the transgenic plants expressing CpTI compared with those containing the control construct. Similar results were obtained in assays against *Heliothis zea* (corn earworm) and *Spodoptera littoralis* (armyworm). In trials run 'to destruction' with *Manduca sexta* (tobacco hornworm) results were visually more spectacular, with transgenic CpTI-expressing plants showing little or no damage, whereas control plants were eaten away to a stump. Representative results are presented in Fig. 10.5.

Subsequent experiments showed that the protection observed in the transgenic plants was stably inherited (as a single Mendelian locus in most cases, although multiple copies of the CpTI gene were incorporated); the 'yield penalty' of expressing large amounts of CpTI was not significant under laboratory conditions, even when the plants were stressed (V. A. Hilder, personal communication). These plants seem to satisfy most of the criteria required for practical viability, although the degree of protection afforded (though statistically significant) would not be sufficient to satisfy most commercial growers. Further development of the system is at present under way; a promising approach is to combine different resistance mechanisms in a single transgenic plant, in a manner analogous to the multi-mechanistic resistance shown by many plants in the wild (Boulter *et al.*, 1990a).

Fig. 10.5. Bio-assay of transformed tobacco plants against *Heliothis virescens* (tobacco budworm). Plants are shown seven days after infestation by first-instar larvae. Left: control plant (see legend to Fig. 10.4); right: CpTI-expressing plant. Expression of the foreign protein in transgenic plants gives substantial protection against insect attack. (From Hilder *et al.*, 1987, with permission.)

Other protease inhibitors besides CpTI may be used in transgenic plants to afford protection; for example, potato inhibitors have been shown to give a degree of protection against attack by *Manduca sexta* in transgenic tobacco (Johnson *et al.*, 1990). However, choice of inhibitor to give maximal protection over as wide a range of pests as possible is clearly necessary.

Use of a bacterial toxin gene

The scope of genetic engineering, in allowing genes from any source to be incorporated into transgenic plants, is well illustrated by the production of insect-resistant plants containing genes encoding the insecticidal crystal protein from a soil bacterium, *Bacillus thuringiensis* (Barton *et al.*, 1987; Fischhoff *et al.*, 1987; Vaeck *et al.*, 1987). Sporulating bacteria produce crystals made up mainly of a toxin precursor; proteolytic cleavage of the non-toxic protoxin polypeptide following ingestion by the insect liberates the active toxin. A high degree of specificity with respect to species is shown by toxins produced by different strains of *B. thuringiensis* (Bt), based both on the toxin and on the susceptibility of the protoxin to cleavage, and toxicity towards mammals is theoretically low.

Bt toxin genes have been isolated from a number of different bacterial strains, but genetic engineering has been carried out only on strains active against lepidopteran pests. The introduction of these genes into transgenic plants follows the same principles as outlined for CpTI above, but more involved engineering proved to be necessary to obtain successful expression of Bt toxins, even at contents much lower than that achieved for CpTI (approx. 0.02% of total soluble leaf protein). The complete protoxin molecule was only expressed at very low amounts and constructs containing the toxin part of the polypeptide (the N-terminal end) had to be used. The engineered toxin polypeptide coding sequence ('truncated toxin') was placed between a constitutive promoter and a terminator, as for CpTI, and was introduced into tobacco plants via *Agrobacterium*-mediated transfer.

Despite the low degree of expression, transgenic plants containing the truncated Bt toxin were effectively protected against *Manduca sexta* when tested in bio-assays. Marginal protection against *Heliothis* spp. was also observed. More impressively, in a later experiment with a different gene construct conducted under field conditions (Fuchs *et al.*, 1989), partial protection of tomato plants containing a truncated Bt toxin gene against *Heliothis zea* was observed, and substantial protection against a fortuitous infestation by *Keiferia lycopersicella* (tomato pinworm) was afforded. The commercial potential of such genetically engineered plants is currently being evaluated.

Conclusions

The production of plants with inherent resistance to insect attack is certain to become an increasingly important part of integrated pest management strategies in future, not only on the grounds of economic savings, but also as a result of legislative and consumer pressure. Previous criticism of inherent resistance, in that it cannot provide a high enough degree of protection for commercial growers and that it can be overcome by pest adaptation, can be addressed by management systems that use other control measures (possibly even a low-level use of insecticides) in concert with resistant plant varieties. Agricultural systems that continue to place an extremely strong selection pressure on a pest, by means of a single pesticidal control measure, will inevitably cause the rapid development of pesticide resistance.

Conventional plant breeding still offers the surest route to generating insect-resistant plant varieties, provided that a suitable source of resistance can be found; new techniques have a role in such breeding programmes in identifying and screening for substances or genes responsible for resistance. In the longer term, techniques such as somaclonal variation and protoplast fusion offer means of generating additional sources of breeding material. Plant genetic engineering is perhaps the most exciting of the new methods, and has a proven track record of actually producing plants with enhanced insect resistance. It remains to be seen whether this initial promise can be translated into practical success, and whether transgenic plants can surmount the hurdles of public and legislative acceptance and commercial viability, to become a reality in the farmer's field.

Acknowledgements

Although the shortcomings of this article are solely the responsibility of the author, I acknowledge with gratitude the help of colleagues, particularly Dr V. A. Hilder, Dr A. M. R. Gatehouse and Professor D. Boulter, in developing and discussing its contents.

References

Applebaum, S. W. & Guez, M. (1972) Comparative resistance of *Phaseolus vulgaris* beans to *Callosobruchus chinensis* and *Acanthoscelides obtectus* (Col. Bruchidae): the differential digestion of soluble heteropolysaccharide. *Entomology Experimental and Applied* 15, 64–74.

Applebaum, S. W., Tadmor, U. & Podoler, II. (1970) The effect of starch and of a heteropolysaccharide fraction from *Phaseolus vulgaris* on development and fecundity of *Callosobruchus maculatus. Entomology Experimental and Applied* 13, 61–70.

Barton, K. A., Whiteley, H. R. & Yang, N.-S. (1987) *Bacillus thuringiensis*-endotoxin expressed in *Nicotiana tabacum* provides resistance to lepidopteran insects. *Plant Physiology* 85, 1103–9.

Belitz, H.-D. & Weder, J. K. P. (1990) Protein inhibitors of hydrolases in plant foodstuffs. *Food Reviews International* 6(2), 151–211.

Berardi, L. C. & Goldblatt, L. A. (1980) Gossypol. In: Liener, I. E. (ed.), *Toxic Constituents of Plant Foodstuffs*, 2nd edn. Academic Press, New York, pp. 183–237.

Birch, A. N. E., Crombie, L. & Crombie, W. M. (1985) Rotenoids of *Lonchocarpus salvadoriensis*; their effectiveness in protecting seeds against bruchid predation. *Phytochemistry* 24, 2881–3.

Birch, A. N. E., Fellows, L. E., Evans, S. V. & Docherty, S. V. (1986) Para-aminophenylalanine in *Vigna*: possible taxonomic and ecological significance as a seed defence against bruchids. *Phytochemistry* 25, 2745–51.

Boughdad, A., Gillon, Y. & Gagnepain, C. (1986) Influence des tannins condensés du tégument de fèves (*Vicia faba*) sur le développement larvaire de *Callosobruchus maculatus*. *Entomology Experimental and Applied* 42, 125–32.

Boulter, D., Edwards, G. A., Gatehouse, A. M. R., Gatehouse, J. A. & Hilder, V. A. (1990a) Additive protective effects of different plant-derived insect resistance genes in transgenic tobacco plants. *Crop Protection* 9, 351–4.

Boulter, D., Gatehouse, J. A., Gatehouse, A. M. R. & Hilder, V. A. (1990b) Genetic engineering of plants for insect resistance. *Endeavour* 14, 185–90.

Broadway, R. M. & Duffy, S. S. (1986) Plant proteinase inhibitors: mechanism of action and effect on the growth and digestive physiology of larval *Heliothis zea* and *Spodoptera exigua*. *Journal of Insect Physiology* 32, 827–33.

Dethier, V. G. (1972) Chemical interactions between plants and insects. In: Sondheimer, E. & Simeone, J. B. (eds), *Chemical Ecology*. Academic Press, New York, pp. 83–102.

Ehrlich, P. R. & Raven, P. H. (1964) Butterflies and plants: a study in co-evolution. *Evolution* 18, 586–608.

Erickson, J. M. & Feeny, P. (1974) Sinigrin: a chemical barrier to the black swallowtail butterfly *Papilio polyxenes*. *Ecology* 55, 103–11.

Evans, S. V., Gatehouse, A. M. R. & Fellows, L. E. (1985) Detrimental effects of 2,5-dihydroxymethyl-3,4-dihydroxypyrrolidine in some tropical legume seeds on larvae of the bruchid *Callosobruchus maculatus*. *Entomology Experimental and Applied* 37, 257–61.

Feeny, P. (1975) Biochemical co-evolution between plants and their insect herbivores. In: Gilbert, L. E. & Raven, P. H. (eds), *Co-evolution of Animals and Plants*. University of Texas Press, Austin, Texas, pp. 3–19.

Fischhoff, D. A., Bowdish, K. S., Perlak, F. J., Marrone, P. G., McCormick, S. M., Niedermeyer, J. G., Dean, D. A., Kusano-Kretzmer, K., Mayer, E. J., Rochester, D. E., Rogers, S. G. & Fraley, R. T. (1987) Insect tolerant transgenic tomato plants. *BioTechnology* 5, 807–13.

Fraenkel, G. (1969) Evaluation of our thoughts on secondary plant substances. *Entomology Experimental and Applied* 12, 474–86.

Fuchs, R. L., MacIntosh, S., Kishore, G., Perlak, F., Dean, D., Stone, T., Sims, S., Hoffman, N., Greenplate, J. T., Marrone, P. & Fischhoff, D. A. (1989) Enhanced expression/efficiency of transgenic plants which express the *Bacillus thuringiensis* insect control protein *Agbiotech '89*, 210.

Garcia-Olmedo, F., Salcedo, G., Sanchez-Monge, R., Gomez, L., Royo, J. & Carbonero, P. (1987) Plant proteinaceous inhibitors of proteinases and alpha-amylases. In: Miflin, B. J. (ed.), *Oxford Surveys of Plant Molecular and Cell Biology,* vol. IV. Oxford University Press, Oxford, pp. 275–334.

Gatehouse, A. M. R., Gatehouse, J. A., Dobie, P., Kilminster, A. M. & Boulter, D. (1979) Biochemical basis of insect resistance in *Vigna unguiculata*. *Journal of the Science of Food and Agriculture* 30, 949–58.

Gatehouse, A. M. R., Gatehouse, J. A. & Boulter, D. (1980) Isolation and characterisation of trypsin inhibitors from cowpea (*Vigna unguiculata*). *Phytochemistry* 19, 751–6.

Gatehouse, A. M. R., Dewey, F. M., Dove, J., Fenton, K. A. & Pusztai, A. (1984) Effect of seed lectin from *Phaseolus vulgaris* on the development of larvae of *Callosobruchus maculatus*: mechanism of toxicity. *Journal of the Science of Food and Agriculture* 35, 373–80.

Gatehouse, A. M. R., Butler, K. J., Fenton, K. A. & Gatehouse, J. A. (1985) Presence and partial characterisation of a major proteolytic enzyme in the larval gut of *Callosobruchus maculatus*. *Entomology Experimental and Applied* 39, 279–86.

Gatehouse, A. M. R., Fenton, K. A., Jepson, I. & Pavey, D. J. (1986) The effects of α-amylase inhibitors on insect storage pests: inhibition of α-amylase *in vitro* and effects on development *in vivo*. *Journal of the Science of Food and Agriculture* 37, 727–34.

Gatehouse, A. M. R., Dobie, P., Hodges, R. J., Meik, J., Pusztai, A. & Boulter, D. (1987) Role of carbohydrates in insect resistance in *Phaseolus vulgaris*. *Journal of Insect Physiology* 33, 843–50.

Gatehouse, A. M. R., Barbieri, L., Stirpe, F. & Croy, R. R. D. (1990a) Effects of ribosome inactivating proteins on insect development – differences between Lepidoptera and Coleoptera. *Entomology Experimental and Applied* 54, 43–51.

Gatehouse, J. A., Hilder, V. A. & Gatehouse, A. M. R. (1990b) Genetic engineering of plants for insect resistance. In: Grierson, D. (ed.), *Plant Genetic Engineering*. Blackie and Son, London, pp. 105–35.

Gregory, P., Ave, D. A., Bouthyette, P. J. & Tingey, W. M. (1986) In: Juniper, B. & Southwood, T. R. E. (eds), *Insects and the Plant Surface*. Edward Arnold, London, pp. 173–84.

Hahlbrook, K. & Schiel, D. (1989) Physiology and molecular biology of phenylpropanoid metabolism. *Annual Review of Plant Physiology* 40, 347–69.

Hamamura, Y., Hayashiya, K., Naito, K., Matsuura, K. & Nishida, J. (1962) Food selection by silkworm larvae. *Nature* 194, 754–5.

Harborne, J. B. (1988) *Introduction to Ecological Biochemistry*, 3rd edn. Academic Press, London.

Hare, J. D. (1990) Ecology and management of the Colorado potato beetle. *Annual Review of Entomology* 35, 81–100.

Herrera-Estrella, L. & Simpson, J. (1988) Foreign gene expression in plants. In: Shaw, C. H. (ed.), *Plant Molecular Biology – a Practical Approach*. IRL Press, Oxford, pp. 131–60.

Hiatt, A., Cafferkey, R. & Bowdish, K. (1989) Production of antibodies in transgenic plants. *Nature* 342, 76–8.

Hilder, V. A., Gatehouse, A. M. R., Sheerman, S. E., Barker, R. F. & Boulter, D. (1987) A novel mechanism of insect resistance engineered into tobacco. *Nature* 330, 160–3.

Hilder, V. A., Barker, R. F., Sammour, R. A., Gatehouse, A. M. R., Gatehouse, J. A. & Boulter, D. (1989) Protein and cDNA sequences of Bowman–Birk inhibitors from the cowpea (*Vigna unguiculata* Walp.). *Plant Molecular Biology* 13, 701–10.

Hill, D. S. (1983) *Agricultural Insect Pests of the Tropics and their Control.* Cambridge University Press, Cambridge.

Hill, D. S. (1987) *Agricultural Insect Pests of Temperate Regions and their Control.* 2nd edn, Cambridge University Press, Cambridge.

Ishii, S. (1952) Studies on the host preference of cowpea weevil (*Callosobruchus chinensis*). *Bulletin of the National Institute of Agricultural Science Japan* 1, 185–256.

Ishimoto, M. & Kitamura, K. (1988) Identification of the growth inhibitor on azuki bean weevil in kidney bean (*Phaseolus vulgaris* L.). *Japanese Journal of Breeding,* 38, 367–70.

Janzen, O. H., Juster, H. B. & Liener, I. E. (1976) Insecticidal action of the phytohaemagglutinin in black beans on a bruchid beetle, *Science* 192, 795–6.

Jefferson, R. A. (1987) Assaying chimaeric genes in plants: the GUS gene fusion system. *Plant Molecular Biology Reports* 5, 387–405.

Johnson, R., Narvaez, J., An, G. & Ryan, C. A. (1990) Expression of potato proteinase inhibitors I and II in transgenic tobacco plants: effects on natural defence against *Manduca sexta* larvae. *Proceedings of the National Academy of Sciences USA* 86, 9871–5.

Klocke, J. A. & Chan, B. G. (1982) Effects of cotton condensed tannin on feeding and digestion in the cotton pest, *Heliothis zea. Journal of Insect Physiology* 28, 911–15.

Lindsey, K. & Jones, M. G. K. (1989) *Plant Biotechnology in Agriculture.* Open University Press, Milton Keynes, UK.

Mann, J. (1987) *Secondary Metabolism,* 2nd edn. Oxford University Press, Oxford.

Marchal, P. (1908) The utilization of auxiliary entomophagous insects in the struggle against insects injurious to agriculture. *Popular Science Monthly* 72, 352–70.

Minney, B. H. P., Gatehouse, A. M. R., Dobie, P., Denedy, J., Cardona, C. & Gatehouse, J. A. (1990) Biochemical bases of seed resistance to *Zabrotes subfasciatus* (bean weevil) in *Phaseolus vulgaris* (common bean): a mechanism for arcelin toxicity. *Journal of Insect Physiology* 36, 757–67.

Moore, D. (1984) The role of silica in protecting Italian ryegrass (*Lolium multiflorum*) from attack by dipterous stem-boring larvae (*Oscinella frit* and other related species). *Annals of Applied Biology* 104, 161–6.

Nash, R. J., Fenton, K. A., Gatehouse, A. M. R. and Bell, E. A. (1986) Effects of the plant alkaloid castanospermine as an antimetabolite of storage pests. *Entomology Experimental and Applied* 42, 71–7.

Oka, I. N. & Pimental, D. (1976) Herbicide (2,4-D) increases insect pathogen pests on corn. *Science* 193, 239–40.

Osborn, T. C., Blake, T., Gepts, P. & Bliss, F. A. (1986) Bean arcelin 2: genetic variation, inheritance and linkage relationships of a novel seed protein of *Phaseolus vulgaris* L. *Theoretical and Applied Genetics* 71, 847–55.

Osborn, T. C., Alexander, D. C., Sun, S. S. M., Cardona, C. & Bliss, F. A. (1988) Insecticidal activity and lectin homology of arcelin seed protein. *Science* 240, 207–10.

Pimental, D., Shoemaker, C., LaDue, E. L., Rovinsky, R. B. & Russell, N. P. (1977) Alternatives for reducing insecticides on cotton and corn: economic and environmental impact. *Environmental Research Laboratory Official Research Development* 145. EPA, Athens, Georgia, USA.

Pusztai, A., Clarke, E. M. W. & King, T. P. (1979) The nutritional toxicity of *Phaseolus vulgaris* lectins. *Journal of the Nutritional Society* 38, 115–20.

Redden, R. J., Dobie, P. & Gatehouse, A. M. R. (1983) The inheritance of seed resistance to *Callosobruchus maculatus* F. in cowpea (*Vigna unguiculata* L. Walp.). I. Inheritance of parental, F1, F2, F3 and backcross seed generations. *Australian Journal of Agricultural Research* 34, 681–95.

Richardson, M. (1977) The proteinase inhibitors of plants and micro-organisms. *Phytochemistry* 16, 159–69.

Romero Andreas, J., Yandell, B. S. & Bliss, F. A. (1986) Bean arcelin 1: inheritance of a novel seed protein of *Phaseolus vulgaris* L. and its effects on seed composition. *Theoretical and Applied Genetics* 72, 123–8.

Rosenthal, G. A. (1982) *Plant Nonprotein Amino and Imino Acids*. Academic Press, New York, London.

Rosenthal, G. A., Dahlman, D. L. & Janzen, D. H. (1976) A novel means for dealing with L-canavanine, a toxic metabolite. *Science* 192, 256–8.

Ryan, C. A. (1983) Insect-induced chemical signals regulating natural plant protection responses. In: Denno, R. F. & McClure, M. S. (eds), *Variable Plants and Herbivores in Natural and Managed Systems*. Academic Press, New York, pp. 43–60.

Ryan, C. A. (1985) Proteinase inhibitors. In: Marcus, A. (ed.), *The Biochemistry of Plants: a Comprehensive Treatise*, vol. 6. Academic Press, New York, pp. 351–70.

Schoonhoven, A. V., Cardona, C. & Valor, J. F. (1983) Resistance to the bean weevil and the Mexican bean weevil (Coleoptera: Bruchidae) in noncultivated common bean accessions. *Journal of Economic Entomology* 76, 1255–9.

Schoonhoven, L. M. (1972) Secondary plant substances and insects. *Recent Advances in Phytochemistry* 5, 197–224.

Shukle, R. H. and Murdoch, L. L. (1983) Lipoxygenase, trypsin inhibitor and lectin from soybeans: effects on larval growth of *Manduca sexta* (Lepidoptera: Sphingidae). *Environmental Entomology* 12, 787–91.

Singh, S. R. (1978) Resistance to insect pests in Nigeria. In: Singh, S. R., Van Emden, H. F. & Taylor, T. A. (eds), *Pests of Grain Legumes: Ecology and Control*. Academic Press, New York, pp. 267–97.

Smith, H., Billett, F. E. & Giles, A. B. (1977) The photocontrol of gene expression in higher plants. In: Smith, H. (ed.), *Regulation of Enzyme Synthesis and Activity in Higher Plants*. Academic Press, London, pp. 93–128.

Steffens, R., Fox, F. R. & Kassel, B. (1978) Effect of trypsin inhibitors on growth and metamorphosis of corn borer larvae *Ostrinia nubialis* (Hubner). *Journal of Agricultural and Food Chemistry* 26, 170–4.

Strong, D. R. (1979) Biogeographic dynamics of insect–host plant communities. *Annual Review of Entomology* 24, 89–119.

Swain, T. (1977) Secondary compounds as protective agents. *Annual Review of Plant Physiology* 28, 479–501.

Vaeck, M., Reynaerts, A., Hofte, H., Jansens, S., De Beuckeleer, M. D., Dean, C., Zabeau, M., van Montagu, M. V. & Leemans, J. (1987) Transgenic plants protected from insect attack. *Nature* 328, 33–7.

Wood, D. L. (1982) The role of pheromones, kairomones and allomones in the host selection and colonization behaviour of bark beetles. *Annual Review of Entomology* 27, 411–46.

Wright, R. J., Dimock, M. J., Tingey, W. M. & Plaisted, R. L. (1985) Colorado potato beetle (Coleoptera: Chrysomelidae): expression of resistance in *Solanum berthaultii* and interspecific potato hybrids. *Journal of Economic Entomology* 78, 576–82.

Xavier-Filho, J., Campos, F. A. B., Ary, M. B., Silva, C. P., Carvalho, M. M. M., Macedo, M. L. R., Lemos, F. J. A. & Grant, G. (1989) Poor correlation between levels of proteinase inhibitors found in seeds of different cultivars of cowpea (*Vigna unguiculata*) and the resistance/susceptibility to predation by *Callosobruchus maculatus*. *Journal of Agricultural and Food Chemistry* 37, 1139–43.

Chapter 11
Resistance to Fungal Diseases

A. K. Chakravorty & K. J. Scott
Department of Biochemistry, The University of
Queensland, Queensland 4072, Australia

Introduction

Plant diseases caused by viral, bacterial, fungal and other pathogens are responsible for enormous economic losses world-wide. The most important pathogens of economically important crop plants are represented by biotrophic and necrotrophic fungi. Necrotrophic fungi cause death of plant cells and grow on nutrients derived from the dead cells. Biotrophic fungi, such as the rust and powdery mildew fungi, can only grow and reproduce on living host plants of susceptible cultivars. This has been recognized for a long time and a considerable amount of scientific endeavour has been devoted towards understanding the molecular basis of resistance to fungal diseases. A vast literature on the physiology and biochemistry now exists but it has contributed little to our understanding of the mechanism of disease resistance. Analyses by techniques of classical genetics of interactions between fungal pathogens and their host plants have led to such proposals as the gene-for-gene hypothesis (Flor, 1956). However, these studies have not elucidated the mechanism of resistance of plants to fungal pathogens.

With the increasing application of molecular biology to problems of plant–microbial interactions (Kerr, 1987; Leong & Holden, 1989), there is a growing expectation that considerable advances will soon be made in our understanding of plant–fungal interactions that lead to either a compatible (host:susceptible) or an incompatible (host:resistant to fungal infection) reaction. This expectation has been bolstered by the discovery of pathogenesis-related (PR) proteins – a group of inducible host plant-encoded proteins whose synthesis is associated with resistance to plant pathogens as well as various forms of physical and chemical stress (reviewed by Carr & Klessig, 1989). In the past four or five years there has been considerable excitement over many new findings that extended our understanding of the structure, synthesis, location and, in some cases, the function of PR proteins in plant tissues (Legrand *et al.,* 1987; White *et al.,* 1987).

In this chapter, we outline recent advances in the biochemical and molecular aspects of the resistance responses of plants to phytopathogenic fungi, concentrating mainly on the active responses in plants triggered by pathogenic fungi. It should be pointed out that whereas the modes of infection by necrotrophic and biotrophic fungi are often quite different, the responses of plants to attack by different phytopathogenic fungi are quite similar. We therefore discuss resistance to fungal diseases and make no distinction between biotrophic and necrotrophic fungi.

Passive and active defence mechanisms

Plants, being stationary organisms, are vulnerable to adverse environmental conditions as well as to attack by plant-eating and pathogenic organisms. The ability of a plant to stop an invasion of pathogenic fungi depends a great deal on the presence of preformed barriers (passive defence) and pathogen-induced active responses. Plants have cell walls containing, in addition to cellulosic and non-cellulosic polysaccharides, interesting proteins that are regarded as protective biopolymers (Carr & Klessig, 1989). These proteins include extensin, which is a hydroxyproline-rich glycoprotein (HRGP), a glycine-rich protein (GRP) and, in some plants, leaf-specific thionins which have antifungal activity and may constitute the first line of defence against fungal pathogens. Some of these proteins are specified by plant gene families that are induced by fungal attack as well as by biotic and abiotic elicitors. These biopolymers are, therefore, part of both active and passive defence mechanisms. Another such biopolymer is lignin, which is formed by random condensation of phenylpropanoid alcohols, and is a component of secondary plant cell walls. Pathogen-induced lignification has been shown to follow the induction of the many genes which specify lignin biosynthetic enzymes, such as cinnamyl alcohol dehydrogenase, peroxidases and polyphenol oxidases (Carr & Klessig, 1989). Callose, a β-1,3-glucan, has also been shown to accumulate in plant cell walls in response to infection and mechanical injury.

Two major types of defence reactions in plants have been recognized. The first is the local defence response, which restricts the spread of the pathogen to the infection site, and which in some cases is accompanied by localized cell death and a hypersensitive reaction (HR). The second is the systemic resistance, in which the plants respond to primary infection by activating general defence mechanisms throughout the plant. Both of these modes of defence reactions are usually part of active defence mechanisms.

Another common form of resistance of plants to fungal pathogens is the so-called non-host resistance (Heath, 1982), in which neither the host plant nor the fungus is capable of recognizing and interacting with the other. This form of resistance is not discussed further in this article. With most other host–fungal combinations, however, survival of the plants in their natural habitat depends on their ability to respond actively and quickly to external stimuli provided by the invading pathogen. It is now known that plants have evolved highly sophisticated mechanisms to respond to external signals. The speed and efficacy of these responses play a pivotal role in determining the resistance of plants to fungal pathogens.

Active defence responses

The phytoalexins

In some plants, as the fungal spores germinate, the resistant cultivars respond by the formation of cytoplasmic aggregates deposited internally on the epidermal cell wall. These appositions result in the formation of structures called papillae (Bryngelsson & Collinge, 1991). These structures form physical barriers, largely preventing hyphae from penetrating the cell wall. This 'papilla resistance' may well be one of the major forms of defence mechanism in the case of biotrophic fungi and form the basis of hypersensitive reaction (HR). In the case of most necrotrophic fungi HR is accompanied by the production of phytoalexins in the surrounding cells.

Phytoalexins are low-molecular-weight antimicrobial compounds which act as broad-spectrum antibiotics and are believed to play an important role in arresting the growth of fungal pathogens in resistant plants. They are products of plant secondary metabolism. In some cases, their production requires the induction of enzymes catalysing reactions in the steps of the phenylpropanoid biosynthetic pathway (Bryngelsson & Collinge, 1991). The expression of these genes may be due to elicitors produced by the partial breakdown of host and fungal cell walls (Tepper & Anderson, 1986). Other biotic elicitors of phytoalexin synthesis include glucans, proteins, glycoproteins and fatty acid derivatives. The induction of the enzymes for the phytoalexin biosynthetic pathway is also caused by abiotic agents such as mechanical injury, ultraviolet irradiation and heavy metals. Whatever the elicitor, the induction of these genes is very rapid and is linked to the expression of genes specifying HRGP precursors and the enzymes for lignin biosynthesis (Carr & Klessig, 1989).

The role of phytoalexins in diseases of cereals caused by biotrophic fungi is quite uncertain (Deverall, 1989; Bryngelsson & Collinge, 1991). However, the first enzyme in the phenylpropanoid biosynthetic pathway, phenylalanine ammonia-lyase (PAL), has been reported to be induced in barley in response to infection by the powdery mildew fungus (Shiraishi *et al.*, 1989).

In other host–fungal interactions, it has been proposed that the inability of a fungal pathogen to detoxify phytoalexins contributes to the resistance of the host plant (Van Etten *et al.*, 1989). The first phytoalexin to be purified and identified was pisatin (Perrin & Bottomley, 1962). Some fungi possess mechanisms of phytoalexin tolerance that do not involve metabolic degradation. The phytoalexins have been given trivial names such as pisatin in pea, medicarpin and maakiain in several species of leguminous plants. The bean pathogen, *Fusarium solani* f. sp. *phaseoli*, can detoxify at least four of the major phytoalexins of bean, namely kievitone, phaseolin, phaseollidin and phaseollinisoflavan. The mechanism of inactivation of many of the phytoalexins has been elucidated (Van Etten *et al.*, 1989). Lubinin and rishitin are two of the major sesquiterpenoid phytoalexins of potato. Although the genes conferring phytoalexin detoxification are often linked to pathogenicity, there are several fungi which are sensitive to phytoalexins even though they can metabolize them (Van Etten *et al.*, 1982).

The key enzyme in phytoalexin synthesis is PAL, which catalyses the deamination of L-phenylalanine to *trans*-cinnamic acid. This is the first step in the biosynthesis of a large class of secondary metabolites based on the phenylpropane skeleton (Jones, 1984). These compounds have important roles in plant development and protection against pathogenic microorganisms. They also act as signal molecules that are responsible for the induction of the virulence genes in *Agrobacterium* and nodulation genes in *Rhizobium* (Downie & Johnston, 1986; Stachel & Zambryski, 1986). In recent years there has been a great deal of interest in the crucifer *Arabidopsis* as a model plant for molecular biological studies because of its small genome size and short life cycle (Meyerowitz & Pruitt, 1985). In *Arabidopsis*, PAL is specified by a small family of genes. Ohl *et al.* (1990) have cloned and partially characterized one member of the multigene family, PAL 1. The promoter of this gene has been found to contain sequence elements homologous to other putative regulatory elements that are conserved among several genes involved in phenylpropanoid biosynthesis. These workers have studied the tissue-specific expression of PAL 1-GUS (β-glucuronidase) gene fusions in transgenic *Arabidopsis* and found that the PAL 1 promoter is involved in photoregulation as well as induction by heavy metals. These observations and the recent discovery that *Arabidopsis thaliana* is susceptible to infection by a downy mildew fungus (Koch & Slusarenko, 1990) mean that it may now be possible to clone resistance determinants from a host plant which is very amenable to genetic and molecular analyses. Further studies along these lines are likely to provide a definitive answer to the long-debated question of the role of phytoalexins in fungal disease resistance.

Several other enzymes of phytoalexin biosynthetic pathways have been studied at the molecular level. These include 4-coumarate:CoA ligase, 6-hydroxychalcone synthetase, chalcone isomerase and cinnamic acid 4-hydroxylase. All these enzymes are induced by various pathogens, including fungi, and by elicitors. In both suspension cell cultures and intact plants, the induction of these enzymes in most cases has been shown to be at the transcriptional level (Carr & Klessig, 1989).

Plant peptides and proteins directly or indirectly involved in resistance to fungal pathogens

During the initial stages of infection of plants by pathogenic fungi, the expression of a number of host genes is induced. Some of these genes are involved in phytoalexin synthesis. Others include the genes specifying toxin receptor proteins, proteinase inhibitors, hydrolytic enzymes and pathogenesis-related (PR) proteins. These disease-response genes are expressed in response to pathogens as well as to biotic and abiotic elicitors.

Toxin receptor proteins

In oat plants, a single dominant locus, Pc2, conditions the sensitivity to the host-specific toxin Victorin (produced by *Helminthosporium victoriae*) but also is a disease-resistance gene against the biotrophic fungus *Puccinia coronata*, conditioning HR to avirulent races (Dixon & Lamb, 1990). Victorin is a basic peptide containing some unusual amino acids. Recently, a large protein of M_r 100 kDa has been shown to contain the Victorin binding site (Wolpert & Macko, 1989). It should now be possible to isolate this protein and to test whether or not it is the Pc2 gene product.

Proteinase inhibitors

The induction of synthesis and accumulation of proteinase inhibitors I and II in plants is triggered by a proteinase inhibitor inducing factor (PIIF), which is a large pectin fragment of approximate M_r 200 kDa released from plant cell walls. The proteinase inhibitors, in contrast to phytoalexins, are synthesized systemically (Ryan, 1987). The accumulation of proteinase inhibitors caused by environmental stress or infection may contribute to disease resistance, or at least be part of an induced resistance mechanism against insects (Chapter 10) and micro-organisms. These proteins, mostly found in plant cell vacuoles, provide limited protection against microbes, by inhibiting their proteinase activities and thus reducing the amount of nutrients available to the parasites.

In plants, activation of defence genes by pathogen attack or mechanical wounding is caused by signal molecules that are released at the infection or wound site. These signal molecules are transported locally by diffusion through intracellular and intercellular fluids and systemically through the vascular system of the plant. Until recently, ethylene was the only chemical signal known to travel through the atmosphere to activate defence genes in other plants. Farmer & Ryan (1990) have reported that methyl jasmonate, a common plant secondary metabolite, when applied to surfaces of tomato plants induces the synthesis and accumulation of proteinase inhibitor proteins, not only in the treated plants, but also in the untreated plants nearby. The presence of methyl jasmonate in the atmosphere is sufficient to induce the synthesis of proteinase inhibitor proteins in three different species of plants belonging to two families. This demonstrates that inter-plant communication to activate the expression of defence genes is possible in the absence of physical contact.

Hydrolytic enzymes

It has been proposed that hydrolytic enzymes involved in the degradation of fungal cell walls are important in both constitutive and inducible defence reactions in plants against fungal pathogens. As chitin and β-1,3-glucan (callose) are important components of the cell walls of many phytopathogenic fungi, two inducible glucanhydrolases, β-1,3-glucanase and chitinase, have been strongly implicated in the antifungal defence reactions of plants (Bell, 1981). These two enzymes occur naturally in healthy plant cells (Hoj *et al.*, 1989). They are co-ordinately induced in several plants by ethylene, which is produced in plants in response to stress caused by infection. The regulation of the genes specifying these enzymes during normal development and in response to several forms of stress, including wounding, pathogen attack or treatment with chitosan or ethylene, has been described by Boller (1985) and Mauch *et al.* (1988). Some chitinases show low lysozyme (peptidoglucanase) activity. The ethylene-induced chitinase of bean is encoded by a small family of genes (Broglie *et al.*, 1986) induced at the transcriptional level in response not only to ethylene, but also to elicitors and fungal pathogens in intact plants, as well as in cell suspension cultures.

In tobacco plants undergoing HR, the mRNAs for both chitinase and β-1,3-glucanases are apparently co-ordinately induced. As discussed in a later section, in many plants β-1,3-glucanase genes are involved in defence against pathogens, and in some cases glucanases may be classified as PR proteins.

The chitinase and β-1,3-glucanase enzymes are thought to be involved in degrading fungal cell walls and release elicitor-active carbohydrate molecules which trigger the synthesis of fungitoxic phytoalexins in the host plant.

Proteins and enzymes involved in cell wall modifications

As mentioned above, parasitically induced changes in the structure and composition of plant cell walls constitute a form of active defence mechanism. Apart from the formation of lignified secondary walls in response to fungal infection, cuticles and suberized cell walls provide additional barriers to penetration by fungal hyphae.

Other changes in plant cell walls in response to fungal infection include the synthesis of HRGP, phenolic compounds, papillae formation and silicon deposition. The main component of papillae is callose, which is synthesized by the Ca^{2+}-dependent enzyme, callose synthase (β-1,3-glucan synthase) located in the plasma membrane. Papillae also contain phenolic compounds as well as cellulose and silica (Heitefuss & Ebrahim-Nesbat, 1986).

The synthesis of plant cell wall-associated proteins of unusual composition such as HRGPs in response to fungal infection is another method of defence. These HRGP molecules are coated with covalently bound sugar residues and these proteins are insolubilized through cross-linking by peroxidases associated with plant cell walls (Cooper & Varner, 1984). The accumulation of HRGPs in response to infection as well as to mechanical injury in many dicotyledonous plants suggests that these cell wall-associated proteins have a role in disease resistance.

Leaf-specific thionins

Leaf-specific thionins are a family of low-molecular-weight sulphur-containing cell wall proteins, which are widespread throughout the plant kingdom (Bohlmann *et al.*, 1988). The thionins of barley have been well characterized. These represent highly abundant polypeptides with antifungal activity found in the cell walls of barley leaves. The leaf-specific thionins of barley are specified by a complex multigene family of 50 100 members, all of which are located on chromosome 6. The toxicity of thionins for phytopathogenic fungi and the observation that their synthesis is triggered by pathogens suggest that these represent inducible plant proteins which are involved in defence against microbial infections.

Peroxidases

The exact role of peroxidases in defence against fungal pathogens is not known. The fact that there is often a substantial increase in the activity of these enzymes during infection, their extracellular location and their role in lignification and suberization and in the polymerization of HRGPs clearly suggest a function for these enzymes in defence against fungal diseases based on cell wall modification (Bryngelsson & Collinge, 1991). There are many peroxidase isozymes found in a wide range of plant tissues. Some of these isozymes are highly acidic, others are basic, and all are glycosylated haemoproteins of M_r 30–35 kDa. Several peroxidase isozymes have also been shown to possess phenoloxidase activity (De Biasi & Badiana, 1990).

Peroxidase activity increases following infection of several different plant species by fungal pathogens (Hislop & Stahmann, 1971; Arora & Bajaj, 1985). Increases in both peroxidase and polyphenol oxidase activities appear to be part of the induced resistance of mung bean to *Rhizoctonia solani*. Specific peroxidase isozymes are induced by ethephon treatment and pathogen attack. Six peroxidase isozymes are induced in cucumber hypocotyls infected with *Cladosporium cucumerinum*, only one of which is present in uninoculated control plants. Interestingly, the same isozymes are also induced by wounding but not by an oligogalacturonide elicitor. By contrast, heptagalacturonide, but not wounding, has been shown to induce lignification (Svalkheim & Robertson, 1990), suggesting that peroxidases are not responsible for lignification in the cucumber–*C. cucumerinum* system.

Pathogenesis-related proteins in tobacco

The PR proteins are a group of host-encoded proteins induced by pathogen infection or by other forms of stress (Antoniw *et al.*, 1980). They were initially discovered in tobacco showing HR to tobacco mosaic virus (TMV) infection, and during the last few years information has been accumulating concerning their biochemical properties, their induction and, to some extent, their role in disease resistance. During the induction of HR by pathogenic micro-organisms a set of defence genes is activated in the host, and a large number of soluble, low-molecular-weight proteins are synthesized and transported to the intercellular spaces. The presence of these PR proteins is correlated with induced resistance. Many of these same PR proteins have been detected in tobacco plants infected with pathogens other than TMV or treated with certain chemicals. Abundant, new, host-encoded proteins, similar to those in tobacco, have been detected in more than 20 species of plants after infection. Therefore the tobacco PR proteins can be regarded as the prototypes for the PR proteins of other plants.

The tobacco PR proteins have been grouped into five main families, termed PR1 to PR5 (Shinishi *et al.*, 1987; Van Loon *et al.*, 1987). The members of these families are serologically related to one another and share homology both of the nucleotide sequence of the genes and of the amino acid sequence of the proteins. Some of these PR proteins are acidic and others are basic in nature.

Family PR1 consists of three acidic members, PR1a, PR1b and PR1c, which differ from one another in their isoelectric points. In addition to these, there are other serologically cross-reacting PR1-type proteins found in different tobacco cultivars. Although this group of proteins has been intensively investigated, no biological function can yet be attributed to any of these proteins. The PR1 genes examined so far do not contain introns but have interesting flanking sequences at their 5' region which may play an important role in the control of their expression.

Family PR2 proteins are known to be β-1,3-glucanases (Kauffman *et al.*, 1987). In the tobacco cultivars Samsun NN and Xanthi-nc there are three acidic members of the PR2 family which are usually called PR2, PRN and PRO, or 2a, 2b and 2c. There is also a basic PR2 protein which resembles the basic, hormone-regulated β-1,3-glucanase produced by cultured cells of tobacco (Shinishi *et al.*, 1987).

Chitinases represent the third family of PR proteins, which consists of both acidic and basic proteins. The PR3 proteins are specified by small gene families and the proteins show approximately 65% homology in amino acid sequence despite differences in M_r values and isoelectric points.

As for the PR4 proteins, their function is unknown. Although some members of the family have been reported to be antigenically similar to some PR2 proteins, the general consensus of opinion is that the PR4 proteins represent a distinct family.

The fifth family of PR proteins in tobacco, as well as in other plant species, consists of the so-called thaumatin-like proteins. Thaumatin is the trivial name given to an intensely sweet protein found in the fruit of a tropical plant *Thaumatococcus danielli*. Thaumatin shares extensive homology with a bifunctional inhibitor from maize which is active, *in vitro*, against trypsin and α-amylase. The bifunctional inhibitor protein in maize is thought to act in defence against herbivorous insects (Richardson *et al.*, 1987). The thaumatin-like PR5 protein(s) of tobacco shares important structural features not only with thaumatin and maize bifunctional inhibitor protein but also with other proteins induced by stress, such as salt-induced stress protein in tomato roots. Thus thaumatin-like proteins, occurring through-out the plant kingdom, may constitute a superfamily of proteins.

The classification of PR proteins outlined above is based on genetic, biochemical, serological or functional similarities. This is only provisional, and as more information about the biological functions of PR proteins becomes available it may be necessary to change the classification. Furthermore, defence-related extracellular peroxidases as well as other extracellular proteins (Shotten-Toma & de Witt, 1988) induced in response to pathogens such as the GRPs could also be considered as PR proteins.

Pathogenesis-related proteins in cereals

White *et al.* (1987) reported the presence of PR1-type proteins in powdery mildew-infected barley. Several PR proteins including four chitinases were found in brome mosaic virus-infected maize as well as powdery mildew-infected barley (Bryngelsson & Collinge, 1991). About ten new proteins in barley were visualized by two-dimensional SDS-polyacrylamide gel electrophoresis and isoelectric focusing. These proteins are induced only during HR, have low M_r values and either very high or very low isoelectric points, thus resembling the PR proteins found in tobacco and other dicotyledonous plants.

Bryngelsson & Collinge (1991) purified several of the barley PR proteins to homogeneity. One of these, a highly acidic monomeric protein, exhibited about 45% homology to osmotin and NP24 proteins, which are synthesized by dicotyledonous plants in response to salinity. A similar degree of amino acid homology to this protein was also found with a maize trypsin/α-amylase bifunctional inhibitor and thaumatin.

The availability of near-isogenic cultivars of barley has made it possible to investigate protein synthesis associated with the expression of resistance to the powdery mildew fungus. Manners et al. (1985) used pairs of susceptible and resistant near-isogenic cultivars of barley, which differed in the M1a, M1k and M1p genes for resistance to the powdery mildew fungus. Patterns of protein synthesis associated with primary infection by this pathogen suggested the enhanced synthesis of five and eight polypeptides at 25 h and 30 h respectively in barley carrying the M1a gene. The enhanced synthesis of these same polypeptides was observed at 48 h and 72 h after inoculation of barley carrying either the M1p or the M1k genes. These enhancements correlated with the onset of resistance, suggesting a possible role of these polypeptides in the resistance reaction. These were termed resistance-related or infection-related (IR) proteins and it was suggested that resistance genes have a regulatory role in determining the synthesis of resistance-related proteins during the infection process.

Davidson et al. (1987) prepared a cDNA library of infection-related mRNAs isolated from leaves of cv M1p at 2 days after inoculation. The library was screened by hybridization to ^{32}P-labelled cDNA prepared from control or infected leaves of M1p at 2 days after inoculation. Six cDNA clones showing greater hybridization to cDNA prepared from inoculated leaves were selected. The translation products of these cDNA clones corresponded to IR proteins in two-dimensional fluorograms.

The induction kinetics of host mRNA in near-isogenic cultivars of barley has been studied using the cDNA clones as hybridization probes (Davidson et al., 1988). The amounts of mRNA corresponding to four of the cDNA clones increased to greater values in cv M1p than in the near-isogenic susceptible cv m1p over a 2-day period following inoculation.

Like the PR proteins of dicotyledonous plants, induction of IR mRNAs in barley is caused by a range of pathogens (Scott et al., 1990). Barley was inoculated with the compatible pathogens chlorostriate mosaic virus, barley stripe virus, Pseudomonas syringae, Pyrenophora teres, Bipolaris sorokiniana and Erysiphe graminis f. sp. hordei, and with the incompatible pathogens E. graminis f. sp. tritici, Puccinia graminis f. sp. tritici and

Pyrenophora tritici-repentis. Wheat was inoculated with the compatible pathogens *B. sorokiniana, P. graminis* f. sp. *tritici* and *E. graminis* f. sp. *tritici* and the incompatible pathogen *E. graminis* f. sp. *hordei*. Induction of IR mRNAs was observed with necrotrophic and biotrophic fungi, but not after these plants were inoculated with either bacteria or viruses. Furthermore, infection by the pathogens *Pyrenophora tritici-repentis* and *B. sorokiniana* resulted in the extremely rapid induction of all IR mRNAs, indicating that IR mRNAs are not synthesized specifically in response to infection by biotrophic fungi alone.

In barley, the possible functions of the proteins encoded by the six IR mRNAs have been investigated by sequencing the cDNAs. The proteins encoded by two of the six cloned IR mRNAs have been identified (Jutidamrongphan *et al.,* 1991; Scott, 1991). One of these encodes β-1,3-glucanase and the other specifies a protein of the PR1 family.

Pathogenesis-related proteins in other plants

In response to fungal, viral and viroid infection, tomato plants (*Lycopersicon esculentum*) synthesize at least one PR1-like protein and several other acidic and basic PR proteins. The PR1-like protein in tomato is a basic protein of M_r 14 kDa and is termed p14. The p14 and three other PR proteins accumulate in resistant tomato leaves infected with *Cladosporium fulvum*, suggesting a possible role in resistance to fungi. Another host-encoded protein associated with resistance to fungal infection has been found in plants undergoing resistance response to *C. fulvum* infection but not in near isogenic plants undergoing a susceptible response. A few other less well-characterized PR proteins have been found in tomato plants in response to pathogen attack or treatment with chemicals (Carr & Klessig, 1989).

Several different PR proteins have been found in the intercellular spaces of potato leaves (*Solanum tuberosum*). These include proteins serologically related to the basic PR1-type, the acidic PR2-type (β-1,3-glucanase) and the thaumatin-like PR5-type, in addition to peroxidases produced in response to infection by TMV or *Phytophthora infestans*, or treatment with a fungal elicitor. In some cultivars of potato, fungal infection also induced six extracellular chitinases (PR3-type).

In bean plants (*Phaseolus vulgaris*) undergoing HR to fungi and other pathogens, several PR proteins are synthesized. Some of these proteins also accumulate in response to treatment with such chemicals as mercuric chloride and salicylic acid. Although all of these proteins have been characterized with respect to their M_r values, they cannot at present be identified with any of the five PR families in tobacco as outlined in the preceding sections.

In cucumber (*Cucumis sativus*) cotyledons infected with a number of viruses and *Colletotrichum lagenarium*, several PR-like proteins are synthesized during HR. One of these proteins, which accumulates in the inter-cellular spaces, is an extracellular chitinase, a PR3-type protein.

Apart from the species of plants mentioned above, PR proteins, of one type or another, have been found in practically every species of plant investigated so far (Andersen, 1989; Carr & Klessig, 1989).

Elicitors and elicitor receptors

In addition to pathogen attack, the synthesis of phytoalexins and defence-related proteins can be induced by wounding and by factors derived from pathogen culture fluids or cell walls. These molecules ('factors') have been termed elicitors. Many different compounds have been shown to elicit defence responses in intact plants as well as cell cultures. These compounds include glycoproteins, simple and complex carbohydrates, certain fatty acids, such as arachidonic acid, and polypeptides (Templeton & Lamb, 1988). These are among the biotic elicitors, which are molecules of either host or pathogen origin that can induce defence responses in plants. In addition to these, there is a vast array of abiotic elicitors that can induce defence responses (such as phytoalexin synthesis and the accumulation of host gene-encoded PR proteins). These include heavy metal salts, detergents, glutathione, autoclaved ribonuclease and many physical agents, such as low temperature and ultraviolet light treatment (Darvill & Albersheim, 1984; Dixon, 1986).

The best characterized elicitor of fungal origin is a β-linked hepta-glucoside from race 3 of *Phytophthora megasperma* f. sp. *glycineae*, which causes root rot of soybean. It is a very strong elicitor, acting in the nano-molar range in the elicitation of glyceollin in soybean cotyledons. Many isomers of this elicitor fragment failed to induce glyceollin synthesis, suggesting that only the active molecule interacts with a specific receptor. It has also been found by Schmidt & Ebel (1987) that the binding sites were located in the plasma membrane. These observations suggested the exist-ence of a cognate plant receptor molecule which specifically interacts with the β-linked heptaglucoside, culminating in the elicitor activity. In contrast, experiments on the effects of the degree of depolymerization and the extent of *N*-acetylation on the activity of chitosan oligosaccharides as elicitors of callose synthesis in *Catharanthus roseus* cell cultures suggest that this elicitor interacts primarily with regularly spaced, negatively charged phospholipid head groups of the plasma membrane rather than with a specific macro-molecular receptor (Kauss *et al.*, 1989).

A cell wall β-1,3-glucan, with an average degree of polymerization of 22, has been used to identify high-affinity binding sites on soybean cell mem-branes. Binding was found to be reversible, and labelled glucan elicitor can be displaced by elicitor-active glucans but not by inactive ones. These obser-vations suggest the existence of a receptor. The identity of the receptor molecule for this elicitor as well as that for the linked heptaglucoside remains to be established.

As plants are known to be capable of responding to developmental and environmental signals as well as elicitors by the regulation of specific sets of genes, there must be signal transduction mechanisms involving specific receptor molecules. In the case of phytoalexin elicitation in bean cells, transcription is stimulated within 5 minutes, suggesting that there are only a few steps in the signal transduction pathway (Lawton & Lamb, 1987). The only plant receptor that has been cloned and characterized so far is the photoreceptor phytochrome (Hershey *et al.*, 1984; Schmidt & Ebel, 1987).

A number of factors have been considered as components of the stress signal transduction pathways in plants. Some of these are briefly outlined below.

Infection or exposure of plant cells to fungal elicitors is known to cause rapid changes in membrane potential and proton transport. Modulation of membrane-associated ATPase activity may account for some of these effects (Anderson, 1989). Harper *et al.* (1989) have recently cloned the proton translocation ATPase of *Arabidopsis* and it should now be possible to ascertain by gene transfer technique whether or not the H^+ ATPase has a role in the internalization of elicitor signals.

Disturbances in oxidative metabolism have also been implicated as a component of the signal transduction process (Dixon & Lamb, 1990). Elicitors from *Verticillium dahliae* induce a striking decrease in the fluorescence of membrane potential-sensitive and pH-sensitive dyes in cotton, tobacco and soybean cells due to fluorescence quenching by the rapid production of H_2O_2. Various elicitors are known to induce these changes within 5 min. An elicitor-stimulated vectorial NADH oxidase located at the plasma membrane is considered to be responsible for the production of H_2O_2 at the cell wall. Exogenous H_2O_2 is itself an elicitor, and phytoalexin induction by fungal elicitor is inhibited by catalase, suggesting a pivotal role of H_2O_2, which has also been suggested to be involved in elicitor-stimulated insolubilization of a proline-rich cell wall protein. Dixon & Lamb (1990) have suggested that stress-induced extracellular H_2O_2 probably has a dual function: firstly as a substrate for the very fast initial localized defence response, and secondly as a component of a signal transduction pathway leading to the activation of plant defence genes. Glutathione, which is a powerful inducer of plant defence genes, can act as a substrate in a complex oxygenase reaction to generate H_2O_2. Dixon & Lamb (1990) convincingly argued that the powerful action of glutathione on rapid and selective plant defence gene activation may be H_2O_2-mediated.

Other signal transduction mechanisms in plants are thought to involve calcium- and protein kinase-mediated pathways. The stimulation of callose synthesis by chitosan involves Ca^{2+}-mediated activation of callose synthetase in the plasma membrane. This finding led to the speculation that

Ca^{2+} may be involved in the internalization of signals for defence gene activation. In pea pod tissues infected by *Fusarium solani* or treated with chitosan, phytoalexin accumulation is unaffected by high concentrations of Ca^{2+} channel blockers or calmodulin inhibitors (Kendra & Hadwiger, 1987). However, a role for plasma membrane Ca^{2+} fluxes in the stimulation of phytoalexin synthesis by *Phytophthora megasperma* f. sp. *glycineae* elicitor in soybean cell suspension culture has been proposed on the basis of the following two findings: elicitor activity is inhibited by external Ca^{2+} depletion and by the Ca^{2+} channel blocker verapamil, and phytoalexin synthesis is stimulated by the Ca^{2+} ionophore A23187 (Stab & Ebel, 1987). Similarly, external Ca^{2+} is required for elicitor induction of several genes of the phytoalexin biosynthetic pathway. Calmodulin antagonists strongly inhibit the accumulation of terpenoid phytoalexins if added prior to the elicitor. These and other studies (Templeton & Lamb, 1988) have suggested, but have not conclusively demonstrated, that Ca^{2+} phosphoinositols and cyclic AMP molecules, which take part in signal transduction in animal cells, play a similar role in elicitation in plant cells.

Inducers of pathogenesis-related proteins

BIOTIC INDUCERS. The expression of genes specifying various PR proteins in a number of plants is affected by plant hormones. Ethephon (2-chloroethyl-phosphonic acid), which breaks down to release ethylene, induces the accumulation of PR1 and PR2 proteins in tobacco. The resistance response in tobacco is often accompanied by ethylene production, suggesting that this hormone may be part of the natural induction pathway for PR proteins in tobacco. Van Loon (1983) reported that the administration of aminoethoxyglycine, which blocks the synthesis of the natural ethylene precursor 1-amino-cyclopropane-1-carboxylic acid, prevents both ethylene synthesis and PR protein accumulation. Other growth regulators inducing PR1 proteins in tobacco callus culture include benzyl adenine, indoleacetic acid, 2,4-dichlorophenoxyacetic acid and kinetin. Gibberellic acid was also shown to induce PR1 proteins in tobacco cell cultures (Carr & Klessig, 1989).

Cell-free extracts of the bacteria *Pseudomonas solanacearum* and *Nocardia asteroides* as well as the fungus *Stachybotrys chartarum* are capable of inducing tobacco PR proteins. Application of a fungal elicitor derived from *Phytophthora infestans* induced the synthesis of PR2 (β-1,3-glucanase) and PR3 (chitinase) proteins in potato (Kombrink *et al.*, 1988). Biotic microbial elicitors that induce the synthesis and accumulation of PR proteins have been isolated from many other fungi, including *Phytophthora megasperma*, and there are also inducers of PR proteins in extracts that are derived from plants. Several investigations have shown that oligosaccharides released from plant cell walls constitute important elicitors of plant defence responses (Darvill & Albersheim, 1984; Templeton & Lamb, 1988).

ABIOTIC INDUCERS. Polyacrylic acid, an inducer of interferon synthesis in animal systems, induces the accumulation of about six PR proteins. Salicylic acid and many of its derivatives have been shown to be powerful inducers of PR1 proteins, not only in tobacco, but also in a range of both dicotyledonous and monocotyledonous plants (White *et al.,* 1987). Other chemicals have been tested for PR protein induction and many, including methylbenz-imidazol-2-carbonate, phytic acid, manganese and barium salts, have been found to induce PR1 proteins.

Pathogenesis-related proteins are also induced by tissue damage, probably due to the wound hormone ethylene or to the release of endogenous elicitors (Carr & Klessig, 1989).

Mechanism of induction of pathogenesis-related proteins

It is believed that the induction of PR1 proteins (PR1a, b and c) is controlled at the transcriptional level. The isolation and sequencing of TMV- or salicylic acid-responsive genes have allowed the examination of putative promoter regions (Carr & Klessig, 1989). These workers identified some sequence motifs in the promoter regions of PR genes which show strong homology to sequences in other plant genes (e.g. light- and stress-induced genes) which are required for transcriptional control. They suggested that patterns of expression of genes which are regulated by light, pathogen attack, stress or in a tissue-specified manner may require combinations of a small repertoire of conserved sequence elements.

The *in vivo* response of plant cells to attack by pathogenic fungi is probably a highly co-ordinated network of events involving recognition of signals that lead to the activation of defence genes in the host plant. The components of several different signalling systems in plants are known. These include, in addition to polysaccharides, phospholipase C, inositol, phospholipids, protein kinase C, calmodulin and annexins (Bowles, 1990)

Signal transduction in plant–bacterial interactions

In a number of instances, signal transduction in bacteria is mediated by sensor and regulator proteins (Stock *et al.*, 1990). The N-terminal domain of the sensor protein receives an environmental signal and transduces it to its C-terminal domain. The C-domain of the sensor phosphorylates the N-domain of the regulator. Phosphorylation of the regulator protein changes the activity of its C-domain, which ultimately carries out the appropriate response. In most cases, the C-domain of the sensor and the N-domain of the regulator are homologous. In this way, different environmental signals are transduced, using a common mechanism to regulate a wide range of processes in bacteria. These include response to starvation, osmolarity changes, chemotaxis, sporulation, symbiosis and bacterial pathogenesis.

There is currently considerable excitement over the discovery of co-ordinate regulation and sensory transduction in bacteria in response to different signal molecules and there is high expectation that similar mechanisms will be discovered in eukaryotic pathogens, such as fungi, that infect plants. In the following paragraphs, we briefly describe two of these systems, both of which are involved in interactions between bacteria and plants. The bacteria involved are members of the family Rhizobiaceae: *Rhizobium* spp. and *Agrobacterium tumefaciens*.

Signal transduction in Rhizobium–*legume interactions*

Bacteria belonging to the genus *Rhizobium* are capable of infecting legumes and inducing the formation of symbiotic nitrogen-fixing nodules. In bacterial nitrogen metabolism the key enzyme is glutamine synthetase (GS), which mediates the entry of ammonia (the primary product of nitrogen fixation) into anabolic metabolism. The expression of the glutamine synthetase gene, *gln A*, is rapidly induced in cells by nitrogen starvation.

In the nitrogen regulation (Ntr) system, the response regulator is a transcriptional activator protein, NR_I. When phosphorylated, NR_I binds to the *gln A* gene promoter and activates its transcription. The state of phosphorylation of the regulator protein (NR_I) is controlled by the enzyme, histidine kinase (NR_{II}). The sensor in the Ntr system is a bifunctional enzyme (UTase/UR) that uridylates (UTase) or deuridylates tyrosine residues in a control protein, P_{II}, which, in turn, controls the histidine kinase (NR_{II}) activity. The signal is generated by the balance between carbon and nitrogen metabolism which is reflected in the relative concentrations of α-ketoglutarate (2KG) and glutamine (Gln). High ratios of Gln:2KG favour UR over UTase. The unmodified form of P_{II} activates phosphatase activity associated with NR_{II}, the histidine kinase. This provides a mechanism for regulating the degree of phosphorylation of the regulator, NR_I. At low ratios of Gln:2KG, NR_I is phosphorylated and it binds to DNA sequences upstream of the promoter for the *gln A* and activates the transcription of the operon that contains genes that specify GS, NR_{II} and NR_I, which occur in that order in the *gln ALG* operon.

Signal transduction in Agrobacterium-*induced crown gall tumour formation*

Agrobacterium tumefaciens has been described as containing a model sensory transduction system (Miller *et al.*, 1989). *A. tumefaciens* is a soil bacterium that can genetically transform some plants and cause malignant tumours called crown gall tumours (Chapter 3). The virulent strains of the bacterium contain a large plasmid, the Ti (tumour-inducing) plasmid. During transformation, a part of the Ti plasmid, the T-DNA, is transferred to and integrated into the nuclear genome of the infected plant. In the plant cells, genes in T-DNA are expressed, resulting in the synthesis of opines (used as carbon and nitrogen sources by the resident *A. tumefaciens*; see Table 3.1), and of auxin and cytokinin, which cause neoplastic growth.

Apart from the eight eukaryotic genes in the T-DNA, Ti plasmids also contain a cluster of operons, known as the *vir* region. The products of these prokaryotic genes are required for T-DNA transfer from the bacterial to the plant cell. The *vir* region contains six separate transcriptional units, *vir* A, B, D, G, C and E. The transcription of all these *vir* loci is induced by signal molecules which are phenolic compounds such as acetosyringone and hydroxyacetosyringone present in the exudates of wounded plants.

The *vir* operons of Ti plasmids are subjected to positive transcriptional control by the products of *vir* A (sensor) and *vir* G (regulator) proteins. The sensor responds to signal molecules provided by wounded plants and transduces these to the regulator, and the result is activation of the virulence gene expression.

The existence of signal transduction mechanisms similar to those outlined above in host plants and pathogenic fungi is difficult to establish, primarily due to the genetic complexity of eukaryotes and the lack of highly defined elicitors to trigger the system.

It is conceivable that the induction of plant genes specifying PR proteins in response to wounding, elicitor treatment or pathogen attack involves signal transduction mechanisms as seen in a wide range of bacteria. This view is strengthened by the striking overall similarity between signal transduction in mammalian cells and some of those seen in bacteria (Bishop, 1987). At the surface of mammalian cells, there are membrane receptors, such as the epidermal growth factor (EGF) receptor, which recognize peptide hormones and transduce a signal to the nucleus, where the signal causes changes in gene expression. These receptors exhibit kinase activity and, just like prokaryotic signal transduction, such as the Ntr system in *Rhizobium*, signal transduction in these eukaryotic cells may involve autophosphorylation and communication with cytoplasmic proteins by phosphotransfer to control gene expression. Thus, bacterial signal transduction may be an archetype for understanding responses of plants to various signals.

Summary and conclusions

In summary, there are several classes of defence-related genes in plants the expression of which is triggered by pathogen attack, by environmental stress and by biotic or abiotic elicitors. These have been grouped into three classes (Bowles, 1990). The first of these specifies products that directly change the properties of the extracellular matrix, thus strengthening and repairing the cell walls. This class of defence genes includes those specifying HRGPs and GRPs as well as a wide range of enzymes involved in cell wall modifications.

The second class of genes encodes defence-related proteins that act as deterrents, exhibit antimicrobial activities or catalyse the synthesis of products that retard microbial growth. These include amylase-proteinase inhibitors, thionins, hydrolytic enzymes, such as β-1,3-glucanases, chitinases and enzymes involved in the synthesis of phytoalexins. The third class of genes encodes the so-called PR proteins, whose appearance is correlated with defence responses. The function of many of these PR proteins is not known. An understanding of how the expression of these plant genes is controlled during infection by fungal and other pathogens is of paramount importance in defining the molecular basis of disease resistance in plants.

References

Andersen, J. B. (1989) 'Studies of PR proteins in barley inoculated with barley powdery mildew'. Ph.D. thesis, Royal Veterinary and Agricultural College, Copenhagen, Denmark.

Anderson, A. J. (1989) The biology of glycoproteins as elicitors. In: Kosuge, T. & Nester, E. W. (eds), *Plant–Microbe Interactions: Molecular and Genetic Perspective*. McGraw-Hill, New York, 3, pp. 87–130.

Antoniw, J. F., Ritter, C. E., Pierpoint, W. S. & van Loon, L. C. (1980) Comparison of three pathogenesis-related proteins from two cultivars of tobacco infected with TMV. *Journal of General Virology* 47, 79–87.

Arora, Y. K. & Bajaj, K. L. (1985) Peroxidase and polyphenoloxidase associated with induced resistance of mung bean to *Rhizoctonia solani* Kuhn. *Phytopathologische Zeitschrift* 114, 325–31.

Bell, A. A. (1981) Biochemical mechanisms of disease resistance. *Annual Review of Plant Physiology* 32, 21–81.

Bishop, J. M. (1987) The molecular genetics of cancer. *Science* 235, 305–11.

Bohlmann, H., Clausen, S., Behnke, S., Giese, H., Hiller, C., Reimann-Phillip, U., Schrader, G., Barkholt, V. & Apel, K. (1988) Leaf-specific thionins of barley – a novel class of cell wall proteins toxic to plant pathogenic fungi and possibly involved in the defence mechanism of plants. *EMBO Journal* 7, 1559–65.

Boller, T. (1985) Induction of hydrolases as a defence reaction against pathogens. In: Key, J. L. & Kosuge, T. (eds), *Cellular and Molecular Biology of Plant Stress*. Alan R. Liss, New York, pp. 247–62.

Bowles, D. J. (1990) Defence-related proteins in higher plants. *Annual Review of Biochemistry* 59, 873–907.

Broglie, K. E., Gaynor, J. J. & Broglie, R. M. (1986) Ethylene-regulated gene expression: molecular cloning of the genes encoding an endochitinase from *Phaseolus vulgaris*. *Proceedings of the National Academy of Sciences USA* 83, 6820–24.

Bryngelsson, T. & Collinge, D. B. (1991) Biochemical and molecular analyses of the response of barley to infection by powdery mildew. In: Shewry, P. R. (ed.), *Barley: Genetics, Biochemistry, Molecular Biology and Biotechnology*. CAB International, Wallingford, Oxon, pp. 452–73.

Carr, J. P. & Klessig, D. F. (1989) The pathogenesis related proteins of plants. In: Setlow, J. K. (ed.), *Genetic Engineering: Principles and Methods*. Plenum, New York, pp. 65–109.

Cooper, J. B. & Varner, J. E. (1984) Cross-linking of soluble extensin in isolated cell walls. *Plant Physiology* 76, 414–17.

Darvill, A. G. & Albershcim, P. (1984) Phytoalexins and their elicitors: a defence against microbial infection in plants. *Annual Review of Plant Physiology* 35, 243–75.

Davidson, A. D., Manners, J. M., Simpson, R. S. & Scott, K. J. (1987) cDNA cloning of mRNAs induced in resistant barley during infection by *Erysiphe graminis* f. sp. *hordei*. *Plant Molecular Biology* 8, 77–85.

Davidson, A. D., Manners, J. M., Simpson, R. S. & Scott, K. J. (1988) Altered host gene expression in near-isogenic barley conditioned by different genes for resistance during infection by *Erysiphe graminis* f.sp. *hordei*. *Physiological and Molecular Plant Pathology* 32, 127–39.

De Biasi, M. G. & Badiani, M. (1990) The phenoloxidase-like activity of purified peroxidase from *Triticum aestivum* L. seedling leaves. *Plant Science* 67, 29–37.

Deverall, B. J. (1989) Mechanisms of resistance and pathogenic specialization in rust–wheat interactions. *New Phytologist* 113, 233–44.

Dixon, R. A. (1986) The phytoalexin response: elicitation, signalling and control of host gene expression. *Biological Reviews* 61, 239–91.

Dixon, R. A. & Lamb, C. J. (1990) Molecular communication in interactions between plants and microbial pathogens. *Annual Review of Plant Physiology and Plant Molecular Biology* 41, 339–67.

Downie, J. A. & Johnston, A. W. B. (1986) Modulation of legume by *Rhizobium*: the recognised root? *Cell* 47, 153–4.

Farmer, E. E. & Ryan, C. A. (1990) Interplant communication: airborne methyl jasmonate induces synthesis of proteinase inhibitors in plant leaves. *Proceedings of the National Academy of Sciences USA* 87, 7713–16.

Flor, H. H. (1956) The complementary genic systems in flax and flax rust. *Advances in Genetics* 8, 29–54.

Harper, J. F., Surowy, T. K. & Sussman, M. R. (1989) Molecular cloning and sequence of cDNA encoding the plasmamembrane proton pump (H$^+$ ATPase) of *Arabidopsis thaliana*. *Proceedings of the National Academy of Sciences USA* 86, 1234–8.

Heath, M. C. (1982) Host defense mechanisms against infection by rust fungi. In: Scott, K. J. & Chakravorty, A. K. (eds), *The Rust Fungi*. Academic Press, London, pp. 223–45.

Heitefuss, R. & Ebrahim-Nesbat, F. (1986) Ultrastructural and histochemical studies on mildew of barley (*Erysiphe graminis* DC. f. sp. *hordei* Marchal). III. Ultrastructure of different types of papillae in susceptible and adult plant resistant leaves. *Journal of Phytopathology* 116, 358–73.

Hershey, H. P., Colbert, J. T., Lissemore, J. L., Barner, R. F. & Quail, P. H. (1984) Molecular cloning of cDNA for *Avena* phytochrome. *Proceedings of the National Academy of Sciences USA* 81, 2332–6.

Hislop, E. C. & Stahmann, M. A. (1971) Peroxidase and ethylene production by barley leaves infected with *Erysiphe graminis* f.sp. *hordei*. *Physiological Plant Pathology* 1, 297–312.

Hoj, P. B., Hartman, D. J., Morrice, N. A., Doan, D. N. P. & Fincher, G. B. (1989) Purification of 1-3-beta-glucan endohydrolase isoenzyme II from germinated barley and determination of its primary structure from a cDNA clone. *Plant Molecular Biology* 13, 31–42.

Jones, D. H. (1984) Phenylalanine ammonia-lyase: regulation of its induction and its role in plant development. *Phytochemistry* 23, 1349–59.

Jutidamrongphan, W., Andersen, J. B., Mackinnon, G., Manners, J. M., Simpson, R. S. & Scott, K. J. (1991) Induction of β-1, 3-glucanase in barley in response to infection by the fungal pathogens. *Molecular Plant–Microbe Interactions*, 4, 234–8.

Kauffmann, S., Legrand, M., Geoffroy, P. & Fritig, B. (1987) Biological function of 'Pathogenesis-related' proteins: four PR-proteins of tobacco have 1,3-beta-glucanase activity. *EMBO Journal* 6, 3209–12.

Kauss, H., Jeblick, W. & Domard, A. (1989) The degree of polymerization and N-acetylation of chitosan determine its ability to elicit callose formation in suspension cells and protoplasts of *Catharanthus roseus*. *Planta* 178, 385–92.

Kendra, D. F. & Hadwiger, L. A. (1987) Calcium and calmodulin may not regulate the disease resistance and pisatin formation responses of *Pisum sativum* to chitosan or *Fusarium solani*. *Physiological and Molecular Plant Pathology* 31, 337–48.

Kerr, A. (1987) The impact of molecular genetics on plant pathology. *Annual Review of Phytopathology* 25, 87–110.

Koch, E. & Slusarenko, A. (1990) *Arabidopsis* is susceptible to infection by a downy mildew fungus. *Plant Cell* 2, 437–45.

Kombrink, E., Schroder, M. & Hahlbrock, K. (1988) Several 'pathogenesis-related' proteins in potato are 1,3-beta-glucanases and chitinases. *Proceedings of the National Academy of Sciences USA* 85, 782–6.

Lawton, M. A. & Lamb, C. J. (1987) Transcriptional activation of plant defence genes by fungal elicitor, wounding and infection. *Molecular and Cellular Biology* 7, 335–41.

Legrand, M., Kauffman, S., Geoffrey, P. & Fritig, B. (1987) Biological function of pathogenesis-related proteins: four tobacco pathogenesis-related proteins are chitinases. *Proceedings of the National Academy of Sciences USA* 84, 6750–4.

Leong, S. A. & Holden, D. W. (1989) Molecular genetic approaches to the study of fungal pathogenesis. *Annual Review of Phytopathology* 27, 463–81.

Manners, J. M., Davidson, A. D. & Scott, K. J. (1985) Patterns of post-infectional protein synthesis in barley carrying different genes for resistance to the powdery mildew fungus. *Plant Molecular Biology* 4, 275–83.

Mauch, F., Hadwiger, L. A. & Boller, T. (1988) Antifungal hydrolases in pea tissue. 1. Purification and characterization of two chitinases and two β-1,3-glucanases differentially regulated during development and in response to fungal infection. *Plant Physiology* 87, 325–33.

Meyerowitz, E. M. & Pruitt, R. E. (1985) *Arabidopsis thaliana* and plant molecular genetics. *Science* 229, 1214–18.

Miller, J. F., Mexalanos, J. J. & Falkow, S. (1989) Coordinate regulation and sensory transduction in the control of bacterial virulence. *Science* 243, 916–22.

Ohl, S., Hedrick, S. A., Chory, J. & Lamb, C. J. (1990) Functional properties of a phenylalanine ammonia-lyase promoter from *Arabidopsis*. *Plant Cell* 2, 837–48.

Perrin, D. R. & Bottomley, W. (1962) Studies on phytoalexins v: the structure of pisatin from *Pisum sativum* L. *Journal of American Chemical Society* 84: 1919–22.

Richardson, M., Valdes-Rodriguez, S. & Blanco-Labra, A. (1987) A possible function for thaumatin and a TMV induced protein suggested by homology to a maize inhibitor. *Nature* 327, 432–4.

Ryan, C. A. (1987) Oligosaccharide signalling in plants. *Annual Review of Cell Biology* 3, 294–317.

Schmidt, W. E. & Ebel, J. (1987) Specific binding of a fungal glucan phytoalexin elicitor to membrane fractions from soybean *Glycine max*. *Proceedings of the National Academy of Sciences USA* 84, 4117–21.

Scott, K. J. (1991) The molecular analysis of barley resistance to powdery mildew. In: Shewry, P. R. (ed.), *Barley: Genetics, Biochemistry, Molecular Biology and Biotechnology*. CAB International, Wallingford, Oxon, pp. 474–88.

Scott, K. J., Davidson, A. D., Jutidamrongphan, W., Mackinnon, G. & Manners, J. M. (1990) The activation of genes in wheat and barley by fungal pathogens. *Australian Journal of Plant Physiology* 17, 229–38.

Shottens-Toma, I. M. J. & de Witt, P. J. (1988) Purification and primary structure of a necrosis-inducing peptide from the apoplastic fluids of tomato infected with *Cladosporium fulvum* (syn. *Fulvia fulva*). *Physiological and Molecular Plant Pathology* 33, 59–67.

Shinishi, H., Mohnen, D. & Meins, Jr, F. (1987) Regulation of a plant pathogenesis-related enzyme: inhibition of chitinase mRNA accumulation in cultured tobacco tissues by auxin and cytokinin. *Proceedings of the National Academy of Sciences USA* 84, 89–93.

Shiraishi, T., Yamaoka, N. & Kunoh, H. (1989) Association between increased phenylalanine ammonia-lyase activity and cinnamic acid synthesis and the induction of temporary inaccessibility caused by *Erysiphe graminis* primary germ tube penetration of the barley leaf. *Physiological and Molecular Plant Pathology* 34, 75–83.

Stab, M. R. & Ebel, J. (1987) Effects of Ca^{2+} on phytoalexin induction by fungal elicitor in soybean cells. *Archives of Biochemistry and Biophysics* 257, 416–23.

Stachel, S. E. & Zambryski, P. C. (1986) *Agrobacterium tumefaciens* and the susceptible plant cell: a novel adaptation of extra-cellular recognition. *Cell* 47, 155–7.

Stock, J. B., Stock, A. M. & Mottonem, J. M. (1990) Signal transduction in bacteria. *Nature* 344, 395–400.

Svalkheim, O. & Robertson, B. (1990) Induction of peroxidases in cucumber hypocotyls by wounding and fungal infection. *Physiologia Plantarum* 78, 261–7.

Templeton, M. D. & Lamb, C. J. (1988) Elicitors and defence gene activation. *Plant Cell and Environment* 11, 395–401.

Tepper, C. S. & Anderson, A. J. (1986) Two cultivars of bean display a differential response to extracellular components from *Colletotrichum lindemuthianum Physiological and Molecular Plant Pathology* 29, 411–20.

Van Etten, H. D., Matthews, D. E. & Smith, D. A. (1982) Metabolism of phytoalexins. In: Bailey, J. A. & Mansfield, J. W. (eds), *Phytoalexins from the Leguminosae*. Blackie, London, pp. 181 217.

Van Etten, H. D., Matthews, D. E. & Matthews, P. S. (1989) Phytoalexin detoxification: importance for pathogenicity and practical implications. *Annual Review of Phytopathology* 27, 143–64.

Van Loon, L. C. (1983) The induction of pathogenesis-related proteins by pathogens and specific chemicals. *Netherland Journal of Plant Pathology* 89, 265–73.

Van Loon, L. C., Gerritsen, Y. A. M. & Ritter, C. E. (1987) Identification, purification and characterization of pathogenesis-related proteins from virus-infected samsun-NN tobacco leaves. *Plant Molecular Biology* 9, 593–609.

White, R. F., Rybicki, E. P., Von Wechmar, M. B., Dekker, J. L. & Antonew, J. F. (1987) Detection of PRI-type proteins in Amaranthacae, Chenopodiaceae, Gramineae and Solanaceae by immunoelectroblotting. *Journal of General Virology* 68, 2043–8.

Wolpert, T. J. & Macko, V. (1989) Specific binding of victorin to a 100 kDa protein from oats. *Proceedings of the National Academy of Sciences USA* 86, 4092–6.

Chapter 12
Advances in Breeding for Resistance to Bacterial Pathogens

Maureen C. Whalen
Department of Biology, Colby College, Waterville, ME 04901, USA

Introduction

Genetic resistance is one means of controlling plant diseases caused by bacterial pathogens, and in some situations, the only means. All strategies for breeding for resistance include identifying the disease and the causal agent, devising appropriate tests for resistance, detecting resistant germplasm in breeding lines or wild species, incorporating this resistance into a crop line, and selecting maximally resistant plants that have other desirable characters (Helgeson, 1989). Traditional methods by which resistance has been incorporated into crop lines include pedigree selection, use of resistant inbred lines for F_1 hybrid selection, recurrent selection, transgressive selection, backcross breeding and interspecific hybridization (Coyne & Schuster, 1983). The specific method selected by an individual breeder depends on the state of the available germplasm for resistance, which naturally depends on the breeding history of a species.

The terminology of genetic resistance has proliferated to produce a bewildering array of operational categories (see Table 5.3 in Crute, 1985). Thus, resistance, which can be determined by major or minor genes, can be qualified as vertical or horizontal, qualitative or quantitative, host or non-host, and race-specific or race-non-specific. The terms for these categories have become associated with judgements regarding whether the type of resistance each defines is durable (Buddenhagen & de Ponti, 1984; Simmonds, 1984). The minimalist approach reduces the categories to two: resistance against bacterial pathogens may be conditioned by a single gene or by more than one gene. This reduction is justified because the technology is at hand to approach an understanding of the genetic basis of resistance against most bacterial pathogens. In some cases, the distinctions between categories begin to fade when knowledge about the genetics or molecular genetics of the resistance is gained. In two examples, artificial distinctions regarding genetic control created by the use of the terms host and non-host resistance appear to be unfounded. In bean and soybean, non-host resistance against bacterial pathogens may be in part controlled similarly to host resistance (Whalen *et al.*, 1988; Kobayashi *et al.*, 1989; Keen & Buzzell, 1991).

Durable resistance, defined as resistance that remains effective while a cultivar possessing it is widely cultivated in an environment favouring the disease (Johnson, 1983), is the goal of all breeding programmes. Durable resistance against plant pathogens may be conditioned by a single gene or by many genes (Johnson, 1983; Sharp, 1983). Resistance conditioned by single genes may be overcome by changes in the pathogen population resulting from influx of new virulent strains into the population, or mutation of existing strains to virulence (Kearney *et al.*, 1988). Similarly, resistance conditioned by several genes may be overcome when environmental conditions change (French & De Lindo, 1982). Attempting to understand the mechanistic basis of a particular resistance may prove more fruitful than attempting to classify it into existing 'types'. Furthermore, learning about the potential for variability in virulence of a certain pathogen and the field conditions under which deployment of resistance will take place will allow rational predictions about durability of a particular source of resistance. This knowledge may prove to be the key to breeding for durable resistance.

A recent study on the molecular basis of resistance in pepper illustrates the potential of attempting to understand the resistance mechanism (Kearney & Staskawicz, 1990). Resistance against *Xanthomonas campestris* pv. *vesicatoria* in pepper cultivar ECW20R is conditioned by the genetic interaction between a single gene in the host, *Bs2*, and a single gene in the pathogen, *avrBs2* (Kearney & Staskawicz, 1990; Minsavage *et al.*, 1990). Avirulence gene *avrBs2* confers a selective advantage on *X. campestris* pv. *vesicatoria* strains; when *avrBs2* is mutated, growth of mutant strains is inhibited in susceptible hosts (Kearney & Staskawicz, 1990). Moreover, all races of *X. campestris* pv. *vesicatoria* tested to date contain an active copy of *avrBs2*. *Bs2* is therefore likely to provide durable resistance, since not only does it condition resistance against all races of *X. campestris* pv. *vesicatoria*, but if *avrBs2* is rendered inactive by a mutation, the mutant strains, in which growth will be inhibited, are unlikely to persist (Kearney & Staskawicz, 1990). Incorporation of *Bs2* in pepper cultivars may provide resistance that is extremely durable. Using a particular source of resistance for breeding based on knowledge of the molecular mechanism now appears to be possible. Resistance conditioned by a single gene, if it is potentially durable, is worth characterizing for possible deployment in the field since it would be easier to incorporate into agronomically important lines than that conditioned by multiple genes. As often as possible, deployment of resistant germplasm should be based on sound characterization of the resistance mechanism.

Wild species that interbreed with domestic species have traditionally been good sources of disease resistance and other economically important characters (Harlan, 1976). Members of the wild species are crossed with breeding lines and progeny with desirable traits are selected and then further crossed with breeding lines until a cultivar is found with superior characteristics in addition to the desired character from the wild species. In many cases, however, incompatibilities that prevent viable crosses or fertile progeny make this standard breeding pathway impossible. Some researchers are pursuing somatic hybridization using protoplast fusion as a way to overcome barriers to sexual hybridization for incorporating resistance to bacterial pathogens (see section on *Pseudomonas solanacearum*). Protoplasts isolated from two related lines are fused and then experimentally encouraged to regenerate into mature plants. Fertile regenerants with the selected phenotype are assessed for genetic stability in subsequent generations. This method (and any method that uses tissue culture) has problems related to the high frequency of somatic and germline mutations that plants regenerated from callus tissue may exhibit. In fact, standard tissue culture has been explored as a means of generating useful mutants (Larkin & Scowcroft, 1983). Resistant plants derived from protoplast fusion or tissue culture must be carefully bred to ensure genotype stability. The time required for this additional breeding must be considered when assessing these methods.

Many of the future requirements for improving resistance to bacterial pathogens can probably be met by traditional breeding methods. Success will always be limited by the prolonged effort generally required, and in extreme cases, by the lack of interfertile, resistant germplasm. Genetic engineering technologies may have a significant impact on breeding for resistance against bacterial pathogens in addition to their important contribution to understanding the underlying molecular genetic basis of resistance. Since many resistance genes that have been identified are single, dominant Mendelian traits, current technologies provide an approach for their isolation. The potential exists for introduction of isolated resistance genes to previously susceptible lines. In addition, some of the limitations associated with traditional approaches to breeding may be overcome. The isolated genes can be transferred into agronomically important lines with relative speed compared with most traditional breeding programmes. Genes from diverse genetic backgrounds can potentially be introduced into heterologous backgrounds and thereby enable otherwise impossible interspecific and intergeneric hybridizations. The first transformation of a susceptible line to resistance is still in the future, however, considering that the first step – isolating a resistance gene – has not yet been realized. Once a resistant transformant is produced, a minimal breeding programme will most probably be necessary to ensure that somaclonal variation from tissue culture is avoided. Whether a non-traditional breeding programme is successful at keeping up with the market's elusive standards remains to be seen.

A novel approach to engineering resistance against bacterial pathogens producing toxins that are responsible for development of disease symptoms involves *P. syringae* pv. *tabaci*, which causes wildfire disease of tobacco and produces a toxin called tabtoxin. Anzai *et al.* (1989) introduced a tabtoxin resistance gene with the appropriate expression features, the acetyl-transferase gene *ttr* isolated from *P. syringae* pv. *tabaci*, into tobacco. The transgenic tobacco plants showed high expression of the *ttr* gene and no symptoms upon infection with *P. syringae* pv. *tabaci*. This strategy could be widely applied to other diseases caused by bacteria producing pathogenic toxins.

Efforts to isolate resistance genes in the hopes of transforming susceptible lines to resistance are currently under way (see sections on *P. syringae* pv. *tomato*, *X. campestris* pv. *vesicatoria*, *X. oryzae* and *X. campestris* pv. *campestris*). At least three isolation strategies are being pursued employing: restriction fragment length polymorphism (RFLP) mapping followed by chromosomal walking, gene tagging with transposable elements, and genomic subtraction. These strategies are chosen to compensate for the fact that there are no known functional assays for the presence of resistance genes against most bacterial pathogens.

To be isolated using chromosomal walking techniques, a gene must first be genetically mapped, and then mapped in reference to RFLPs (Chapter 2). Genetic maps were first made with morphological markers and later with isozyme markers. Recombinant DNA technology has allowed the production of molecular maps using RFLP markers. To localize a resistance gene (using any type of genetic marker) the co-segregation of markers with resistance against a certain pathogen is studied. An example of mapping a disease locus with RFLPs is work by Young *et al.* (1988), who mapped the *Tm2ᵃ* gene in tomato for resistance against tobacco mosaic virus. Using RFLP markers the chromosomal position of a locus can be closely mapped. The two nearest flanking markers are then used as starting-points for chromosomal walking following detailed physical mapping using restriction enzyme digests of high-molecular-weight DNA. In chromosomal walking a previously isolated DNA fragment is used to isolate the next overlapping fragment in a directional manner (Bender *et al.*, 1983). Chromosomal walking is facilitated by use of yeast artificial chromosome vectors (YAC) which can carry inserts of up to 700 kb (Burke *et al.*, 1987). YAC libraries have been produced for tomato with inserts of approximately 150 kb (S. D. Tanksley, personal communication) and for *Arabidopsis* (Guzmán & Ecker, 1988). Verification of isolation of a resistance gene clone from a walk is by transformation of a susceptible line to resistance.

Another strategy actively being followed is gene tagging, which is based on work showing that the maize transposable element *Ac* is active in tobacco (Baker *et al.*, 1986). Additional work suggests that *Ac* is active in other members of the Solanaceae (Yoder *et al.*, 1988; Yoder, 1990) and in the Cruciferae (Van Sluys *et al.*, 1987). Gene tagging involves transforming a resistant cultivar with *Ac*, followed by screening primary transformants and their progeny for loss of the resistant phenotype that would occur when gene expression is disrupted by the random insertion of the transposable element (see Shepherd, 1988). The element can then be used as a probe to isolate the flanking DNA from a genomic library of the mutant. The flanking DNA is in turn used as a probe to isolate the wild-type gene. For this strategy to be considered, transformation with the transposable element and final verification of isolation of the resistance gene clone must both be possible.

A third strategy, genomic subtraction, is based on subtraction hybridization and use of the polymerase chain reaction (PCR) (Straus & Ausubel, 1990; Wieland *et al.*, 1990). The starting plant material should ideally be two lines that are isogenic except for a resistance gene. These lines may be existing isolines or created using deletion mutagenesis. DNA sequences that the two lines do not have in common may represent the resistance gene. Common sequences are eliminated after several rounds of denaturation and reassociation and the remaining DNA is multiplied by PCR. Verification of the isolation of the resistance gene clone again rests on transformation of a susceptible line to resistance.

Some of the recent work on breeding for resistance against the following bacterial pathogens will be reviewed: *P. syringae* pv. *tomato*, *P. solanacearum*, *X. campestris* pv. *vesicatoria*, *X. oryzae*, *X. citri*, *X. campestris* pv. *citrumelo* and *X. campestris* pv. *campestris*. For each disease caused by these pathogens, depending on the availability of information, I will address useful sources of germplasm for resistance, the genetics of resistance, the success of breeding for resistance to date, and alternative approaches currently being developed for breeding for resistance against these pathogens.

Pseudomonas syringae pv. *tomato*

P. syringae pv. *tomato* is the causal agent of bacterial leaf speck of tomato. Resistance to *P. syringae* pv. *tomato* in three different lines appears to be conferred by three distinct genes, *Pto* (Kerr & Cook, 1983), *Pto2* (Pilowsky & Zutra, 1986) and an unnamed gene originating in Rehovot-13 (Fallik *et al.*, 1984). *Pto* provides resistance against *P. syringae* pv. *tomato* race 0 but not against race 1 (Lawton & MacNeill, 1986). Resistance allelic to *Pto* also occurs in some wild species including *Lycopersicon pimpinellifolium* and *L. hirsutum* f. *glabratum* (Lawson & Summers, 1984b); indeed, *Pto* is thought to be derived from *L. pimpinellifolium* (Pitblado & MacNeill, 1983).

Pto maps to chromosome 5 at position 30 based on linkage to the morphological markers sundwarf and macrocalyx (Pitblado *et al.*, 1984). Recent work, however, may reassign *Pto* to another position on chromosome 5 based on linkage to the markers trifoliate and anthocyanin free (F. Carland & B. Staskawicz, personal communication). In addition to the linkage map with over 200 morphological and isozyme markers produced by the long history of genetic studies, there is an extensive molecular map that includes over 400 RFLP markers (Zamir & Tanksley, 1988). Other features besides the availability of fine genetic maps make tomato an attractive candidate for molecular genetic studies of disease resistance. Tomato is transformable by *Agrobacterium tumefaciens* and transformants can be regenerated into fertile plants (McCormick *et al.*, 1986).

Recently several laboratories have begun to conduct molecular genetic studies on *Pto* (Pautot *et al.*, 1989; Martin *et al.*, 1990; Salmeron *et al.*, 1990). At least three strategies for isolating *Pto* are currently being pursued: gene tagging, genomic subtraction and chromosomal walking.

Attempts are under way to isolate *Pto* using the maize transposable element *Ac*, which has been shown to actively transpose in tomato (Yoder *et al.*, 1988). Transposition and insertion of *Ac* into *Pto* should result in loss of resistance against race 0 of *P. syringae* pv. *tomato*. A population of *Pto* lines containing *Ac* would be screened to isolate one that has lost resistance. Once the insertion is shown to map to the *Pto* locus, *Ac* provides a molecular probe for flanking sequences, which could then be used to isolate clones containing the wild-type allele. Putative clones of *Pto* are transferred into a susceptible line for identification of the gene. To begin this process, *Ac* has been introduced by *Agrobacterium*-mediated transformation into the tomato line 76R containing *Pto* (Ronald *et al.*, 1989; Salmeron *et al.*, 1990). Progeny of primary transformants are currently being screened to find plants in which T-DNA insertions (as assayed by kanamycin resistance) are linked to *Pto* resistance (F. Carland, J. Salmeron & B. Staskawicz, personal communication). Since the *Ac* element carried within the T-DNA is more likely to transpose to closely linked loci, plants in which the T-DNA is linked to *Pto* will be further characterized.

A second strategy for isolating *Pto* involves finding closely linked RFLP markers as the starting place for chromosomal walking techniques. Uncovering RFLPs with tight linkage to *Pto* and to morphological markers known to be close to *Pto* is the first step (Tanksley *et al.* 1989; F. Carland & B. Staskawicz, personal communication). RFLP markers that flank *Pto* provide anchorage (starting sites) for chromosomal walking strategies. High-resolution, physical mapping of the region lying between two RFLP markers that are closely linked to *Pto* will help ensure directionality of a chromosomal walk. Since chromosomal walking is facilitated by libraries with large inserts, libraries of tomato genomic DNA using yeast artificial chromosome vectors with average insert sizes of 150 kb are currently being developed (S. D. Tanksley, personal communication).

Isolation of a gene such as *Pto* will not only allow deployment of resistance against *P. syringae* pv. *tomato* race 0 into susceptible lines, but also allow characterization of the molecular basis of resistance. The resistance afforded by *Pto* will only be important in regions where race 0 of *P. syringae* pv. *tomato* is present but race 1 is absent.

Pseudomonas solanacearum

P. solanacearum is the causal agent of bacterial wilt or brown rot of a vast number of natural hosts in over 25 families (Bradbury, 1986). Resistant members of *Solanum* spp. and *Solanum tuberosum* are scarce despite the large number screened (Martin, 1979). Since *S. phureja* shows a consistently high level of resistance to *P. solanacearum*, its resistance was bred into *S. tuberosum* (Thurston & Lozano, 1968; Sequeira & Rowe, 1969; Rowe & Sequeira, 1970). Resistance found in other *Solanum* species does not appear to be complete (Martin, 1979). A high degree of specificity is found between host lines and bacterial strains, yet a physiological race structure has not been uncovered. Estimation of the number of resistance genes is thus difficult. Resistance from *S. phureja* against the three strains of *P. solanacearum* was found to be controlled by several independent genes, none of which appeared to be common (Sequeira, 1979). In addition, resistance from *S. phureja* was found to be profoundly affected by temperature and light intensity (Sequeira, 1979; French & De Lindo, 1982), making this source of resistance inadequate for warm, lowland regions. Resistance in tomato cultivars Venus and Saturn (Henderson & Jenkins, 1972) is also inadequate at high temperatures.

At least two experimental approaches may help to solve the problem of incorporating resistance against *P. solanacearum* in potato: somatic hybridization (Chapter 8) and use of quantitative trait loci (QTL) mapping to screen for incorporation of quantitatively inherited traits.

Wild *Solanum* species have been known to be sources of germplasm with resistance against many diseases (see Table 1 in Helgeson, 1989). Many of these wild species, however, are not interfertile with tetraploid North American potato lines or even with diploid breeding lines. Interspecific somatic cell hybridization, which may provide access to genes for resistance in these wild species, would be employed as follows. Protoplasts from the two species are fused, and then hybrid cells capable of regenerating fertile plants that have incorporated the traits of interest are selected. Hybrids are then crossed with cultivars in conventional sexual crosses to transfer desirable agronomic traits from the wild species to a commercial cultivar. The possibility that resistance against *P. solanacearum* may be incorporated into commercial potato lines using this technique is currently being tested (J. P. Helgeson, personal communication). Success with this technique was apparent in incorporating resistance to soft rot, *Erwinia* spp. (Austin *et al.*, 1988). Although the genetic basis of resistance to *Erwinia* soft rot is considered to be complex, hexaploid somatic hybrids derived from *S. brevidens* (diploid) and *S. tuberosum* (tetraploid) were resistant to *Erwinia* spp. and furthermore, this high degree of resistance was sexually transferred (Austin *et al.*, 1988). As the first step in using somatic hybridization to incorporate resistance to *P. solanacearum* in potato, sexually incompatible accessions and somatic hybrid progeny are being screened for genetically stable, temperature-insensitive resistance.

Because the resistance against *P. solanacearum* in *Solanum* spp. may be governed by several genes, and because a molecular RFLP map exists for potato (Bonierbale *et al.*, 1988), a technique such as QTL mapping (Paterson *et al.*, 1988) may allow detection of the incorporation of these loci into potato. A segregating population from an interspecific cross or a population of somatic hybrids can be analysed for the co-segregation of the quantitative trait with several RFLP markers on each chromosome. Using statistical analysis, the loci encoding the trait are individually mapped within the interval between two linked RFLP markers. The flanking RFLP markers can subsequently be used as probes to detect incorporation of the QTL into a particular line.

Xanthomonas campestris pv. *vesicatoria*

X. campestris pv. *vesicatoria* causes bacterial leaf spot disease on pepper and tomato. The disease is a serious problem in regions where humidity and temperature are high such as Florida and parts of Australia.

There are several sources of resistance conditioned by independent, single genes to *X. campestris* pv. *vesicatoria* in pepper: *Bs1*, *Bs2*, *Bs3*, and *Bs4* genes (Cook & Stall, 1963; Cook & Guevara, 1984; Kim & Hartmann, 1985; Hibberd *et al.*, 1987, 1988). A gene-for-gene relationship appears to control the outcome of interactions between *X. campestris* pv. *vesicatoria* and pepper, with three corresponding avirulence genes in *X. campestris* pv. *vesicatoria*, *avrBs1*, *avrBs2* and *avrBs3*, having been cloned and characterized (Ronald & Staskawicz, 1988; Swanson *et al.*, 1988; Bonas *et al.*, 1989; Minsavage *et al.*, 1990). The gene *Bs1* proved to be easily overcome (Dahlbeck & Stall, 1979) because an insertion element carried by a copper resistance plasmid disrupted expression of *avrBs1* (Kearney *et al.*, 1988). *Bs4* resistance in pepper, although determined by a single gene, is quantitatively assessed resistance, depressing lesion size and number (Hibberd *et al.*, 1988). The *Bs4* gene was incorporated into a commercial line in Florida, but an influx of race 1 strains from Asia rendered the resistance ineffective (R. E. Stall, personal communication).

As discussed in the introduction, resistance encoded by *Bs2* may prove to be a good source of durable, single-gene resistance. Attempts to isolate the *Bs2* gene by genomic subtraction will use mutagenized ECW20R deleted for the *Bs2* gene or the ECW isoline as subtractor lines (B. Staskawicz, personal communication). Analysis of co-segregation of RFLP markers (Tanksley *et al.*, 1988) with *Bs2* resistance in a segregating population should allow discovery of tightly linked RFLP markers which can be used both as a starting-point for chromosomal walking techniques and as a way to monitor deletion of *Bs2*. Since verification of isolation of *Bs2* depends on transforming a susceptible line to resistance, pepper presents a technological problem because although it has been transformed, fertile plants have not been regenerated from tissue culture.

For many years there was no known source of resistance against *X. campestris* pv. *vesicatoria* in tomato (Crill *et al.*, 1972; Lawson & Summers, 1984a). After screening about 300 tomato accessions, Scott & Jones (1986) reported a source of resistance. The line Hawaii 7998 was shown to have hypersensitive resistance (HR) to *X. campestris* pv. *vesicatoria* (Jones & Scott, 1986). Field tests of the inheritance of this resistance indicated quantitative gene action with moderate to high heritability (Scott & Jones, 1989). Recent evidence suggests, however, that the HR resistance in Hawaii 7998 may be controlled in a simpler fashion. The avirulence gene *avrRxv* induces an HR on Hawaii (M. C. Whalen, unpublished data), suggesting that a single gene in Hawaii 7998 corresponding to the single avirulence gene may control resistance. Tests of the inheritance of resistance associated with *X. campestris* pv. *vesicatoria* have been complicated by several factors. The resistance in Hawaii 7998 is temperature-sensitive, active at 24°C but not at 30°C (M. C. Whalen, F. Carland & B. Staskawicz, unpublished data). In addition, analysis of inheritance patterns is problematic because of the potential action of various modifying genes (R. E. Stall, personal communication). Knowing the mechanism of inheritance is crucial for future work concerning isolation of the gene and subsequent incorporation into commercial cultivars.

Once genetic engineering technologies allow isolation of resistance genes followed by transfer to any plant species, other sources of resistant germplasm for tomato may be forthcoming from pepper. Pepper (ECW) is resistant to the tomato race of *X. campestris* pv. *vesicatoria*, producing an HR upon infection (Minsavage *et al.*, 1990). This HR resistance appears to be controlled by a gene-for-gene relationship mediated by an avirulence gene from *X. campestris* pv. *vesicatoria* named *avrBsT*. The corresponding resistance gene may provide useful resistance in tomato. In addition, as described earlier, the resistance gene *Bs2* from pepper appears to recognize all members of *X. campestris* pv. *vesicatoria* and in most cases responds with a resistant reaction (Kearney & Staskawicz, 1990). It may also provide a durable source of resistance against *X. campestris* pv. *vesicatoria* in tomato.

Xanthomonas oryzae

X. oryzae, the causal agent of bacterial leaf blight, is one of the most serious diseases of rice in Asia (Mew, 1987). Since the introduction of high-yielding cultivars, which are highly susceptible, and the application of nitrogen fertilizers, which favours disease development in susceptible cultivars, the problem has worsened over time (Mew, 1987). Use of resistant varieties is considered the most effective means of control.

Moderate resistance is common in rice germplasm. Sixteen independent genes for resistance against *X. oryzae* have been described (*Xa*1–12, Mew, 1987; *xa*13, Sahu & Khush, 1989; *Xa*16, Noda & Ohuchi, 1989). Twelve are dominant (*Xa*1–4, *Xa*6, *Xa*7, *Xa*10–12 and *Xa*14–16), four are recessive (*xa*5, *xa*8, *xa*9 and *xa*13) and one is allelic (*Xa*4^a and *Xa*4^b). Depending on the bacterial strain and the resistance gene or combination of genes present, the resistant response conditioned by each gene differs in the developmental stage at which it is expressed and in the phenotype of resistance (Parry & Callow, 1986; Nayak *et al.*, 1987). Linkage among the *Xa* series of genes is present: *Xa*4 is linked to *Xa*6, *xa*9 and *Xa*10; *Xa*2 is linked to *Xa*1 and *Xa*12 (*Xa*-kg); and *xa*5 is linked to *xa*13 (Yoshimura *et al.*, 1984; Sahu & Khush, 1989). A source of race-non-specific resistance that appears to be quantitatively inherited has been described (Yamada, 1984, 1986). Susceptible cultivars have been mutated for increased resistance (Miah *et al.*, 1981; Nakai *et al.*, 1985, 1990). In-depth genetic characterization of one of these mutant genes, *xa*-nm (t) showed that the resistance may indeed be novel since the strain-specific resistance it confers has not been previously described (Nakai *et al.*, 1990). An avirulence gene, *avr*10, has been isolated from *X. oryzae* race 2 which induces a resistance response from Cas 209 carrying *Xa*10 (Kelemu & Leach, 1990), implying that the interaction between the race 2 strain and Cas 209 is controlled according to the gene-for-gene model of resistance.

Because of the extensive variability in strains of *X. oryzae*, a single resistance gene, unless it is effective against all races and also confers a selective advantage upon the parent strain, may not prove to be agriculturally important. Molecular analysis of an isolated resistance gene, however, may be useful for directing efforts to design alternative means of control. Strategies employing chromosomal walking and genomic subtraction are two that may be fruitful in isolating one of the *Xa* genes. Near-isogenic lines differing only in the presence of one of the *Xa* genes are available. These lines provide a starting-point for genomic subtraction. To facilitate chromosomal walking, an extensive RFLP map of rice is under construction (McCouch *et al.*, 1988; Tanksley *et al.*, 1989) and a YAC vector library from a rice isoline containing one of the *Xa* genes is being produced. RFLP mapping of resistance gene *Xa*4 in an isoline has not yet produced any confirmed linkages (S. D. Tanksley, personal communication).

One of the obstacles for genetic engineering in rice is transformation and regeneration of fertile plants from all lines of rice. Although successful transformation and regeneration of fertile plants have been obtained from two particular lines of rice (Hayashimoto *et al.*, 1990), many genotypes appear to be recalcitrant. Perhaps the protocol successfully used in maize (Gordon-Kamm *et al.*, 1990) can be adapted for rice transformation and regeneration.

Xanthomonas campestris pv. *citrumelo* and *X. citri*

X. citri causes citrus bacterial canker and *X. campestris* pv. *citrumelo* causes citrus bacterial spot. Strains on which pv. *citrumelo* is based were originally identified as *X. citri* (as *X. campestris* pv. *citri*) but many studies have demonstrated that they are distinct (Hartung & Civerolo, 1987, 1989; Schoulties *et al.*, 1987; Gabriel *et al.*, 1988, 1989; Dienelt & Lawson, 1989; Lawson *et al.*, 1989; Graham & Gottwald, 1990b; Vauterin *et al.*, 1990).

Differences in the extent of resistance of citrus scions to *X. citri*, which are thought to be heritable, have been reported (Koizumi & Kuhara, 1982). Screening of scion and rootstock lines with resistance to *X. campestris* pv. *citrumelo* and *X. citri* is presently under way (Graham & Gottwald, 1990a; T. R. Gottwald, personal communication). Several *Citrus* species have been found to be resistant to *X. campestris* pv. *citrumelo*. In general, *X. campestris* pv. *citrumelo* appears to attack only rootstock cultivars and does not damage commercial scion varieties. Control measures such as removal of diseased leaves from the rootstock seedlings, spraying with copper, budding with a resistant scion variety and removing susceptible tissue from the rootstock appear to allow the scion to remain disease-free (Graham & Gottwald, 1990a).

Although adequate genetic variability for resistance to *X. citri* and *X. campestris* pv. *citrumelo* may be found within *Citrus* and related genera, conventional breeding is limited because of the reproductive biology of citrus. Problems that plague breeding programmes include high levels of heterozygosity among cultivars with consequent inbreeding depression, pollen and ovule sterility, apomixis and a long juvenility phase (Grosser & Gmitter, 1990).

An alternative to traditional breeding that may circumvent some of these problems is somatic hybridization. Problems associated with sexual incompatibility, sterility and apomixis can be overcome (Grosser & Gmitter, 1990). Allotetraploids formed from somatic hybridization appear to exhibit no loss of vigour and express traits in an additive fashion (Grosser & Gmitter, 1990). The potential of somatic hybridization to surmount obstacles in breeding for resistance in citrus against bacterial pathogens may be good. In addition to somatic hybridization, there is the possibility of introducing genes by transformation. Both particle-gun delivery methods and *Agrobacterium*-mediated transformation are under development for citrus (J. L. Schell, personal communication). Techniques in which plant cells go through tissue culture have problems associated with somaclonal variation which must be accounted for.

Another limiting factor in breeding for resistance to citrus bacterial disease is the low priority afforded such breeding programmes. Resistance to diseases caused by *Citrus* tristeza virus, *Phytophthora* spp. and the nematode *Tylenchulus semipenetrans* demand breeding programmes with higher priority (Grosser & Gmitter, 1990). Moreover, there exist control measures involving resistant scions that reduce the possibility of a major outbreak of *X. campestris* pv. *citrumelo* (Graham & Gottwald, 1990a).

Xanthomonas campestris pv. *campestris*

X. campestris pv. *campestris* causes black rot disease of crucifers. Williams *et al.* (1972) reported that resistance to *X. campestris* pv. *campestris* in *Brassica oleracea* Early Fuji was determined by a single, recessive gene influenced by one dominant and one recessive modifier gene. Later tests found that this resistance may not be expressed in young seedlings (Hunter *et al.*, 1987). Resistance from a line from China had a high degree of both seedling and adult resistance and was inherited as a single, recessive gene that interacted with a recessive modifier (Dickson & Hunter, 1987; Hunter *et al.*, 1987).

With the availability of genetic engineering technologies, another source of resistance may be forthcoming from *Arabidopsis thaliana*, the weedy crucifer that has proved to be amenable to molecular genetic manipulations. Tsuji & Somerville (1988) and Simpson & Johnson (1990) reported that *A. thaliana* ecotypes responded differentially to infection with *X. campestris* pv. *campestris*. These results suggest that there is genetic variation in resistance to *X. campestris* pv. *campestris*. In the future when resistance genes are identified and then cloned from *A. thaliana*, they can potentially be transferred to various crucifers, which would thereby be transformed to resistance. Sources of resistance would no longer be limiting. To ensure optimal durability, the resistance should not be race-specific and the corresponding avirulence gene should confer a selective advantage on *X. campestris* pv. *campestris*.

Novel sources of resistance

Investigations into the molecular mechanisms of non-host resistance have presented the possibility of novel sources of resistant germplasm against bacterial diseases. In a study of non-host resistance of bean against *X.*

campestris pv. *vesicatoria*, resistance was shown to be genetically controlled – in part in a gene-for-gene manner. A single gene in the pathogen, *avrRxv*, interacted with a single gene in bean, *Rxv*, to condition the resistant response (Whalen *et al.*, 1988). This suggests that bean may potentially be a source of resistance against *X. campestris* pv. *vesicatoria*. In addition, the resistance against *X. campestris* pv. *vesicatoria* in certain lines of soybean, alfalfa, cotton, corn and tomato line Hawaii 7998 appeared also to be controlled by *avrRxv*. These findings suggest that a resistance gene analogous to *Rxv* may function in diverse genetic backgrounds. In a similar study, Kobayashi *et al.* (1989) showed that *avrD*, a gene in the tomato pathogen *P. syringae* pv. *tomato*, induced a resistance reaction in soybean in a cultivar-specific manner. The resistance in soybean associated with *avrD* is conditioned by a single gene *Rpg4* (Keen & Buzzell, 1991). Again in this case, non host resistance appeared to function similarly to host resistance. Once resistance genes can be routinely isolated from any species and effectively transferred to others, potential sources of germplasm for resistance will be expanded.

In light of these studies, the possibility that *Arabidopsis thaliana* may be a potential source of resistance genes is under study. *A. thaliana* has many features that make it amenable to molecular genetic manipulations (Meyerowitz, 1989). It has a small genome size with a low number of repeated sequences and it is transformable by *Agrobacterium*. It has a long history of genetic studies that have resulted in a genetic map with many morphological markers and a molecular map with many RFLP markers. Work is under way to produce a restriction map and YAC libraries have been created (Guzmán & Ecker, 1988). All of these characteristics make *A. thaliana* a potentially good source of genes.

A model system for studying *A. thaliana* as a source of resistance genes has been developed (Whalen *et al.*, 1990). After screening many different *P. syringae* pv. *tomato* strains on *A. thaliana* ecotype Col-0, two were found that induced different responses, *P. syringae* pv. *tomato* 1065 and DC3000. *P. syringae* pv. *tomato* 1065 induces a resistant reaction on *A. thaliana* which appears to be in part controlled by an avirulence gene, *avrRpt2*. There is variation in response of ecotypes to a virulent strain carrying *avrRpt2*, suggesting that there is a single resistance gene corresponding to the avirulence gene. Studies of the inheritance of resistance associated with the avirulence gene are under way (Bent *et al.*, 1990). After mapping the resistance gene with RFLP markers, chromosomal walking will be employed to isolate the gene. Gene tagging with *Ac* is also being pursued (Innes *et al.*, 1989). Once it is isolated, the identity of the gene will be confirmed by transforming a susceptible *A. thaliana* ecotype to resistance. Moreover, it will be transferred to a susceptible tomato line to test whether *A. thaliana* can, indeed, provide resistance genes to other species.

Acknowledgements

I thank my colleagues for sharing their unpublished results. I am grateful to B. J. Staskawicz for introducing me to the field of plant–pathogen interactions and for many stimulating discussions. The expert editorial assistance of R. L. Moe and the scientific literature database searching performed by S. W. Cole are gratefully acknowledged. The author is supported by the Clare Booth Luce Foundation.

References

Anzai, H., Yoneyama, K. & Yamaguchi, I. (1989) Transgenic tobacco resistant to a bacterial disease by the detoxification of a pathogenic toxin. *Molecular and General Genetics* 219, 492–4.

Austin, S., Lojkowska, E., Ehlenfeldt, M. K., Kelman, A. & Helgeson, J. P. (1988) Fertile interspecific somatic hybrids of *Solanum*: a novel source of resistance to *Erwinia* soft rot. *Phytopathology* 78, 1216–20.

Baker, B., Schell, J., Lörz, H. & Federoff, N. (1986) Transposition of the maize controlling element 'Activator' in tobacco. *Proceedings of the National Academy of Science USA* 83, 4844–8.

Bender, W., Spierer, P. & Hogness, D. S. (1983) Chromosomal walking and jumping to isolate DNA from the *Ace* and *Rosy* loci and the bithorax complex in *Drosophila melanogaster*. *Journal of Molecular Biology* 168, 17–33.

Bent, A. F., Innes, R. W. & Staskawicz, B. J. (1990) Molecular genetic analysis of *Arabidopsis thaliana* resistance to the bacterial pathogen *Pseudomonas syringae* pv. *tomato*. *Fifth International Symposium on the Molecular Genetics of Plant–Microbe Interactions*, 173.

Bonas, U., Stall, R. E. & Staskawicz, B. (1989) Genetic and structural characterization of the avirulence gene *avrBs3* from *Xanthomonas campestris* pv. *vesicatoria*. *Molecular and General Genetics* 218, 127–36.

Bonierbale, M. W., Plaisted, R. L. & Tanksley, S. D. (1988) RFLP maps based on a common set of clones reveal modes of chromosomal evolution in potato and tomato. *Genetics* 120, 1095–103.

Bradbury, J. F. (1986) *Guide to Plant Pathogenic Bacteria*. CAB International, Wallingford, Oxon.

Buddenhagen, I. W. & de Ponti, O. M. B. (1984) Crop improvement to minimize future losses to diseases and pests in the tropics. In: FAO (ed.), *Breeding for Durable Disease and Pest Resistance*. Food and Agriculture Organization of the United Nations, Rome, pp. 23–47.

Burke, D. T., Carle, G. F. & Olson, M. V. (1987) Cloning of large segments of exogenous DNA into yeast by means of artificial chromosome vectors. *Science* 236, 806–12.

Cook, A. A. & Guevara, Y. G. (1984) Hypersensitivity in *Capsicum chacoense* to race 1 of the bacterial spot pathogen of pepper. *Plant Disease* 68, 329–30.

Cook, A. A. & Stall, R. E. (1963) Inheritance of resistance in pepper to bacterial spot. *Phytopathology* 53, 1060–2.

Coyne, D. P. & Schuster, M. L. (1983) Genetics of and breeding for resistance to bacterial pathogens in vegetable crops. *HortScience* 18, 30–6.

Crill, P., Jones, J. P. & Burgis, D. S. (1972) Relative susceptibility of some tomato genotypes to bacterial spot. *Plant Disease Reporter* 56, 504–7.

Crute, I. R. (1985) The genetic bases of relationships between microbial parasites and their hosts. In: Fraser, R. S. S. (ed.), *Mechanisms of Resistance to Plant Diseases.* Martinus Nijhoff/Dr W. Junk Publishers, Dordrecht, Netherlands, pp. 80–142.

Dahlbeck, D. & Stall, R. E. (1979) Mutations for change of race in cultures of *Xanthomonas campestris. Phytopathology* 69, 634–6.

Dickson, M. D. & Hunter, J. E. (1987) Inheritance of resistance in cabbage seedlings to black rot. *HortScience* 22, 108–9.

Dienelt, M. M. & Lawson, R. H. (1989) Histopathology of *Xanthomonas campestris* pv. *citri* from Florida and Mexico in wound-inoculated detached leaves of *Citrus aurantifolia:* transmission electron microscopy. *Phytopathology* 79, 336–48.

Fallik, E., Bashan, Y., Okon, Y. & Kedar, N. (1984) Genetics of resistance to bacterial speck of tomato caused by *Pseudomonas syringae* pv. *tomato. Annals of Applied Biology* 104, 321–5.

French, E. R. & De Lindo, L. (1982) Resistance to *Pseudomonas solanacearum* in potato: specificity and temperature sensitivity. *Phytopathology* 72, 1408–12.

Gabriel, D. W., Hunter, J. E., Kingsley, M. T., Miller, J. W. & Lazo, G. R. (1988) Clonal population structure of and genetic diversity among citrus canker strains. *Molecular Plant–Microbe Interactions* 1, 59–65.

Gabriel, D. W., Kingsley, M. T., Hunter, J. E. & Gottwald, T. (1989) Reinstatement of *Xanthomonas citri* (ex Hasse) and *X. phaseoli* (ex Smith) to species and reclassification of all *X. campestris* pv. *citri* strains. *International Journal of Systematic Bacteriology* 39, 14–22.

Gordon-Kamm, W. J., Spencer, T. M., Mangano, M. L., Adams, T. R., Daines, R. J., Start, W. G., O'Brien, J. V., Chambers, S. A., Adams, W. R., Jr, Willets, N. G., Rice, T. B., Mackey, C. J., Krueger, R. W., Kausch, A. P. & Lemaux, P. G. (1990) Transformation of maize cells and regeneration of fertile transgenic plants. *Plant Cell* 2, 603–18.

Graham, J. H. & Gottwald, T. R. (1990a) Susceptibility of *Citrus, Poncirus* and their hybrids to citrus bacterial spot. *Citrus Industry* 71, 48–9.

Graham, J. H. & Gottwald, T. R. (1990b) Variation in aggressiveness of *Xanthomonas campestris* pv. *citrumelo* associated with citrus bacterial spot in Florida citrus nurseries. *Phytopathology* 80, 190–6.

Grosser, J. W. & Gmitter, F. G. J. (1990) Protoplast fusion and citrus improvement. *Plant Breeding Reviews* 8, 339–74.

Guzmán, P. & Ecker, J. R. (1988) Development of large DNA methods for plants: molecular cloning of large segments of *Arabidopsis* and carrot DNA into yeast. *Nucleic Acids Research* 16, 11091–105.

Harlan, J. R. (1976) Genetic resources in wild relatives of crops. *Crop Science* 16, 329–33.

Hartung, J. S. & Civerolo, E. I. (1987) Genomic fingerprints of *Xanthomonas campestris* pv. *citri* strains from Asia, South America, and Florida. *Phytopathology* 77, 282–5.

Hartung, J. S. & Civerolo, E. I. (1989) Restriction fragment length polymorphisms distinguish *Xanthomonas* strains isolated from Florida citrus nurseries from *X. c.* pv. *citri*. *Phytopathology* 79, 793–9.

Hayashimoto, A., Li, Z. & Murai, N. (1990) A polyethylene glycol-mediated protoplast transformation system for production of fertile transgenic rice plants. *Plant Physiology* 93, 857–63.

Helgeson, J. P. (1989) Postharvest resistance through breeding and biotechnology. *Phytopathology* 79, 1375–7.

Henderson, W. R. & Jenkins, S. F. Jr (1972) Venus and Saturn. *North Carolina Agricultural Experiment Station. Bulletin* 444, 13 pp.

Hibberd, A. M., Bassett, M. J. & Stall, R. E. (1987) Allelism tests of three dominant genes for hypersensitive resistance to bacterial spot of pepper. *Phytopathology* 77, 1304–7.

Hibberd, A. M., Stall, R. E. & Bassett, M. J. (1988) Qualitatively assessed resistance to bacterial leaf spot in pepper that is simply inherited. *Phytopathology* 78, 607–12.

Hunter, J. E., Dickson, M. H. & Ludwig, J. W. (1987) Source of resistance to black rot of cabbage expressed in seedlings and adult plants. *Plant Disease* 71, 263–6.

Innes, R. W., Baker, B. & Staskawicz, B. J. (1989) A two element transposon tagging system for *Arabidopsis thaliana*. *Journal of Cellular Biochemistry* Supplement 13C, 320.

Johnson, R. (1983) Genetic background of durable resistance. In: Lamberti, F., Waller, J. M. & Van der Graaff, N. A. (eds), *Durable Resistance in Crops*. Plenum Press, New York and London, pp. 5–26.

Jones, J. B. & Scott, J. W. (1986) Hypersensitive response in tomato to *Xanthomonas campestris* pv. *vesicatoria*. *Plant Disease* 70, 337–9.

Kearney, B. & Staskawicz, B. (1990) Widespread distribution and fitness contribution of *Xanthomonas campestris* avirulence gene *avrBs2*. *Nature* 346, 385–6.

Kearney, B., Ronald, P. C., Dahlbeck, D. & Staskawicz, B. J. (1988) Molecular basis for evasion of plant host defence in bacterial spot disease of pepper. *Nature* 332, 541–3.

Keen, N. T. & Buzzell, R. I. (1991) New disease resistance genes in soybean against *Pseudomonas syringae* pv. *glycinea*: evidence that one of them interacts with a bacterial elicitor. *Theoretical and Applied Genetics* 81, 133–8.

Kelemu, S. & Leach, J. E. (1990) Cloning and characterization of an avirulence gene from *Xanthomonas campestris* pv. *oryzae*. *Molecular Plant–Microbe Interactions* 3, 59–65.

Kerr, E. A. & Cook, F. I. (1983) Ontario 7710 – a tomato breeding line with resistance to bacterial speck, *Pseudomonas syringae* pv. *tomato*. *Canadian Journal of Plant Science* 63, 1107–9.

Kim, B.-S. & Hartmann, R. W. (1985) Inheritance of a gene (Bs_3) conferring hypersensitive resistance to *Xanthomonas campestris* pv. *vesicatoria* in pepper (*Capsicum annuum*). *Plant Disease* 69, 233–5.

Kobayashi, D. Y., Tamaki, S. J. & Keen, N. T. (1989) Cloned avirulence genes from the tomato pathogen *Pseudomonas syringae* pv. *tomato* confer cultivar specificity on soybean. *Proceedings of the National Academy of Sciences USA* 86, 157–61.

Koizumi, M. & Kuhara, S. (1982) Evaluation of citrus plants for resistance to bacterial canker disease in relation to the lesion extension. *Bulletin of the Fruit Tree Research Station. D* 4, 73–92.

Larkin, P. J. & Scowcroft, W. R. (1983) Somaclonal variation and crop improvement. In: Kosuge, T. & Meredith, C. P. (eds), *Genetic Engineering of Plants*. Plenum Press, New York, pp. 289–314.

Lawson, R. H., Dienelt, M. M. & Civerolo, E. L. (1989) Histopathology of *Xanthomonas campestris* pv. *citri* from Florida and Mexico in wound-inoculated detached leaves of *Citrus aurantifolia*: light and scanning electron microscopy. *Phytopathology* 79, 329–35.

Lawson, V. F. & Summers, W. L. (1984a) Disease reaction of diverse sources of *Lycopersicon* to *Xanthomonas campestris* pv. *vesicatoria* pepper strain race 2. *Plant Disease* 68, 117–19.

Lawson, V. F. & Summers, W. L. (1984b) Resistance to *Pseudomonas syringae* pv. *tomato* in wild *Lycopersicon* species. *Plant Disease* 68, 139–41.

Lawton, M. B. & MacNeill, B. H. (1986) Occurrence of race 1 of *Pseudomonas syringae* pv. *tomato* on field tomato in southwestern Ontario. *Canadian Journal of Plant Pathology* 8, 85–8.

McCormick, S., Niedermeyer, J., Fry, J., Barnason, A., Horsch, R. & Fraley, R. (1986) Leaf disc transformation of cultivated tomato (*Lycopersicon esculentum*) using *Agrobacterium tumefaciense*. *Plant Cell Reports* 5, 81–4.

McCouch, S. R., Kochert, G., Yu, Z. H., Wang, Z. Y., Khush, G. S., Coffman, W. R. & Tanksley, S. D. (1988) Molecular mapping of rice chromosomes. *Theoretical and Applied Genetics* 76, 815–29.

Martin, C. (1979) Sources of Resistance to *Pseudomonas solanacearum*. *Report of the Planning Conference on the Developments in the Control of Bacterial Diseases of Potatoes*. International Potato Center (CIP), Lima, Peru, pp. 49–54.

Martin, G., Ganal, M., Messeguer, R. & Tanksley, S. (1990) Towards the isolation of disease resistance genes from tomato using linked RFLP markers. *Fifth International Symposium on the Molecular Genetics of Plant–Microbe Interactions*, 223.

Mew, T. W. (1987) Current status and future prospects of research on bacterial blight of rice. *Annual Review of Phytopathology* 25, 359–82.

Meyerowitz, E. M. (1989) *Arabidopsis*, a useful weed. *Cell* 56, 263–9.

Miah, A. J., Mansur, M. A. & Jalal Uddin, M. (1981) Improvement of rice through induced mutations. *Indian Journal of Agricultural Science* 51, 145–6.

Minsavage, G. V., Dahlbeck, D., Whalen, M. C., Kearney, B., Bonas, U., Staskawicz, B. & Stall, R. E. (1990) Gene-for-gene relationships specifying disease resistance in *Xanthomonas campestris* pv. *vesicatoria*–pepper interactions. *Molecular Plant–Microbe Interactions* 3, 41–7.

Nakai, H., Kobayashi, M. & Saito, M. (1985) Induction and selection of mutations for resistance against bacterial leaf blight in rice. *Euphytica* 34, 577–85.

Nakai, H., Nakamura, K., Kuwahara, S. & Saito, M. (1990) A new gene, developed through mutagenesis, for resistance of rice to bacterial leaf blight (*Xanthomonas campestris* pv. *oryzae*). *Journal of Agricultural Science. Cambridge* 114, 219–24.

Nayak, P., Suriya Rao, A. V. & Chakrabarti, N. K. (1987) Components of resistance to bacterial blight disease of rice. *Journal of Phytopathology* 119, 312–18.

Noda, T. & Ohuchi, A. (1989) A new pathogenic race of *Xanthomonas campestris* pv. *oryzae* and inheritance of resistance of differential rice variety, Te-tep, to it. *Annals of the Phytopathology Society of Japan* 55, 201–7.

Parry, R. W. H. & Callow, J. A. (1986) The dynamics of homologous and heterologous interactions between rice and strains of *Xanthomonas campestris*. *Plant Pathology* 35, 380–9.

Paterson, A. H., Lander, E. S., Hewitt, J. D., Peterson, S., Lincoln, S. E. & Tanksley, S. D. (1988) Resolution of quantitative traits into Mendelian factors by using a complete linkage map of restriction fragment length polymorphisms. *Nature* 335, 721–6.

Pautot, V., Gaut, B., Holzer, F. & Walling, L. (1989) Gene expression during *Pseudomonas syringae* pv. *tomato* infection. *Journal of Cellular Biochemistry* Supplement 13D, 285.

Pilowsky, M. & Zutra, D. (1986) Reaction of different tomato genotypes to the bacterial speck pathogen (*Pseudomonas syringae* pv. *tomato*). *Phytoparasitica* 14, 39–42.

Pitblado, R. E. & MacNeill, B. H. (1983) Genetic basis of resistance to *Pseudomonas syringae* pv. *tomato* in field tomatoes. *Canadian Journal of Plant Pathology* 5, 251–5.

Pitblado, R. E., MacNeill, B. H. & Kerr, E. A. (1984) Chromosomal identity and linkage relationships of *Pto*, a gene for resistance to *Pseudomonas syringae* pv. *tomato* in tomato. *Canadian Journal of Plant Pathology* 6, 48–53.

Ronald, P. C. & Staskawicz, B. J. (1988) The avirulence gene *avrBs₁* from *Xanthomonas campestris* pv. *vesicatoria* encodes a 50-kD protein. *Molecular Plant–Microbe Interactions* 1, 191–8.

Ronald, P. C., Kearney, B., Dahlbeck, D. & Staskawicz, B. (1989) The *Pto–avrPto* interaction: cloning of genes involved in disease resistance of tomato. *Journal of Cellular Biochemistry* Supplement 13D, 342.

Rowe, P. R. & Sequeira, L. (1970) Inheritance of resistance to *Pseudomonas solanacearum* in *Solanum phureja*. *Phytopathology* 60, 1499–501.

Sahu, R. K. & Khush, G. S. (1989) Genetics of resistance to four races of *Xanthomonas campestris* pv. *oryzae* in some rice cultivars. *Plant Breeding* 102, 232–6.

Salmeron, J., Carland, F., Kearney, B., Ronald, P., Dahlbeck, D. & Staskawicz, B. (1990) Molecular genetics of resistance of tomato to *Pseudomonas syringae*. *Fifth International Symposium on the Molecular Genetics of Plant–Microbe Interactions*, 223.

Schoulties, C. L., Civerolo, E. L., Miller, J. W., Stall, R. E., Krass, C. J., Poe, S. R. & Du Charme, E. P. (1987) Citrus canker in Florida. *Plant Disease* 71, 388–95.

Scott, J. W. & Jones, J. B. (1986) Sources of resistance to bacterial spot in tomato. *HortScience* 21, 304–6.

Scott, J. W. & Jones, J. B. (1989) Inheritance of resistance to foliar bacterial spot of tomato incited by *Xanthomonas campestris* pv. *vesicatoria*. *Journal of the American Society for Horticultural Science* 114, 111–14.

Sequeira, L. (1979) Development of Resistance to Bacterial Wilt Derived from *Solanum phureja*. *Report of the Planning Conference on the Developments in the Control of Bacterial Diseases of Potatoes*. International Potato Center (CIP), Lima, Peru, pp. 55–62.

Sequeira, L. & Rowe, P. R. (1969) Selection and utilization of *Solanum phureja* clones with high resistance to different strains of *Pseudomonas solanacearum*. *American Potato Journal* 46, 451–62.

Sharp, E. L. (1983) Experience of using durable resistance in the USA. In: Lamberti, F., Waller, J. M. & Van der Graaff, N. A. (eds), *Durable Resistance in Crops*. Plenum Press, New York and London, pp. 385–99.

Shepherd, N. S. (1988) Transposable elements and gene-tagging. In: Shaw, C. H. (ed.), *Plant Molecular Biology*. IRL Press, Oxford and Washington, pp. 187–220.

Simmonds, N. W. (1984) Strategy of disease resistance breeding. In: FAO (ed.) *Breeding for Durable Disease and Pest Resistance*. Food and Agriculture Organization of the United Nations, Rome, pp. 11–22.

Simpson, R. B. & Johnson, L. J. (1990) *Arabidopsis thaliana* as a host for *Xanthomonas campestris* pv. *campestris*. *Molecular Plant–Microbe Interactions* 3, 233–7.

Straus, D. & Ausubel, F. M. (1990) Genomic subtraction for cloning DNA corresponding to deletion mutations. *Proceedings of the National Academy of Science USA* 87, 1889–93.

Swanson, J., Kearney, B., Dahlbeck, D. & Staskawicz, B. (1988) Cloned avirulence gene of *Xanthomonas campestris* pv. *vesicatoria* complements spontaneous race-change mutants. *Molecular Plant-Microbe Interactions* 1, 5–9.

Tanksley, S., Bonierbale, M., Fulton, T., Ganal, M., Lapitan, N. S. M., Messeguer, R., Paterson, A., Prince, J., Vincente, C., Wang, Z., Yound, N. & Yu, Z. (1989) Genetic and physical mapping of tomato and rice chromosomes. *Journal of Cellular Biochemistry* Supplement 13C, 235.

Tanksley, S. D., Bernatzky, R., Lapitan, N. I. & Prince, J. P. (1988) Conservation of gene repertoire but not gene order in pepper and tomato. *Proceedings of the National Academy of Science USA* 85, 6419–23.

Thurston, H. D. & Lozano, J. C. (1968) Resistance to bacterial wilt of potatoes in Columbian clones of *Solanum phureja*. *American Potato Journal* 45, 51–5.

Tsuji, J. & Somerville, S. C. (1988) *Xanthomonas campestris* pv. *campestris*-induced chlorosis in *Arabidopsis*. *Arabidopsis Information Service* 26, 1–8.

Van Sluys, M. A., Tempé J. & Federoff, N. (1987) Studies on the introduction and mobility of the maize *Activator* element in *Arabidopsis thaliana* and *Daucus carota*. *EMBO Journal* 6, 3881–9.

Vauterin, L., Swings, J., Kersters, K., Gillis, M., Mew, T. W., Schroth, M. N., Palleroni, N. J., Hildebrand, D. J., Stead, D. E., Civerolo, E. L., Hayward, A. C., Maraite, H., Stall, R. E., Vidaver, A. K. & Bradbury, J. F. (1990) Towards an improved taxonomy of *Xanthomonas*. *International Journal of Systematic Bacteriology* 40, 312–16.

Whalen, M. C., Stall, R. E. & Staskawicz, B. J. (1988) Characterization of a gene from a tomato pathogen determining hypersensitive resistance in non-host species and genetic analysis of this resistance in bean. *Proceedings of the National Academy of Science USA* 85, 6743–7.

Whalen, M. C., Innes, R. I., Bent, A. F. & Staskawicz, B. J. (1991) Identification of *Pseudomonas syringae* pathogens of *Arabidopsis thaliana* and a bacterial locus determining avirulence on both *Arabidopsis* and soybean. *Plant Cell* 3, 49–59.

Wieland, I., Bolger, G., Asouline, G. & Wigler, M. (1990) A method for difference cloning: gene amplification following subtractive hybridization. *Proceedings of the National Academy of Science USA* 87, 2720–4.

Williams, P. H., Staub, T. & Sutton, J. C. (1972) Inheritance of resistance in cabbage to black rot. *Phytopathology* 62, 247–52.

Yamada, T. (1984) Studies on genetics and breeding of resistance to bacterial leaf blight in rice. VI. Inheritance of quantitative resistance of the variety IR 28 to bacterial groups II, III and IV of *Xanthomonas campestris* pv. *oryzae* of Japan. *Japanese Journal of Breeding* 34, 181–90.

Yamada, T. (1986) Estimates of genetic parameters of quantitative resistance to bacterial leaf blight of rice variety IR26 and their implications in selection. *Japanese Journal of Breeding* 36, 112–21.

Yoder, J. I. (1990) Rapid proliferation of the maize transposable element *Activator* in transgenic tomato. *Plant Cell* 2, 723–30.

Yoder, J. I., Palys, J., Alpert, K. & Lassner, M. (1988) *Ac* transposition in transgenic tomato plants. *Molecular and General Genetics* 213, 291–6.

Yoshimura, A., Mew, T. W., Khush, G. S. & Omura, T. (1984) Genetics of bacterial blight resistance in a breeding line of rice. *Phytopathology* 74, 773–7.

Young, N. D., Zamir, D., Ganal, M. W. & Tanksley, S. D. (1988) Use of isogenic lines and simultaneous probing to identify DNA markers tightly linked to the *Tm-2a* gene in tomato. *Genetics* 120, 579–85.

Zamir, D. & Tanksley, S. D. (1988) Tomato genome is comprised largely of fast-evolving, low copy-number sequences. *Molecular and General Genetics* 213, 254–61.

Chapter 13
Genetic Engineering for Resistance to Viruses

Gail M. Timmerman
DSIR Crop Research, Private Bag, Christchurch, New Zealand

Introduction

Plant virus diseases can affect both yield and quality of crops and even result in complete crop failure. Various means of controlling viral disease have been practised, including using virus-free seed, avoiding or controlling insect vectors, controlling weed species that act as virus hosts and using cultivars bred for virus resistance. Two other virus control methods that have been applied are cross-protection (Hamilton, 1985) and the use of satellite viruses or RNAs (Francki, 1985).

Plant genetic engineering provides new methods for producing virus-resistant plants and these are the subject of this review. A number of approaches have been taken, but to date most of these involve the integration of viral sequences into plant genomes. Powell-Abel *et al.* (1986) first described the use of a plant viral coat protein gene to produce virus-resistant plants. Subsequently, similar experiments have been described for a number of plant viral coat protein genes. More recently, a non-structural coding region from tobacco mosaic virus (TMV) has been used to produce virus-resistant plants (Golemboski *et al.*, 1990). Resistant plants have also been produced by expressing satellite RNAs or antisense RNAs complementary to viral positive-sense sequences. Other untested strategies exist and have been discussed in the literature. For example, ribozymes may be useful for controlling viral replication. Alternatively, virus disease may be controlled by manipulating plant nuclear genes encoding resistance or involved in defence mechanisms using genetic engineering technology. These genes are currently used by plant breeders to develop genetically virus-resistant cultivars by more conventional methods.

The process used to produce virus-resistant transgenic plants in all studies published to date has been *Agrobacterium*-mediated transformation of chimeric genes into plant cells (Chapter 3), followed by selection and regeneration of these cells in culture. It is well documented that plant cell culture itself may produce genetic changes, a phenomenon termed 'somaclonal variation' (Larkin & Scowcroft, 1981). Indeed, somaclonal variation has been used to select heritable virus resistance from tomato cultures (Barden *et al.*, 1986). Therefore, somaclonal variation could be an unexpected source of the resistance phenotype in any experiment to produce virus-resistant plants by methods involving cell culture. Fortunately, somaclonal variation generally occurs at low frequency for a given character. If necessary, the origin of the virus resistance obtained can be clarified by monitoring co-segregation of the resistance phenotype and the inserted gene in sexual progeny of transgenic resistant plants.

The emphasis in this chapter is on the experiments performed to produce virus-resistant plants by genetic engineering. The characteristics of the resulting plants are examined. These plants provide valuable model systems for testing the usefulness of genetically engineered virus resistance and for dissecting mechanisms by which virus resistance is conferred by the inserted genes. The prospects and possible directions for a number of strategies are discussed.

Resistance using virus-encoded genes

Recent experiments have shown that virus resistance can be produced in plants transformed with and expressing viral coding regions. Viral coat protein (CP) genes have been used to produce plants with complete or nearly complete resistance to potato virus X (PVX) (Hemenway *et al.*, 1988; Hoekema *et al.*, 1989; Lawson *et al.*, 1990), potato virus Y (PVY) (Lawson *et al.*, 1990), TMV (Powell-Abel *et al.*, 1986), alfalfa mosaic virus (AlMV) (Van Dun *et al.*, 1987; Loesch-Fries *et al.*, 1987), tobacco streak virus (TSV) (Van Dun *et al.*, 1988) and tobacco rattle virus (TRV) (Van Dun *et al.*, 1987), as well as lines partially resistant to cucumber mosaic virus (CMV) (Cuozzo *et al.*, 1988). Cell lines showing resistance to beet necrotic yellow vein virus (BNYVV) have also been reported (Kallerhoff *et al.*, 1990). Expression of CP genes appears to be a method that may be generally applicable for controlling positive-sense, ssRNA plant virus infections. It may also be effective for controlling plant viruses from other groups, although this possibility remains untested to date.

In addition, a viral coding region that does not encode a coat protein has been shown to confer resistance on transgenic plants. In a surprising discovery, TMV-resistant plants have been produced by genetically engineering tobacco with a TMV gene sequence for an open reading frame (ORF) whose putative gene product, a 54 kDa protein, has never been observed in virus-infected plants (Golemboski *et al.*, 1990). The resulting transgenic plants displayed a high degree of resistance to TMV.

Coat protein-mediated virus resistance

At present the most promising way to produce virus-resistant plants by genetic engineering is to insert and express a viral CP gene or genes. One of the reasons for the attractiveness of this approach is the relative accessibility of these genes. Single-stranded RNA plant viruses are genetically simple, most having genomes of less than 10,000 nucleotides. Furthermore, the cistron encoding the coat protein is often located at the 3' end of the encoding RNA. This simplifies the synthesis of cDNA molecules once a suitable primer sequence is identified. The technologies for cDNA synthesis, cloning and sequencing are well established and widely practised. Likewise, once a CP coding region has been cloned and sequenced, readily accessible technology exists to construct suitable vectors with a chimeric form of the gene for transfer and expression in model plant systems. Plant transformation systems have been and are continually being developed for important crop plants (Gasser & Fraley, 1989).

The extent of the resistance achieved by genetically engineering for CP gene expression varies (see below). This variation may depend on a number of factors relating to the accumulation of the CP gene product, including its degree of expression, its cellular localization, its aggregation state and the tissue and/or developmental specificity of its expression. In addition, resistance may be affected by other unidentified factors inherent to the biology of the specific virus, vector, host plant or their interaction.

Coat protein gene expression

Transgenic plants expressing a coat protein gene are actually expressing a number of new molecules that could mediate virus resistance, including the coat protein, the CP mRNA, and other viral RNA sequences that are transcribed from the chimeric gene construct, usually 3' untranslated sequences. The available evidence indicates that, at least for TMV, the coat protein confers protection, and the mRNA and 3' untranslated sequences do not contribute to the observed virus resistance (Powell *et al.*, 1990). Researchers have, therefore, focused on attaining reasonable synthesis of coat protein. The accumulation of CP mRNA must also be considered as this is an important and regulated step in the gene expression pathway.

Expression of CP mRNA

The accumulation of any mRNA is affected by factors such as the rates of transcription, processing and degradation. The promoter sequence used in chimeric gene constructs can have a significant effect on mRNA accumulation. So far, most of the CP genes have been expressed from the cauliflower mosaic virus (CaMV) 35S promoter. In addition the CaMV 19S promoter has been used to express AlMV CP (Loesch-Fries *et al.*, 1987) and the 'enhanced' CaMV 35S promoter dimer has also been used (Lawson *et al.*, 1990). The other transcriptional regulatory signal required for expression of viral CP genes in plant tissues is the 3' termination signal. Most of the chimeric genes constructed for CP expression use the nos 3' sequences; however, the petunia rbcS E9 (Cuozzo *et al.*, 1988; Hemenway *et al.*, 1988), CaMV 19S (Loesch-Fries *et al.*, 1987) and soybean conglycinin 7S α' (Lawson *et al.*, 1990) 3' signals have also been used. Reasonable amounts of CP mRNA have been obtained with all these except the CaMV 19S 3' signals; however, in this instance the less efficient 19S promoter was used as well (Loesch-Fries *et al.*, 1987).

Some estimates have been made of the amount of viral CP mRNA accumulating in transgenic plants, and these suggest that promoter strength may be the major factor affecting CP mRNA concentrations. The amounts of viral CP mRNA shown to accumulate from various chimeric gene constructs using the CaMV 35S promoter were similar. For example, Lawson *et al.* (1990) found that 50–250 pg of PVX CP mRNA and 100–250 pg of PVY CP mRNA accumulated per µg total RNA in transgenic potato plants. TMV CP mRNA accumulated to 100 pg/µg total RNA in transgenic tobacco line 3404 (Powell *et al.*, 1990). Slightly higher amounts of TMV CP mRNA were observed to accumulate when the enhanced CaMV 35S promoter 'dimer' was used, resulting in up to 400 pg of TMV CP mRNA/µg total RNA (Powell *et al.*, 1990). In contrast, use of the weaker CaMV 19S promoter resulted in accumulation of only 10 pg of AlMV CP mRNA/µg total RNA (Loesch-Fries *et al.*, 1987). These comparisons suggest that mRNA instability may be a relatively unimportant problem for these sequences since the amounts of CP mRNA that accumulate are fairly consistent with our notions of relative promoter strength.

Expression of the coat protein

The accumulation of viral coat protein in transgenic CP+ plants (plants expressing CP gene) is summarized in Table 13.1. The efficiency of CP mRNA translation and the stability of the CP gene product are factors likely to have a significant effect on accumulation. The 5' leader sequences of the subgenomic RNAs for TMV and AlMV CP expression have been shown to act as translational enhancers (Gallie *et al.*, 1987; Jobling & Gehrke, 1987); therefore, incorporation of these elements into chimeric CP gene constructs may improve protein accumulation. Expression of AlMV CP up to 0.8% of total soluble protein was achieved using constructs that contained the full AlMV CP 5' leader sequence (Tumer *et al.*, 1987).

It is notable that the maximum extent of expression (as percent of total soluble protein) attained for the CP genes of some viruses is quite low, i.e. TRV (0.05%), CMV (0.002%) and PVY (0.05%) (Table 13.1). The mRNA encoding the PVY CP was shown to accumulate to reasonable concentrations (up to 250 pg/µg total RNA) (Lawson *et al.*, 1990), and Cuozzo *et al.* (1988) stated that high concentrations of CMV CP mRNA accumulated in the CP+ transgenic plants. Therefore, the low degree of protein accumulation may be due to the rate of translation of the message, a post-translational event, or rapid turnover of the protein. It is possible that some plant viral coat proteins may be unstable when not assembled with viral RNA into virions.

Table 13.1. Extent of coat protein expression in transgenic plants.

CP gene	Host plant	Expression[a]	Gene construct[b]	Reference
AlMV	*N. tabacum*	0.1–0.4%	p35S–CP–nos 3′	Tumer *et al.*, 1987
AlMV	*L. esculentum*	0.1–0.8%	p35S–CP–nos 3′	Tumer *et al.*, 1987
AlMV	*N. tabacum*	0.01–0.05%	p35S–CP–nos 3′	Van Dun *et al.*, 1987
AlMV	*N. tabacum*	0.004–0.08%	p19S–CP–19S 3′	Loesch-Fries *et al.*, 1988
CMV	*N. tabacum*	0.001–0.002%	p35S–CP–E9 3′	Cuozzo *et al.*, 1988
PVX	*N. tabacum*	0.02–0.1%	p35S–CP–E9 3′	Hemenway *et al.*, 1988
PVX	*S. tuberosum*	< 0.01–0.3%	p35S–CP–nos 3′	Hoekema *et al.*, 1989
PVX	*S. tuberosum*	0.05–0.2%	p35S–CP–7Sα′ 3′	Lawson *et al.*, 1990
PVY	*S. tuberosum*	0.01–0.05%	p35S²–CP–7Sα′ 3′	Lawson *et al.*, 1990
SMV	*N. tabacum*	0.01–0.21%	p35S–CP–nos 3′	Stark & Beachy, 1989
TMV	*N. tabacum*	0.002%	p35S–CP–nos 3′	Bevan *et al.*, 1985
TMV	*N. tabacum*	0.04–0.1%	p35S–CP–nos 3′	Powell-Abel *et al.*, 1986
TMV	*L. esculentum*	0.05%	p35S–CP–nos 3′	Nelson *et al.*, 1988
TRV	*N. tabacum*	0.01–0.05%	p35S–CP–nos 3′	Van Dun *et al.*, 1987
TRV–PLB	*N. tabacum*	< 0.01%	p35S–CP–nos 3′	Angenent *et al.*, 1990
TRV–TCM	*N. tabacum*	0.05%	p35S–CP–nos 3′	Angenent *et al.*, 1990

[a]As percent of total soluble protein.
[b]Abbreviations used for promoter and terminator regions: p35S, CaMV 35S promoter; nos 3′, nopaline synthase terminator region; p19S, CaMV 19S promoter; E9, petunia rbcs E9 terminator region; P35S², CaMV 35S promoter dimer; 7Sα′ 3′, soybean conglycinin 7Sα′ terminator region.

Resistance of CP+ plants to infection

The foremost question is whether CP gene expression provides a high enough degree of resistance to be of practical value in agriculture. The results obtained from field trials of TMV CP+ tomatoes are very encouraging (Nelson *et al.*, 1988). Most published studies to date are the results of laboratory or glasshouse experiments, which are also very encouraging. A number of transgenic plant lines have been produced in which no virus is detectable in any plants even after inoculation with high concentrations of the appropriate virus. These are described below. Other plant lines show partial resistance. This may be demonstrated as a reduction in the percent of plants developing symptoms, in symptom severity or in virus accumulation, and/or as a delay in the appearance of symptoms. Since the appearance of symptoms is often delayed in CP-expressing plants, experiments that monitor disease development for short times may overestimate the degree of the virus resistance phenotype. For these reasons, the most sensitive experiments will test the susceptibility of a reasonably large number of plants at multiple inoculum concentrations. Furthermore, the plants should be monitored for a sufficiently long time and virus accumulation determined using a sensitive assay such as ELISA or Northern blot.

PVX-RESISTANT PLANTS. Transgenic CP+ potatoes (cv. Russett Burbank) have been produced that are resistant to PVX (Lawson *et al.*, 1990). Four independent CP+ lines were produced and the contents of PVX CP in three of these lines were measured. Line 303 (which also contains the PVY CP gene; see below) has the highest PVX resistance, completely withstanding challenge with PVX at an inoculum concentration of 5 μg/ml. In this line, PVX CP is expressed at a low concentration which is consistent from plant to plant (0.2 ng PVX CP/mg extractable protein). Plants from the other two lines are more susceptible to PVX inoculation and yet contain more PVX CP/mg protein. Furthermore, the PVX CP concentration in one of these lines varies 50-fold between clonally propagated plants. It is possible that this line shows a large variation in degree of expression due to the original transgenic plant being chimeric for more than one transformation event. Nevertheless, these results suggest that there are undetermined factors affecting the extent of PVX CP expression, and the interaction between expression and resistance remains uncharacterized.

There are two other reports of transgenic plant lines expressing PVX CP and showing partial resistance to PVX. The first report is of transgenic tobacco plants (Hemenway *et al.*, 1988), and the second is of transgenic potato plants (Hoekema *et al.*, 1989).

POTATO VIRUS Y-RESISTANT PLANTS. The same research group (Lawson *et al.*, 1990) has produced two transgenic Russett Burbank lines expressing the PVY CP gene and showing complete resistance to mechanical inoculation by PVY at virus concentrations of 20 μg/ml. The susceptibility of these two CP+ lines to aphid inoculation was also tested. Only one of these lines (once again line 303) was resistant to inoculation by this insect vector, suggesting that insect inoculation may be a more sensitive test of resistance for insect-transmitted viruses than mechanical inoculation. Since aphid inoculation is the natural route to infection in the field, it is a very important test of the usefulness of the resistance phenotype.

SUSCEPTIBILITY TO MIXED INFECTION BY PVX AND PVY. Since many cultivated plant species are susceptible to several viral diseases, it may be desirable to genetically engineer with multiple coat protein genes from different viruses. For example, viruses infecting potato include PVX, PVY, potato virus S and potato leafroll virus (PLRV). Furthermore, dual plant virus infections can result in synergistic increases in disease severity. In potato, one such combination is simultaneous infection by PVX and PVY, which produces a disease termed 'rugose mosaic' (Delgado-Sanchez & Grogan, 1970). Lawson *et al.* (1990) produced transgenic Russett Burbank plants expressing both the PVX and PVY CP genes. The transgenic line 303 also showed complete resistance to PVX and PVY in mixed infection, demonstrating that multiple resistances can be achieved by genetically engineering with coat protein genes.

This study by Lawson *et al.* (1990) is the first examining the effectiveness of CP-mediated protection against mixed virus infections. The partially resistant transgenic CP+ lines that were tested provided interesting clues about the possible application of this technology. Specifically, susceptibility to PVX increased markedly in those transgenic plants expressing the PVX CP gene that became infected with PVY, indicating in some instances that CP-mediated resistance may be less effective in the presence of a second infection caused by an unrelated virus.

ALFALFA MOSAIC VIRUS-RESISTANT PLANTS. Three laboratories have reported using the AlMV CP gene to produce plants with improved resistance to this economically important virus (Loesch-Fries *et al.*, 1987; Tumer *et al.*, 1987; Van Dun *et al.*, 1987). Van Dun *et al.* (1987, 1988) developed a transgenic tobacco line (S40-7) showing resistance to infection by AlMV at inoculum concentrations up to 2.5 μg/ml. In these experiments, relatively small numbers of S40-7 plants were infected, and therefore a low degree of susceptibility might have been overlooked, especially if partial resistance were seen as disease development in a small percentage of inoculated plants as is often the case with CP-mediated virus resistance. It is relevant that the two other reports (Loesch-Fries *et al.*, 1987; Tumer *et al.*, 1987) described plants with improved but not complete resistance to AlMV infection. Higher inoculum concentrations were used in these two studies than by Van Dun *et al.* (1987) and under these inoculation pressures some of the CP-expressing plants developed disease symptoms.

TOBACCO STREAK VIRUS-RESISTANT PLANTS. Van Dun *et al.* (1988) have engineered tobacco plants for resistance to tobacco streak virus (TSV), with the result that line STSV-1 appears to be resistant to TSV infection. These transgenic plants (STSV-1) showed no sign of viral disease caused by TSV or of infectious virus particles after challenge inoculation with 10 μg virus/ml.

TOBACCO RATTLE VIRUS-RESISTANT PLANTS. Transgenic plant lines resistant to TRV strains TCM and PLB have been developed (Van Dun & Bol, 1988; Angenent *et al.*, 1990). Neither line develops symptoms when challenged at 10 μg/ml with the strain of TRV from which the CP gene originated, nor do these plants accumulate viral RNA when challenged with the appropriate strains as shown by Northern blot analysis.

SYMPTOM EXPRESSION AND DISEASE DEVELOPMENT IN PARTIALLY RESISTANT PLANTS. Many of the plant lines expressing viral CP genes showed improved but not complete resistance to viral infection. Transgenic CP+ plant lines that are partially virus-resistant have been developed for TMV (Powell-Abel *et al.*, 1986), CMV (Cuozzo *et al.*, 1988), AlMV (Loesch-Fries *et al.*, 1987; Tumer *et al.*, 1987), PVX (Hoekema *et al.*, 1989; Hemenway *et al.*, 1988; Lawson *et al.*, 1990) and PVY (Lawson *et al.*, 1990). Careful studies of these plants should enhance our understanding of the progress of the diseases caused by these viruses. When disease does develop, it is often less severe and may result in reduced virus accumulation in inoculated and/or systemically infected leaves, delay in the timing of appearance of systemic disease symptoms and reduction in symptom severity. The progress of viral disease in these plants is best characterized for the TMV CP+ lines, discussed below. Since most of the transgenic CP+ plant lines produced so far show incomplete resistance, this is currently a likely outcome of any experiment to produce virus resistance using CP genes. It will be important to determine the usefulness of partially resistant lines for controlling virus disease in the field.

Partially resistant transgenic plant lines expressing the TMV CP have been characterized in *N. tabacum* (lines 3404 and 3646) (Powell-Abel *et al.*, 1986) and *Lycopersicon esculentum* cv. VF36 (line 306–98) (Nelson *et al.*, 1988). When these CP+ plants become infected, it is typical for the appearance of symptoms to be delayed by at least 2–4 days compared with control CP− plants. In addition, the symptoms that develop on CP+ plants that do become infected are less severe than on control CP− plants (Powell *et al.*, 1989), and the extent of virus accumulation in both the inoculated and the systemically infected leaves of transgenic CP+ plants is lower than in control CP− plants (Nelson *et al.*, 1987; Nejidat & Beachy, 1989).

A factor that has a significant effect on the resistance of TMV CP-expressing tobacco plants is the temperature at which the plants are incubated (Nejidat & Beachy, 1989). Transgenic CP+ tobacco plants are nearly as susceptible to TMV infection when incubated at 30°C or 35°C as CP− plants are at any temperature (22°C, 30°C or 35°C). Furthermore, the numbers of virions that accumulate in infected transgenic CP+ plants incubated at 35°C are the same as in identically treated CP− plants. Plants retain some resistance when subjected to a 35°C day/25°C night, a regime more closely resembling the conditions that would occur in the field. Nejidat & Beachy

(1989) showed that the endogenous TMV CP concentrations fall in unin-fected transgenic CP+ plants incubated at 35°C, and suggested that this may be due to instability of the TMV CP at high temperature.

The breakdown of resistance at high temperature may not be a general phenomenon. Tomato plants expressing TMV CP retained resistance when incubated at 35°C, even though the amount of the TMV CP expressed from the transgene dropped significantly in these plants. In contrast, the amounts of potyviral soybean mosaic virus (SMV) CP accumulating in transgenic tobacco plants expressing this gene do not drop in plants incubated at 35°C (Nejidat & Beachy, 1989). The temperature sensitivity of CP-mediated resistance therefore appears to depend upon both the viral coat protein gene and the plant species.

RESISTANCE TO RELATED STRAINS AND VIRUSES. Classical cross-protection is characterized by the ability of mild strains to protect plants against super-infection by severe strains of the same virus. This phenomenon is used in plant virus classification as a test of relatedness. Although coat protein-mediated virus resistance may function differently from cross-protection, it has been shown to work for strains or viruses that are closely related to the virus from which the CP gene was derived. This has been demonstrated for viruses from four groups: alfalfa mosaic virus group, tobraviruses, potyviruses and tobamoviruses. The extent of the protection obtained is generally related to the sequence similarity between the virus whose CP is in use and the challenging virus. For example, the CP from AlMV strain 425 protects against AlMV strains McKinney and YSMV (Loesch-Fries *et al.*, 1988; Van Dun *et al.*, 1987), but not against the ilarvirus TSV (Van Dun *et al.*, 1988). This is not surprising since there is no significant sequence simi-larity between the AlMV and TSV coat proteins. Similarly, the coat protein from TRV strain TCM protects plants against pea early browning virus (PEBV), which is highly homologous to strain TCM, but not against TRV strain PLB, whose coat protein only shares 38% amino acid sequence identity with the TCM strain (Angenent *et al.*, 1990). In the potyvirus group, wide-range but partial protection was conferred upon transgenic tobacco plants expressing the soybean mosaic virus (SMV) CP gene. These plants showed some resistance to infection by two other potyviruses that are only distantly related to SMV, tobacco etch virus (TEV) and PVY (Stark & Beachy, 1989).

In some cases the protection conferred by endogenously expressed coat protein is wider than would be predicted from cross-protection experiments. For example, plant lines engineered to express the coat protein from TMV strain U1 show good resistance to strain U1 itself and to the highly virulent strain PV230. They also show some resistance to strain L and to tomato mosaic virus (ToMV), which is related to TMV (Nelson *et al.*, 1988). The protection against ToMV is wider than is observed in classical cross-protection experiments, since TMV has been reported not to cross-protect against ToMV (Broadbent, 1964).

Further interesting observations were made by Anderson *et al.* (1989), who showed that expression of TMV CP or AlMV CP interferes with infection by heterologous viruses, but only under specific experimental conditions. In plants expressing the TMV U1 coat protein, disease development was retarded when challenged with PVX, PVY and CMV; and plants expressing the AlMV CP behaved similarly when infected with PVX and CMV. However, this 'heterologous protection' was only observed at very low challenge inoculum concentrations, typically 0.01 μg/ml, with the primary effect being to delay disease development.

Mechanism of coat protein-mediated resistance

The mechanism by which endogenous coat protein expression confers resistance to virus infection is not understood; however, some experiments have been reported that begin to pin-point where the block(s) to disease development may occur.

Genetically engineered protection against TMV is the best studied example. Studies based on protoplast infection suggest that the initial block occurs early in replication, possibly at the stage of virion uncoating (Register & Beachy, 1988, 1989). The block to virus infection in CP+ protoplasts can be overcome by inoculating with either viral RNA or virions pretreated at pH 8.0, forms of inoculum that do not or may not require uncoating, respectively. Pretreatment of TMV virions at pH 8.0 produces an unspecified change that results in stimulation of *in vitro* translation (Wilson, 1984), possibly due to a limited amount of uncoating occurring at the 5' end of the virus rod.

Coat protein-mediated protection also acts to inhibit both localized and systemic spread of virus (Wisniewski *et al.*, 1990). This was studied by inoculating CP+ plants with TMV RNA, which is able to initiate TMV infections. Localized spread between nearby cells was studied by point-inoculating with TMV RNA. In CP+ plants lesion size was significantly smaller than in CP− plants. Systemic spread into stem and upper leaves was also inhibited. In addition, grafting experiments established that TMV was able to move uninhibited through leafless stem sections derived from CP+ plants, but that TMV movement through fully or partially leafed CP+ stem sections was inhibited. This suggests that movement into leaf tissue may normally occur during systemic spread of TMV, and that the leaf may contain the site at which TMV CP acts to inhibit systemic viral spread in CP+ plants.

These experimental results indicate that TMV CP blocks all stages of infection. At the cellular level, the block may occur very early in infection, at or before virion uncoating, which may account for the inhibition seen at all stages. Alternatively, CP may interfere with other events that are specifically involved in TMV movement within the plant. The form that TMV takes while being transported through the plant is unknown; however, a number of studies have shown that CP is required for efficient long-distance translocation of TMV (Takamatsu *et al.*, 1987; Culver & Dawson, 1989).

Although TMV CP-mediated virus resistance is the best studied, it is unlikely to provide a universally applicable model for understanding CP-mediated virus resistance. Indeed, the block caused by PVX CP may occur at a different stage from that caused by TMV CP. This is indicated by comparing the susceptibility of PVX CP+ and TMV CP+ plants to RNA inoculation. The PVX CP+ plants tested were resistant to inoculation with PVX RNA (Hemenway *et al.*, 1988), in contrast to TMV CP+ plants, which are susceptible to TMV RNA inoculation (Nelson *et al.*, 1987). One speculation is that this difference in susceptibility to RNA inocula may be due to differences in virion uncoating mechanisms (Hemenway *et al.*, 1988). Whereas TMV has been shown to uncoat from the 5' end of the virion RNA, *in vitro* disassembly studies suggest potexvirus uncoating begins at the 3' end of the virion (Lok & Abouhaidar, 1981).

The block to TRV infection caused by TRV CP may act after primary infection, affecting events involved in local and systemic viral movement. Angenent *et al.* (1990) showed that TRV CP+ and CP− protoplasts were equally susceptible to infection by TRV virions or RNA, even though good resistance was observed in CP+ whole plants. TRV CP may still act at the cellular level; however, a greater degree of CP expression may be required to protect protoplasts than whole plants.

In summary, experiments on the mechanisms of CP-mediated virus resistance indicate that the block(s) operate at the cellular and/or whole plant levels. These effects may be due to a single site of action or there may be multiple mechanisms at work to confer resistance. Further studies using protoplasts, grafted plants or RNA inoculations of whole plants should help elucidate the nature of the blocks caused by viral coat protein, as well as the characteristics of different viral systems. Studying the mechanism of CP-mediated virus resistance is likely to lead to greater understanding of the normal processes of virus infection and spread.

Resistance conferred by a non-structural viral gene

Highly virus-resistant transgenic plants have been produced by engineering *N. tabacum* with a non-structural gene from TMV strain U1 (Golemboski *et al.*, 1990). The chimeric gene that was inserted into these plants contains the CaMV 35S promoter followed by the coding region of an ORF with the potential to encode a 54 kDa protein. The 54 kDa ORF is interesting because its protein product has never been detected in infected plants or protoplasts, even though it is a large ORF and a subgenomic RNA carrying it has been detected in TMV-infected plants (Sulzinski *et al.*, 1985). In fact, the initial rationale for engineering this ORF into tobacco plants was to examine its function, which may be related to RNA replication, since the 54 kDa ORF contains the conserved Gly–Asp–Asp sequence that is characteristic of viral RNA polymerase domains (Habili & Symons, 1989).

When challenged with TMV-U1, transgenic plants containing the chimeric 54 kDa ORF showed complete resistance, even when inoculated with very high concentrations of virus (up to 500 μg/ml) or viral RNA (up to 300 μg/ml). Furthermore, no symptoms, viral RNA or infectious virions were detectable for the 30 days that the plants were monitored. The resistance that was obtained was specific to strain U1, since the plants were susceptible to the U2 and L strains and also to CMV. Analysis of the chimeric gene's expression in these plants demonstrated that the mRNA was produced, but the 54 kDa protein product was not detectable. As a result, it is unclear whether the resistance is due to the protein or its mRNA; however, this question can be answered with *in vitro* mutagenesis experiments.

This unexpected finding presents the possibility that plants resistant to viruses from other groups of positive-sense, single-stranded RNA viruses may be produced by genetic engineering with the ORF containing the conserved Gly–Asp–Asp sequence, but this remains to be tested. The durability of the resistance conferred by the 54 kDa ORF when challenged by very high inoculum concentrations is a strong advantage in practical applications, but the apparent strain specificity is not.

Other strategies for virus resistance

The remaining strategies that have been attempted for producing virus-resistant plants either show less promise or have serious drawbacks when compared with the two methods discussed above. These remaining methods, the use of satellite RNAs and of antisense RNAs, are discussed below. Another two strategies discussed remain untested to date. These are the introduction by genetic engineering of ribozymes or of plant genes encoding resistance to viruses.

Satellite RNA expression

A group of biological entities that are naturally able to modify plant viral disease are the satellite RNAs. These have potential uses in controlling viral disease since many of these agents result in amelioration of virus symptoms. Satellite RNAs are amenable to genetic engineering, as DNA copies inserted into plant genomes in a form that enables them to be transcribed give rise to unit-length satellite RNAs which behave in all respects like naturally occurring satellite RNAs (Gerlach *et al.*, 1987; Harrison *et al.*, 1987). When the transgenic tobacco plants described in these reports were infected with the appropriate virus, proliferative replication of unit-length, transmissible satellite RNA resulted. Replication of the infecting virus and symptom formation were contained in a few leaves in the transgenic, satellite RNA-expressing plants. In these plants lesions formed on the inoculated leaves and some symptoms appeared on the first to third systemic leaves, but upper leaves were mostly symptom-free and growth of the later leaves was reported as being nearly normal.

There are a number of disadvantages or potential risks to using plants genetically engineered with satellite RNA copies. These relate to the fact that satellite RNAs are transmissible agents and are similar to the problems encountered with classical cross-protection (Hamilton, 1985). These include:

1. satellite RNAs that are ameliorative in one species may endanger other crop species. An example is CMV D-sat RNA which is ameliorative in tobacco, yet lethal in tomato (Waterworth *et al.*, 1979);
2. satellite RNAs mutate rapidly, and sequence variants have been observed within as few as two to four serial passages (Kurath & Palukaitis, 1990). Furthermore, a single nucleotide change introduced by *in vitro* mutagenesis can alter an ameliorative satellite so that it becomes necrogenic (Sleat & Palukaitis, 1990);
3. recombination has recently been observed between satellite RNAs (Cascone *et al.*, 1990).

Once there is a more advanced understanding of the molecular biology of satellite RNAs, it may be possible to manipulate the sequences of these agents so that they lose their transmissibility from the transgenic plant and yet retain their efficacy in ameliorating plant viral disease. In this case, the use of satellite RNA sequences as virus-resistance genes may become more attractive.

Use of antisense RNA

Another strategy that has been tried for producing virus-resistant plants is the expression of antisense RNA against plant viral CP gene sequences (Cuozzo *et al.*, 1988; Hemenway *et al.*, 1988; Rezaian *et al.*, 1988; Powell *et al.*, 1989). Although agriculturally useful protection has not been obtained with this approach, plants that show some resistance to low inoculum concentrations have been produced for CMV (Cuozzo *et al.*, 1988), PVX (Hemenway *et al.*, 1988) and TMV (Powell *et al.*, 1989). The resistance is most often expressed as an increase in the proportion of plants escaping infection, but delayed symptom appearance and reduced severity are also observed.

There may be other viral sequences whose expression as antisense RNA would result in useful plant resistance, particularly sequences that may be binding sites for RNA replicases or sequences that are present in infected cells in relatively low concentrations and yet are crucial for viral replication. Since we have little understanding of the exact mechanisms of gene expression or replication for most plant viruses, the best target sequences are difficult to identify at present. Attempts have been made to use antisense RNAs complementary to non-coat protein sequences to control TMV (Powell *et al.*, 1989) and CMV (Rezaian *et al.*, 1988). The experiments with TMV showed that expression of an antisense RNA for the CP gene and tRNA-like 3' end of the genomic RNA resulted in resistance to inocula of low concentration, while CP antisense alone was ineffective in controlling virus at these concentrations. This suggests that an antisense RNA targeted for the tRNA-like 3' end alone might have an inhibitory effect on TMV replication.

The experiments with CMV (Rezaian *et al.*, 1988) presented a less optimistic prospect for the use of antisense RNAs as viral control agents. These workers produced transgenic plant lines expressing antisense RNA complementary to three regions of the CMV genome, and tested these plants for resistance to CMV. None of the three regions encoded coat protein. One transgenic line showed increased resistance to CMV replication; however, eleven other lines expressing the same RNA at equal or higher concentrations were just as susceptible as control plants. These researchers interpreted this result cautiously and concluded that the differences in the susceptibility of the various lines were most probably due to unknown factors unrelated to the antisense RNA expression, such as somaclonal variation.

Final success in using antisense RNA to disrupt viral replication and therefore disease may require that the antisense RNAs expressed are complementary to key regions involved in the regulation of replication or gene expression. It is probably also important for them to accumulate to reasonably high concentrations in the cell, and to be efficiently transported from the nucleus to the cellular compartment where replication takes place. A more sophisticated understanding of plant RNA metabolism as well as viral life cycles is needed to succeed in designing such antisense RNA molecules.

Ribozymes

An ingenious adjunct to the antisense RNA strategy that has the potential to control plant viral disease is the use of RNA enzymes or 'ribozymes'. Recently, Haseloff & Gerlach (1988) described the design of an RNA enzyme that cleaves a specific target RNA sequence, the bacterial chloramphenicol acetyl transferase (CAT) mRNA, and demonstrated its function *in vitro*. This synthetic ribozyme only requires that the substrate contain a conserved sequence of three nucleotides, GUC (a valine codon), and therefore a ribozyme could be designed to target virtually any RNA. The potential result is that the purpose-designed ribozyme binds to its complementary sequence on the target RNA, and the nuclease activity of the active domain cleaves the substrate RNA.

At present there are no reports of the application of this strategy to control plant viral disease in transgenic plants; however, we can speculate as to how it might be applied. The ribozyme domain must be engineered into an RNA complementary to the target RNA, therefore it could be engineered into an antisense RNA in an attempt to cleave plus-strand RNAs, or conversely into a plus-strand RNA in order to cleave minus-strand RNAs. The most likely difficulty with this strategy for controlling plant virus replication is that the ribozyme-containing control agent will be synthesized in the nucleus, while replication of plus-strand RNA viruses occurs in the cytoplasm. Hence the stability and transport of the ribozyme reagent may limit its potential for plant virus control, and therefore ribozymes may not be any more effective than antisense RNAs for controlling viral disease.

Virus-resistance genes

Another strategy for producing virus-resistant plants by genetic engineering that may be applicable in the future is to incorporate naturally occurring plant genes that protect against plant viral disease. These genes fall into two general categories: virus-resistance genes and host defence genes. The genes in the first category are either constitutively expressed or induced upon infection and are specific for a particular plant virus or strain (Fraser, 1990). Genes in the second category are induced upon infection and limit the plant's susceptibility to subsequent infection on other leaves. Some of the pathogenesis-related proteins that have been described (Kauffmann *et al.*, 1990; Chapter 11) may be products of genes belonging to this latter category.

Plant breeders routinely take advantage of naturally occurring genes conferring resistance to plant viruses when developing new or improved cultivars. Many of these genes have been exploited in plant breeding, as described in two recent reviews (Fraser, 1985, 1990). A limitation to the future exploitation of these genes by genetic engineering is the poor knowledge of the biochemical basis for the resistance that they confer. It is clear from examining the patterns of inheritance of these genes as well as the phenotypes they produce that plant genes for virus resistance act by a number of mechanisms, which are largely unknown at present. In part this is due to our lack of understanding of the precise biology of plant virus replication and spread, and also to a paucity of *in vitro* experimental systems that are suitable for dissecting the biochemistry of resistance. A possible single exception exists. Resistance to cowpea mosaic virus (CPMV) strain SB is controlled by a dominant allele found in cowpea cultivar Arlington. Ponz *et al.* (1988) have identified a protein that inhibits CPMV polyprotein processing in Arlington protoplasts. This protein may also inhibit translation.

At present, this is the sole case where a putative gene product of a plant virus resistance gene has been identified. The purification and characterization of this protein have not yet been reported.

Before the wealth of naturally occurring virus-resistance genes can be used to modify plants by genetic engineering, further advances in plant gene cloning technology and the understanding of plant virus resistance must be made. Promising technologies are developing rapidly. For example, advances are being made in genome mapping and the development of linkage maps saturated with molecular markers (Chapter 2). Molecular markers linked to major disease-resistance genes have been found, such as for the *Pisum* gene for resistance to pea seed-borne mosaic virus (G. M. Timmerman, J. E. Lancaster & N. F. Weeden, unpublished). Complementary developments are also being made in genome library construction (Burke *et al.*, 1987; Anand *et al.*, 1990), chromosome walking (Collins & Weissman, 1984; Fors *et al.*, 1990) and automation of nucleic acid sequencing. Application of these technologies may permit direct isolation of plant genes, including those for virus resistance, without any knowledge of the nature of the gene product. Because of the complexity of many higher plant genomes, further advances over the current state of the art are still required before such projects are feasible.

The other approach that will ultimately yield virus-resistance genes is to take the more traditional biochemical approach and characterize the putative gene product conferring resistance and then clone a cDNA copy of its mRNA or the gene encoding it. This was the approach taken by Ponz *et al.* (1988), as discussed above, and it is feasible for any virus–host combination that is amenable to biochemical dissection, such as CPMV and cowpea. A further consequence will be a much improved understanding of the nature of virus-resistance gene products.

Success in genetically modifying plants with virus-resistance genes will also depend on the genes themselves. Plant virus-resistance genes show three patterns of inheritance: dominant, incompletely dominant, and recessive. It is a straightforward matter to envisage how dominant or incompletely dominant genes, once available, may be added to transgenic plants to produce virus resistance. However, about one-third of plant genes for virus resistance are recessive (Fraser, 1990). These genes may be null alleles of 'susceptibility genes' or may actually encode a functional gene product. In either case, the simple insertion of such genes into a genetic background that is dominant for susceptibility seems an unlikely way to produce virus-resistant plants by genetic engineering.

Summary

Genetic engineering now offers novel sources of genes for virus resistance and the technology for their stable insertion into crop species. Several viral coat protein genes have been shown to produce virus resistance when expressed in transgenic plants, and this technology is advancing well. Throughout the world, field trials have been conducted on transgenic plants containing coat protein genes from eight different plant viruses (OECD, 1990). Most of these trials are on solanaceous plants, which are readily transformed. Genetic engineering for virus resistance will become more widespread as the protocols for transforming more crop species become available.

The genes that are introduced by genetic engineering and successfully confer resistance will serve as valuable adjuncts to the naturally occurring plant genes introduced by conventional breeding. By applying both genetic engineering and plant breeding methodologies, new cultivars can be developed whose resistance to a particular virus is due to two or more genes that act by different mechanisms, for example a coat protein gene and a plant nuclear resistance gene. In such plants, resistance should not break down as easily as if it were conferred by a single gene. Furthermore, genetic engineering should permit virus-resistant cultivars to be developed for crop species in which effective genes that confer virus resistance by conventional breeding have not been identified. Research to introduce virus resistance by genetic engineering should, therefore, make significant contributions toward the development of disease-resistant cultivars for use in agriculture.

References

Anand, R., Riley, J. H., Butler, R., Smith, J. C. & Markham, A. F. (1990) A 3.5 genome equivalent multi access YAC library: construction, characterisation, screening and storage. *Nucleic Acids Research* 18, 1951–6.

Anderson, E. J., Stark, D. M., Nelson, R. S., Powell, P. A., Tumer, N. E. & Beachy, R. N. (1989) Transgenic plants that express the coat protein genes of tobacco mosaic virus or alfalfa mosaic virus interfere with disease development of some nonrelated viruses. *Phytopathology* 79, 1284–90.

Angenent, G. C., Van Den Ouweland, J. M. W. & Bol, J. F. (1990) Susceptibility to virus infection of transgenic tobacco plants expressing structural and nonstructural genes of tobacco rattle virus. *Virology* 175, 192–8.

Barden, K. A., Schiller Smith, S. & Murakishi, H. H. (1986) Regeneration and screening of tomato somaclones for resistance to tobacco mosaic virus. *Plant Science* 45, 209–13.

Bevan, M. W., Mason, S. E. & Goelet, P. (1985) Expression of tobacco mosaic virus coat protein by a cauliflower mosaic virus promoter in plants transformed by *Agrobacterium*. *EMBO Journal* 4, 1921–6.

Broadbent, L. (1964) The epidemiology of tomato mosaic virus. VII. The effect of TMV on tomato fruit yield and quality under glass. *Annals of Applied Biology* 54, 209–24.

Burke, D. T., Carle, G. F. & Olson, M. V. (1987) Cloning of large segments of exogenous DNA into yeast by means of artificial chromosome vectors. *Science* 236, 806–12.

Cascone, P. J., Carpenter, C. D., Li, X. H. & Simon, A. E. (1990) Recombination between satellite RNAs of turnip crinkly virus. *EMBO Journal* 9, 1709–15.

Collins, F. S. & Weissman, S. M. (1984) Directional cloning of DNA fragments at a large distance from an initial probe: a circularization method. *Proceedings of the National Academy of Sciences USA* 81, 6812–16.

Culver, J. N. & Dawson, W. O. (1989) Tobacco mosaic virus coat protein: an elicitor of the hypersensitive reaction but not required for the development of mosaic symptoms in *Nicotiana sylvestris*. *Virology* 73, 755–8.

Cuozzo, M., O'Connel, K. M., Kaniewski, W., Fang, R.-X., Chua, N.-H. & Tumer, N. E. (1988) Viral protection in transgenic tobacco plants expressing the cucumber mosaic virus coat protein or its antisense RNA. *Bio/Technology* 6, 549–57.

Delgado-Sanchez, S. & Grogan, R. G. (1970) Potato virus Y. In: CMI/AAB (eds) *Descriptions of Plant Viruses*. Commonwealth Mycological Institute and the Association of Applied Biologists, Kew, Surrey, England, 4 pp.

Fors, L., Saavedra, R. A. & Hood, L. (1990) Cloning of the shark Po promoter using a genomic walking technique based on the polymerase chain reaction. *Nucleic Acids Research* 18, 2793–9.

Francki, R. I. B. (1985) Plant virus satellites. *Annual Review of Microbiology* 39, 151–74.

Fraser, R. S. S. (1985) Genes for resistance to plant viruses. *CRC Critical Reviews in Plant Sciences* 3, 257–94.

Fraser, R. S. S. (1990) The genetics of resistance to plant viruses. *Annual Review of Phytopathology* 28, 179–200.

Gallie, D. R., Sleat, D. E., Watts, J. W., Turner, P. C. & Wilson, T. M. A. (1987) A comparison of eucaryotic viral 5′ leader sequences as enhancers of mRNA expression *in vivo*. *Nucleic Acids Research* 15, 8693–711.

Gasser, C. S. & Fraley, R. T. (1989) Genetically engineering plants for crop improvement. *Science* 244, 1293–9.

Gerlach, W. L., Llewellyn, D. & Haseloff, J. (1987) Construction of a disease resistance gene from the satellite RNA of tobacco ringspot virus. *Nature* 328, 802–5.

Golemboski, D. B., Lomonosoff, G. P. & Zaitlin, M. (1990) Plants transformed with a tobacco mosaic virus nonstructural gene sequence are resistant to the virus. *Proceedings of the National Academy of Sciences USA* 87, 6311–15.

Habili, N. & Symons, R. H. (1989) Evolutionary relationship between luteoviruses and other RNA plant viruses based on sequence motifs in their putative RNA polymerases and nucleic acid helicases. *Nucleic Acids Research* 17, 9543–55.

Hamilton, R. I. (1985) Using plant viruses for disease control. *HortScience* 20, 848–52.

Harrison, B. D., Mayo, M. A. & Baulcombe, D. C. (1987) Virus resistance in transgenic plants that express cucumber mosaic virus satellite RNA. *Nature* 328, 799–802.

Haseloff, J. & Gerlach, W. L. (1988) Simple RNA enzymes with new and highly specific endoribonuclease activities. *Nature* 334, 585–91.

Hemenway, C., Fang, R.-X., Kaniewski, W. K., Chua, N.-H. & Tumer, N. E.

(1988) Analysis of the mechanism of protection in transgenic plants expressing the potato virus X coat protein or its antisense RNA. *EMBO Journal* 7, 1273–80.

Hoekema, A., Huisman, M. J., Molendijk, L., van den Elzen, P. J. M. and Cornelissen, B. J. C. (1989) The genetic engineering of two commercial potato cultivars for resistance to potato virus X. *Bio/Technology* 7, 273–8.

Jobling, S. A. & Gehrke, L. (1987) Enhanced translation of chimaeric messenger RNAs containing a plant viral RNA untranslated leader sequence. *Nature* 325, 622–5.

Kallerhoff, J., Perez, P., Bouzoubaa, S., Ben Tahar, S. & Perret, J. (1990) Beet necrotic yellow vein virus coat protein mediated protection in sugarbeet (*Beta vulgaris* L.) protoplasts. *Plant Cell Reports* 9, 224–8.

Kauffmann, S., Legrand, M. & Fritig, B. (1990) Isolation and characterization of six pathogenesis-related (PR) proteins of Samsun NN tobacco. *Plant Molecular Biology* 14, 381–90.

Kurath, G. & Palukaitis, P. (1990) Serial passage of infectious transcripts of a cucumber mosaic virus satellite RNA clone results in sequence heterogeneity. *Virology* 176, 8–15.

Larkin, P. J. & Scowcroft, W. R. (1981) Somaclonal variation – a novel source of variability from cell cultures for plant improvement. *Theoretical and Applied Genetics* 60, 197–214.

Lawson, C., Kaniewski, W., Haley, L., Rozman, R., Newell, C., Sanders, P. & Tumer, N. E. (1990) Engineering resistance to a mixed virus infection in a commercial potato cultivar: resistance to potato virus X and potato virus Y in transgenic russet burbank. *Bio/Technology* 8, 127–34.

Loesch-Fries, L. S., Merlo, D., Zinnen, T., Burlop, L., Hill, K., Krahn, K., Jarvis, N., Nelson, S. & Halk, E. (1987) Expression of alfalfa mosaic virus RNA4 in transgenic plants confers virus resistance. *EMBO Journal* 6, 1845–51.

Lok, S. & Abouhaidar, M. (1981) The polar alkaline disassembly of papaya mosaic virus. *Virology* 113, 637–43.

Nejidat, A. & Beachy, R. N. (1989) Decreased levels of TMV coat protein in transgenic tobacco plants at elevated temperatures reduces resistance to TMV infection. *Virology* 173, 531–8.

Nelson, R. S., Powell-Abel, P. & Beachy, R. N. (1987) Lesions and virus accumulation in inoculated transgenic tobacco plants expressing the coat protein gene of tobacco mosaic virus. *Virology* 158, 126–32.

Nelson, R. S., McCormick, S. M., Delanney, X., Dube, P., Layton, J., Anderson, E. J., Kaniewska, M., Proksch, R. K., Horsch, R. B., Rogers, S. G., Fraley, R. T. & Beachy, R. N. (1988) Virus tolerance, plant growth, and field performance of transgenic tomato plants expressing coat protein from tobacco mosaic virus. *Bio/Technology* 6, 403–9.

OECD (1990) *Database File: Field Release of Genetically Modified Organisms*. Organization for Economic Co-operation and Development, Environmental Directorate, Paris. 94 pp.

Ponz, F., Glascock, C. B. & Bruening, G. (1988) An inhibitor of polyprotein processing with the characteristics of a natural virus resistance factor. *Molecular Plant–Microbe Interactions* 1, 25–31.

Powell-Abel, P., Nelson, R. S., De, B., Hoffman, N., Rogers, S. G., Fraley, R. T. & Beachy, R. N. (1986) Delay of disease development in transgenic plants that express the tobacco mosaic virus coat protein gene. *Science* 232, 738–43.

Powell-Abel, P., Stark, D. M., Sanders, P. R. & Beachy, R. N. (1989) Protection against tobacco mosaic virus in transgenic plants that express tobacco mosaic virus antisense RNA. *Proceedings of the National Academy of Sciences USA* 86, 6949–52.

Powell-Abel, P., Sanders, P. R., Tumer, N., Fraley, R. T. & Beachy, R. N. (1990) Protection against tobacco mosaic virus infection in transgenic plants requires accumulation of coat protein rather than coat protein RNA sequences. *Virology* 175, 124–30.

Register, J. C. & Beachy, R. N. (1988) Resistance to TMV in transgenic plants results from interference with an early event in infection. *Virology* 166, 524–32.

Register, J. C. & Beachy, R. N. (1989) Effect of protein aggregation state on coat protein mediated protection against tobacco mosaic virus using a transient protoplast assay. *Virology* 173, 656–63.

Rezaian, M. A., Skene, K. G. M. & Ellis, J. G. (1988) Anti-sense RNAs of cucumber mosaic virus in transgenic plants assessed for control of the virus. *Plant Molecular Biology* 11, 463–71.

Sleat, D. E. & Palukaitis, P. (1990) Site-directed mutagenesis of a plant viral satellite RNA changes its phenotype from ameliorative to necrogenic. *Proceedings of the National Academy of Sciences USA* 87, 2946–50.

Stark, D. M. & Beachy, R. N. (1989) Protection against potyvirus infection in transgenic plants: evidence for broad spectrum resistance. *Bio/Technology* 7, 1257–62.

Sulzinski, M. A., Gabard, K. A., Palukaitis, P. & Zaitlin, M. (1985) Replication of tobacco mosaic virus. VIII. Characterization of a third subgenomic TMV RNA. *Virology* 145, 132–40.

Takamatsu, N., Ishikawa, M., Meshi, T. & Okada, Y. (1987) Expression of bacterial chloramphenicol acetyltransferase gene in tobacco mediated by TMV-RNA. *EMBO Journal* 6, 307–11.

Tumer, N. E., O'Connell, K. M., Nelson, R. S., Sanders, P. R., Beachy, R. N., Fraley, R. T. & Shah, D. M. (1987) Expression of alfalfa mosaic virus coat protein gene confers cross protection in transgenic tobacco and tomato plants. *EMBO Journal* 6, 1181–8.

Van Dun, C. M. P. & Bol, J. F. (1988) Transgenic tobacco plants accumulating tobacco rattle virus coat protein resist infection with tobacco rattle virus and pea early browning virus. *Virology* 167, 649–52.

Van Dun, C. M. P., Bol, J. F. & van Vloten-Doting, L. (1987) Expression of alfalfa mosaic virus coat protein genes in transgenic tobacco plants. *Virology* 159, 299–305.

Van Dun, C. M. P., Overduin, B., van Vloten-Doting, L. & Bol, J. F. (1988) Transgenic tobacco expressing tobacco streak virus or mutated alfalfa mosaic virus coat protein does not cross-protect against alfalfa mosaic virus infection. *Virology* 164, 383–9.

Waterworth, H. E., Kaper, J. M. & Tousignant, M. E. (1979) CARNA5, the small cucumber mosaic virus-dependent replicating RNA, regulates disease expression. *Science* 204, 845–7.

Wilson, T. M. A. (1984) Cotranslational disassembly increases the efficiency of expression of TMV RNA in wheat germ cell-free extracts. *Virology* 138, 353–6.

Wisniewski, L. A., Powell, P. A., Nelson, R. S. & Beachy, R. N. (1990) Local and systemic spread of tobacco mosaic virus in transgenic tobacco. *Plant Cell* 2, 559–67.

Chapter 14
Breeding for Resistance to Physiological Stresses

Christopher A. Cullis
Dean of Mathematics and Natural Sciences, 7080
Crawford Hall, Case Western Reserve University,
Cleveland, Ohio 44106, USA

Introduction

Stress, in some form, is a highly probable occurrence for almost any plant growing under either natural or cultivated conditions. Stress may be even more severe for crops, including forest trees, planted in marginal areas. Mild stresses are likely to occur frequently and result in losses in growth. Severe stresses, however, can lead to catastrophic losses. Breeders have long addressed such problems, normally selecting for stability of performance over a range of environments, using extensive testing and an intricate biometrical approach. This traditional approach to breeding is becoming limited and new methods are required. Increased understanding of how the interaction of chemical and physical environments reduces plant development and yield opens the door to a combination of breeding, physiological and biotechnological approaches to plant modification in a comprehensive strategy for improving resistance to environmental stress.

World crop production is limited largely by environmental stress (Blum, 1988). For example, Dudal (cited in Blum, 1988) estimated that only 10% of the world's arable land is free from some form of stress. The main factor responsible for the difference between potential and actual yield is environmental stress (Cardwell, 1982). Thus, while a specific genetic solution may not be appropriate or available for all situations, the incorporation of stress resistance must be part of the primary goal of all breeding programmes while potential and actual yields are significantly separated.

This chapter will consider three types of stress: water stress, which will be considered mainly in terms of drought stress; temperature stress; and salt stress. Each of these three stresses represents a broad topic with extensive

340

literature and a wealth of experimental detail. All have common components in the sense of having mechanisms for stress tolerance and avoidance. However, there is no single mechanism by which multiple stresses are alleviated. All indications are that the ability to withstand stressful environments is controlled by a number of genes, and the multigenic character imposes limits to subsequent manipulation.

Current breeding strategies have clearly been successful in incorporating degrees of tolerance in a variety of species. In this chapter these successes will be considered along with the current advances in new technologies. The new methods of gene identification and isolation, the ability to transfer genes within and between species by various transformation schemes, and selection in tissue culture will be evaluated for their potential to contribute to the overall management of environmental stress in both the short and long term.

Temperature stress

Any temperature outside the optimum for growth and development may be considered a stress temperature. The definition of a stress temperature will clearly depend on the stage of growth, and the effects of such stresses will not be equal at all stages of growth. In general, the most severe perturbations occur at the extremes of the growing seasons, namely at germination and at fruit formation. It is at these two extremes that much of the concentration in developing resistance has occurred. Three different types of temperature stress will be considered, namely heat stress, chilling stress and freezing stress. Understanding at the physiological and molecular levels is probably greatest for heat stress, although for all three stresses the basic information is still sparse.

Heat tolerance

Damage caused by high temperatures is common in crop plants. At extremes, such stress results in the death of cells, organs or whole plants. The reproductive process is especially sensitive to heat, with the resulting flower abscission, pollen sterility and/or poor fruit set being responsible for large yield reductions. However, even sublethal heat stress, which increases the rate of plant development, can also result in a loss of yield. Thus thermal tolerance must be viewed in terms of the degree of stress and the duration of exposure, as well as in relation to the stage of development at which the stress occurs.

Genetic variation in heat tolerance for various attributes exists in a number of crop plants. For example, genetic variation has been demonstrated in tolerance at germination (Zeng & Khan, 1984; Soman & Peacock,

1985), for growth under heat stress (Mendoza & Estrada, 1979; Stamp *et al.*, 1983; Shpiler & Blum, 1986), for recovery from heat stress (Coffman, 1957), in flowering and fruit or seed set, and in the functions of photosynthesis, translocation and the stability of membranes. The genetic resources for heat tolerance are apparently abundant in a number of crop species such as rice, potatoes, soybean and tomato, but for a number of other crop plants the extent of existing genetic variation has hardly been explored.

Much molecular characterization has been directed towards the heat-shock response in terms of the heat-shock proteins. This heat-shock response is one of the most highly conserved biological responses known, being present in organisms ranging from bacteria to man. In general, the heat-shock response is characterized by a number of steps which include:

1. a decrease in normal protein synthesis;
2. induction of a new set of mRNAs and the translation of these RNAs into the heat-shock proteins;
3. accumulation of large amounts of heat-shock proteins;
4. acquisition of thermotolerance to otherwise lethal temperatures;
5. gradual decline in heat-shock protein synthesis and a return to normal protein synthesis during prolonged heat treatments.

As stated above, plant breeders have been able to manipulate thermal tolerance as an important and heritable agronomic trait. However, the relationship between the thermal tolerance observed in cultivars and the transient heat-shock response in the laboratory is still unclear. It has been established that certain crop species growing in the field can exhibit the synthesis of heat-shock mRNAs and proteins under conditions of heat stress (Kimpel & Key, 1985; Vierling *et al.*, 1988). However, Edelman & Key (cited in Nagao *et al.*, 1990) did not detect qualitative differences in the profiles of heat-shock response in a comparison of soybean seedlings, which show heritable differences in their field thermal tolerance (Nagao *et al.*, 1990), whereas a positive correlation between the degree of thermal tolerance and the contents of low-molecular-weight heat-shock proteins was found in wheat. Thus the exact role of the heat-shock proteins in the generation of thermal tolerance is still unclear, but a subject of intense study.

Cold tolerance

Tolerance to low temperatures can be divided arbitrarily into tolerance to chilling and tolerance to freezing. As an operational definition, chilling tolerance is that which comes into play at temperatures above 0°C, whereas freezing tolerance is that exhibited below 0°C. In both cases one of the primary sites of damage by low temperatures is thought to be the membrane.

Chilling stress

Chilling stress is probably the most common environmental stress during germination. The low temperature at germination not only reduces the germination rate, but also affects the subsequent growth. Chilling can also be important during flowering, with consequences such as deformed fruit or the failure to set fruit or seed. Genetic variation in chilling tolerance has been demonstrated for various physiological processes such as photosynthesis and plasma membrane function, as well as for whole plant processes such as germination, growth and seed set. In practical terms, most information is available for germination and seedling growth. Genetic resources for chilling tolerance are to be found at the margins of a sensitive species range. Ample genetic variation has been found in maize (Mock & Eberhart, 1972), while in tomato the introgression of chilling tolerance from exotic germplasm has been successful (Patterson & Payne, 1983).

Freezing resistance

Plants are known to differ in their ability to withstand freezing temperatures, but the molecular and genetic bases of this differential are unknown. However, it is known that a prior exposure to low non-freezing temperatures (cold acclimation) increases the tolerance to subsequent freezing. Tolerance to freezing stress has been studied in a wide range of plants, with the most detailed knowledge of the inheritance of the trait coming from wheat. Studies, such as those by Nilsson-Ehle (1912) and others (cited in Thomashow, 1990), have concluded that frost hardiness is a quantitative trait controlled by several genes. In addition to the number of genes involved, the interactions between the genes can be complex. Outside wheat, frost hardiness has also been studied in oats and barley, as well as in a variety of woody plants including fruit trees, tea plants, pines and roses (Sakai & Larcher, 1987). Generally the results indicate a complex interaction of many genes. Frost tolerance of tea stems and winterkill of roses may be exceptions, where only a few genes may be involved.

A large number of biochemical changes take place in a cold-acclimated plant, but exactly which are responsible for the increased tolerance is not known. Much attention has been focused on the plasma membrane since disruption of cellular membranes has been regarded as the primary cause of freezing injury in plants. A characterization of the differences in biophysical properties of plasma membranes from cold-acclimated and non-acclimated plants suggests that the biochemical composition of the membrane is altered during cold hardening. Efforts are being directed to determining a causal relationship between specific membrane changes and freezing tolerance. The results of Steponkus *et al.* (1988) strongly support the view that changes in plasma lipid composition are causally related to increased tolerance to expansion-induced lysis. However, the overall increase in freezing tolerance of membranes probably involves multiple alterations, each having different effects.

The changes in membranes are only one part of a series of biochemical alterations associated with cold acclimation. The appearance of new isozymes and an increase in proline and organic acids, as well as sugar and soluble proteins, also occur. The direct causal relationship between these changes and low-temperature survival is still unclear. Specific changes in gene expression have been demonstrated by the identification of cold-regulated genes. The expression of these genes with the synthesis of their polypeptide products closely parallels freezing tolerance in plants (Thomashow, 1990). The current state of knowledge of the molecular genetics of cold acclimatization is unsatisfactory. However, it is clear that the response is distinct from the heat-shock response in terms of the identities of new proteins synthesized and the length of time over which these proteins are synthesized. Much more will have to be learned before a biotechnological manipulation can be considered for this property, and conventional breeding programmes will have to suffice.

Drought stress

Drought resistance is a complex phenomenon in both its physiology and its genetics. The essential requirement for tolerance to drought is the ability of the plant to continue to function under conditions of water limitation. This can be done by drought avoidance where a short growing season can function as an attribute of drought escape. Such a scenario clearly depends on the environment over the growing season as to whether or not such a mechanism is useful. However, the relative yield advantage of early-maturing genotypes under conditions of water stress has been exploited (Reitz, 1974; Chapter 1). An alternative to drought avoidance is dehydration avoidance, which is the ability of a plant to maintain sufficient tissue hydration for appropriate functioning of the metabolic processes involved in growth and development under conditions of environmental water stress.

Genetic variation for the various components of dehydration avoidance certainly occurs. The ability to maintain high leaf water potential is frequently associated with root characteristics, especially the ability to extract soil moisture. Little work on the genetics of root characteristics has been done, but selection for shoot dehydration may be equivalent to selection for root characteristics.

The role of abscisic acid (ABA) in drought resistance is likely to be crucial for understanding how this index can be used in selection schemes. ABA concentration can affect many processes in addition to its possible effect on yield under water stress conditions (Milborrow, 1981). Selection for ABA accumulation may turn out to be an important selection criterion for drought resistance. It should be noted that some of the cold-regulated genes described in the previous section are also induced by both ABA and drought stress. The role of ABA in many processes may make it a common thread through stress tolerance selection, or because of its ubiquity, make it insensitive as a unique selection index for any stress.

Genetic variations in the components of drought tolerance have been observed and manipulated. For example, a weak association between proline accumulation and yield under stress was observed for *Brassica* spp. (Richards & Thurling, 1979), but a natural mutant of barley which accumulated proline did not show an alteration in the plant's water status. Thus, as with many of the components of stress tolerance, it is not just the alteration of a characteristic, but how that alteration occurs in response to a stress environment that is important in evaluating the role of that particular response.

Salt stress

Over the past two decades increasing attempts have been made to develop salt-resistant varieties able to grow under saline conditions. However, in spite of this effort, there are few instances of crop cultivars that have been bred for salinity resistance and used as an economic solution in saline eco-systems. Saline fields are inherently variable in their salt distribution, so that most of the yield from such areas is from the non-saline patches, with little from the saline patches. Thus the question arises as to the appropriate strategy for dealing with such a distribution; whether it is better to breed for high yields for the unstressed patches, or for lower yields (Blum, 1988).

The effects of salinity on plant processes are threefold, in terms of water stress from the osmotic effects, mineral toxicity of the salt and interruptions to the mineral nutrition of the plant. The multipartite nature of the stress results in difficulty in predicting the extent of stress in a saline environment.

The mechanisms by which plants adapt to saline conditions are typically unique to halophytes and have evolved over a considerable period of time. The lessons learned from a study of the integrated control of halophytes may not be immediately applicable to crop breeding programmes in a piecemeal fashion. Thus, genetic material can be found in saline habitats from which useful salt tolerance may be exploited, but a great deal of additional physiological and genetic characterization needs to be done.

The problem of osmoregulation has received a great deal of attention. Although the role of proline accumulation in response to a variety of environmental stresses is not clear, proline is an important solute in osmoregulation in halophytes (Stewart & Lee, 1974). The roles of individual solutes in osmoregulation of glycophytes are much less clear, although proline (Stewart & Lee, 1974), *myo*inositol (Sacher & Staples, 1985) and betaine (Wynn Jones, 1981) have all been implicated.

The root system is also likely to play an important part in salt tolerance. An increased development of the root system would provide a mechanism to meet the high transpirational demands at low water potential, as well as allow the exploitation of the variability in the soil salt content.

The molecular aspects of salt tolerance are still in the early stages of investigation. It is clear that the profile of proteins synthesized in response to saline stress differs from that in the absence of stress (Ramagopal, 1987). However, the role of any of these proteins in a causal relationship to any tolerance that develops is still not confirmed. A major gene which will control salt tolerance is unlikely in view of the continuous nature of tolerance, although the development of tolerant tissue culture lines suggests that such a search may not be hopeless.

Biotechnological approaches

The preceding sections have illustrated the range of genetic resources available, and some of the uses to which these have been put in the development of crops resistant to physiological stresses. Developments over the past two decades in techniques for tissue culture and regeneration of crop plants, and for the introduction of foreign DNA, have been widely discussed in terms of their potential for accelerating the incorporation of novel tolerances as well as for the discovery of new genetic variation. These will be considered in two sections, related to tissue culture and molecular manipulation.

Tissue culture

The use of cell and tissue culture for developing stress-resistant genotypes employs a cell population as the equivalent of the breeding population. In this way the environment can be maintained in a defined form and the size of the population greatly increased over that possible using whole plants. In addition, there is the phenomenon of somaclonal variation, resulting in novel genetic variants within the culture that were absent from the source material used to initiate the culture.

The selection for cell lines which are able to grow under the appropriate environmental stress conditions is the primary aim. From such a resistant cell population plants can be regenerated and tested for their ability to withstand the same environmental stresses as whole plants. The application of these techniques in the development of tolerant lines is limited in two ways. First, there is the complex genetic control which apparently underlies most tolerance mechanisms. However, in the case of somaclonal variation the constant application of the selection pressure may be able to overcome this by the accumulation of many mutations giving the desired tolerance. Secondly, the use of cultured cells can only be effective for cell-based phenomena, and is likely to be ineffective when applied to problems where the basis of resistance is the integration of plant processes, such as an increased root vigour.

Salinity resistance has been extensively investigated through selection *in vitro*. Cell lines with enhanced resistance to salt have been isolated from many plant species (Tal, 1990). The stability of the resistance during mitosis has been demonstrated in many experiments. More rarely the resistance observed in culture has been associated with resistance at the whole plant level (Orton, 1980; Warren & Gould, 1982). The production of salt-resistant somaclones which transmit this characteristic to their progenies has also been described (Nabors, 1983). A corollary of the ability to accumulate multiple mutations in culture is the accumulation of unwanted changes in the selected lines. Plants regenerated from NaCl-selected alfalfa callus were so abnormal in growth and disease susceptibility that their evaluation for salinity resistance was impossible. However, stable saline-resistant variants selected from culture have been reported for rice, wheat, oats and tobacco (Nabors, 1983). The ability to regenerate important crop species has advanced markedly over the past five years so that selection for stress tolerance is available for many more species. However, the selection in culture of useful variants is likely to be limited to those stress phenomena which are cell-based, can be efficiently selected against a background of non-useful 'escapes', and have a relatively simple genetic basis to shorten the time in culture.

Gene transfer

The development of transformation techniques for many of the major crop species over the past decade has opened many possibilities for developing new varieties incorporating novel genetic material which would not have been available to the traditional breeder. New techniques, such as restriction fragment length polymorphism (RFLP) mapping (Chapter 2), may be used in conjunction with conventional breeding programmes to accelerate the process and to reduce the size of the populations required.

There are three parts to a successful strategy to genetically engineer new stress resistances. First is the isolation of appropriate genes of interest and utility. Second, the newly tailored gene or gene complex must be transferred back into the plant of interest. Finally, the novel plants have to be characterized as to their phenotype. The second and third processes can be done relatively efficiently, especially with the advent of the biolistic transformation schemes which have proved to be useful in many important crop species which had previously been intractable (Chapter 5). However, the first prerequisite, the identification and isolation of the genes for important stress-resistance traits, is likely to prove most difficult. As noted earlier, many of the resistances appear to have complex inheritance patterns, involving genes with unknown functions. Thus the isolation of the genes involved will be difficult. The present transformation schemes are also limited in the amount of information which can be transferred. Although the technology is very powerful, its use is likely to be limited by the quantity and quality of the biological knowledge that is available. Much more fundamental work is needed to understand the biochemical basis of stress resistance and to fully utilize the available technology.

One possible success of the technological approach is seen in the work on superoxide dismutase and catalase. Both of these enzyme systems are needed to protect cells against the damaging effects of active oxygen species, which may be produced during photorespiration (Halliwell, 1984), chilling (Wise & Naylor, 1987) and dehydration (Dhindsa & Matowe, 1981). Representative cDNA clones have been isolated for superoxide dismutase (Bowler *et al.*, 1989) and catalase (Redinbaugh *et al.*, 1988). The coupling of these genes to the appropriate regulatory sequences may well provide some measure of protection against a limited array of stressful environments, but is unlikely to be a generic solution. The functions of the many other stress-related polypeptides need to be known before their usefulness in the generation of stress-resistant transgenic material can be considered.

RFLP Mapping

The digestion of genomic DNA with a restriction enzyme (see Table 2.3; p. 36) and subsequent electrophoresis results in the DNA being separated into a reproducible set of fragments. Comparison of these fragment patterns between genetically distinct individuals can reveal differences in some of these fragments. The term restriction fragment length polymorphism (RFLP) has been coined to describe this variation. The potential usefulness of RFLP markers in basic plant genetics and plant improvement programmes has been reviewed by Beckman & Soller (1983), Tanksley (1988) and Tanksley *et al.* (1989). RFLP maps are being generated for a large number of crop species (Chapter 2). These genetic maps will have extensive utility in breeding programmes and may be very useful in developing stress-resistance material.

RFLP markers have proved to be useful in the mapping of quantitative traits and will be useful in determining the number of chromosomal segments involved in the resistance phenotype. The identification of these segments can be used in a couple of ways. The introgression of alien germplasm can be monitored using RFLP markers both for the segment(s) of interest and for the size of the introgressed segment(s). The initial mapping will reduce the size of the populations needed to be carried through to a selection scheme. The second use for the RFLP markers may be in the isolation of the genes responsible for stress tolerance. This can be done by chromosome walking from the known markers to eventually isolate the desired gene. This technique is time-consuming and involves the isolation of a large number of clones, the identification of the gene of interest and the reintroduction of the gene to demonstrate that it is responsible for the desired phenotype. In spite of the amount of work involved in this approach, it is a potential method for the isolation of a gene that has a phenotypic effect, but for which no other information helpful in its isolation is available.

Conclusions

Breeding for stress resistance has been successful in some cases but not in others. The use of the new methods for molecular manipulation will depend on the identification of specific genes which are capable of making a direct contribution to the adaptation to specific stress environments. These genes may be plant-derived or could come from other sources. However, the complex inheritance of stress resistance suggests a tailoring of any novel combinations to very specific sets of conditions rather than for a generic type of stress resistance. The molecular manipulations are most likely to be added to the present assortment of breeding methods rather than replacing them in the future.

References

Beckman, J. S. & Soller, M. (1983) Restriction fragment length polymorphism in genetic improvement: methodologies, mapping and costs. *Theoretical and Applied Genetics* 67, 35–43.

Blum, A. (1988) *Plant Breeding for Stress Environments*. CRC Press, Boca Raton, Florida.

Bowler, C., Alliotte, T., De Loose, M., van Montagu, M. & Inze, D. (1989) The induction of manganese superoxide dismutase in response to stress in *Nicotiana plumbaginifolia*. *EMBO Journal* 8, 31–8.

Cardwell, V. B. (1982) Fifty years of Minnesota corn production: sources of yield increases. *Agronomy Journal* 74, 984–95.

Coffman, F. A. (1957) Factors influencing heat resistance in oats. *Agronomy Journal* 49, 368–76.

Dhindsa, R. S. & Matowe, W. (1981) Drought tolerance in two mosses: correlated with enzymatic defense against lipid peroxidation. *Journal of Experimental Botany* 32, 79–92.

Halliwell, B. (1984) Oxygen derived species and herbicide action. *Physiologica Plantarum* 15, 21–4.

Kimpel, J. A. & Key, J. L. (1985) Presence of heat shock mRNAs in field grown soybeans. *Plant Physiology* 79, 672–8.

Mendoza, H. A. & Estrada, R. N. (1979) Breeding potatoes for tolerance to stress:

heat and frost. In: Mussell, H. & Staples, R. C. (eds), *Stress Physiology in Crop Plants*. Wiley-Interscience, New York, pp. 227–355.

Milborrow, B. V. (1981) Abscisic acid and other hormones. In: Paleg, L. G. & Aspinall, D. (eds), *The Physiology and Biochemistry of Drought Resistance in Plants*. Academic Press, Sydney, pp. 347–88.

Mock, J. J. & Eberhart, S. A. (1972) Cold tolerance in adapted maize populations. *Crop Science* 12, 466–71.

Nabors, M. W. (1983) Increasing the salt and drought tolerance of crop plants. In: Randall, D. D., Blevins, D. G., Larson, R. L. & Rapp, B. J. (eds), *Current Topics in Plant Biochemistry and Physiology*, vol. 2. University of Missouri, Columbia, pp. 165–84.

Nagao, R. T., Kimpel, J. A. & Key, J. L. (1990) Molecular and cellular biology of the heat shock response. *Advances in Genetics* 28, 235–74.

Orton, T. J. (1980) Comparison of salt tolerance between *Hordeum vulgare* and *H. jubatum* in whole plants and callus culture. *Zeitschrift für Pflanzenphysiologie* 98, 105–18.

Patterson, B. D. & Payne, L. A. (1983) Screening for chilling resistance in tomato seedlings. *HortScience* 18, 340–7.

Ramagopal, S. (1987) Differential messenger RNA transcription during salinity stress in barley. *Proceedings of the National Academy of Sciences USA* 84, 94–8.

Redinbaugh, H. D., Wadsworth, G. J. & Scandalios, J. G. (1988) Characterization of catalase transcripts and their differential expression in maize. *Biochimica et Biophysica Acta* 951, 104–16.

Reitz, L. P. (1974) Breeding for more efficient water-use – is it real or a mirage? *Agricultural Meteorology* 14, 3–6.

Richards, R. A. & Thurling, N. (1979) Genetic analysis of drought stress response in rapeseed (*Brassica campestris* and *B. napus*). III. Physiological characters. *Euphytica* 28, 755–60.

Sacher, R. F. & Staples, R. C. (1985) Inositol and sugars in adaptation of tomato to salt. *Plant Physiology* 77, 206–10.

Sakai, A. & Larcher, W. (1987) *Frost Survival of Plants: Responses and Adaptations to Freezing Stress*. Springer Verlag, Berlin.

Shpiler, L. & Blum, A. (1986) Differential reaction of wheat cultivars to hot environments. *Euphytica* 35, 483–92.

Soman, P. & Peacock, J. M. (1985) A laboratory technique to screen seedling emergence of sorghum and pearl millet at high soil temperature. *Experimental Agriculture* 21, 335–42.

Stamp, P., Geisler, G. & Thiraporn, R. (1983) Adaptation to sub and supraoptimal temperatures of inbred maize lines differing in origin with regard to seedling development and photosynthetic traits. *Physiologica Plantarum* 58, 62–8.

Steponkus, P. L., Uemura, M., Balsamo, R. A., Arvinte, T. & Lynch, D. V. (1988) Transformation of the cryobehavior of rye protoplasts by modification of the plasma membrane lipid composition. *Proceedings of the National Academy of Sciences USA* 86, 9026–30.

Stewart, G. R. & Lee, J. A. (1974) The role of proline accumulation in halophytes. *Planta* 120, 279–89.

Tal, M. (1990) Somaclonal variation for salt resistance. In: Bajaj, Y. P. S. (ed.), *Somaclonal Variation in Crop Improvement*, vol. I. Springer-Verlag, Berlin, pp. 236–57.

Tanksley, S. D. (1988) Resolution of quantitative traits into Mendelian factors using a complete RFLP linkage map. *Nature* 335, 721–6.

Tanksley, S. D., Young, N. D., Paterson, A. H. & Bonierbale, M. W. (1989) RFLP mapping in plant breeding: new tools for an old science. *Bio/Technology* 7, 257–64.

Thomashow, M. F. (1990) Molecular genetics of cold acclimation in higher plants. *Advances in Genetics* 28, 99–131.

Vierling, E., Nagao, R. T., DeRocher, A. E. & Harris, L. M. (1988) A heat shock protein localized to chloroplasts is a member of a eukaryotic super family of heat shock proteins. *EMBO Journal* 7, 575–81.

Warren, R. S. & Gould, A. R. (1982) Salt tolerance expressed as a cellular trait in suspension cultures developed from the haplophytic grass *Distichlis spicata*. *Zeitschrift für Pflanzenphysiologie* 107, 347–56.

Wise, R. R. & Naylor, A. W. (1987) Chilling enhanced photooxidation: the peroxidative destruction of lipids during chilling injury to photosynthesis and ultrastructure. *Plant Physiology* 83, 272–7.

Wynn Jones, R. G. (1981) Salt tolerance. In: Johnson, C. B. (ed.), *Environmental Factors Limiting Plant Productivity*. Butterworths, London, pp. 271–322.

Zeng, G. W. & Khan, A. A. (1984) Alleviation of high temperature stress by preplant permeation of phthalimide and other growth regulators into lettuce seed via acetone. *Journal of the American Society for Horticultural Science* 109, 782–5.

Index